Essential Calculus-based

PHYSICS

Study Guide Workbook

Volume 3: Waves, Fluids, Sound, Heat, and Light

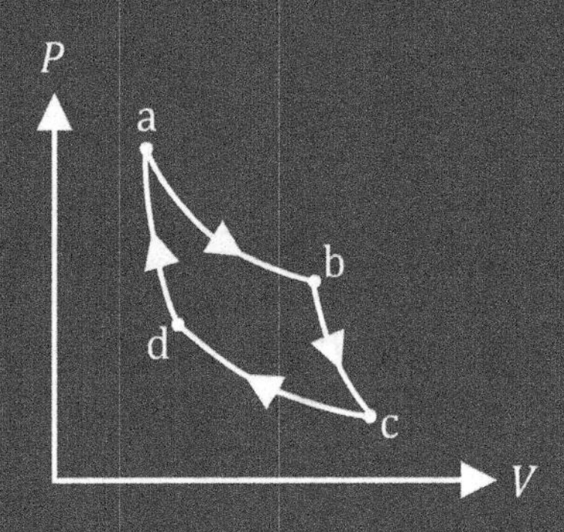

$$W = \int P\, dV$$

Chris McMullen, Ph.D.

Essential Calculus-based Physics Study Guide Workbook
Volume 3: Waves, Fluids, Sound, Heat, and Light
Learn Physics with Calculus Step-by-Step

Chris McMullen, Ph.D.
Physics Instructor
Northwestern State University of Louisiana

www.monkeyphysicsblog.wordpress.com
www.improveyourmathfluency.com
www.chrismcmullen.wordpress.com

Zishka Publishing

ISBN: 978-1-941691-19-9

Textbooks > Science > Physics
Study Guides > Workbooks> Science

CONTENTS

EXPECTATIONS

Prerequisites:

- As this is the third volume of the series, the student should already have studied the material from first-semester physics, including motion, applications of Newton's second law, conservation of energy, Hooke's law, and rotation.
- The student should have learned basic calculus skills, including how to find the derivative of a polynomial or trig function and how to integrate over a polynomial or trig function. Essential integration skills were reviewed in Volumes 1-2.
- The student should know some basic algebra skills, including how to combine like terms, how to isolate an unknown, how to solve the quadratic equation, and how to apply the method of substitution. Needed algebra skills were reviewed in Volume 1.
- The student should have prior exposure to trigonometry. Essential trigonometry skills were reviewed in Volume 1.

Use:

- This book is intended to serve as a supplement for students who are attending physics lectures, reading a physics textbook, or reviewing physics fundamentals.
- The goal is to help students quickly find the most essential material.

Concepts:

- Each chapter reviews relevant definitions, concepts, laws, or equations needed to understand how to solve the problems.
- This book does not provide a comprehensive review of every concept from physics, but does cover most physics concepts that are involved in solving problems.

Strategies:

- Each chapter describes the problem-solving strategy needed to solve the problems at the end of the chapter.
- This book covers the kinds of fundamental problems which are commonly found in standard physics textbooks.

Help:

- Every chapter includes representative examples with step-by-step solutions and explanations. These examples should serve as a guide to help students solve similar problems at the end of each chapter.
- Each problem includes the main answer(s) on the same page as the question. At the **back of the book**, you can find hints, intermediate answers, directions to help walk you through the steps of each solution, and explanations regarding common issues that students encounter when solving the problems. It's very much like having your own **physics tutor** at the back of the book to help you solve each problem.

INTRODUCTION

The goal of this study guide workbook is to provide practice and help carrying out essential problem-solving strategies that are standard in waves, fluids, sound, heat, and light. The aim here is not to overwhelm the student with comprehensive coverage of every type of problem, but to focus on the main strategies and techniques with which most physics students struggle.

This workbook is not intended to serve as a substitute for lectures or for a textbook, but is rather intended to serve as a valuable supplement. Each chapter includes a concise review of the essential information, a handy outline of the problem-solving strategies, and examples which show step-by-step how to carry out the procedure. This is not intended to teach the material, but is designed to serve as a time-saving review for students who have already been exposed to the material in class or in a textbook. Students who would like more examples or a more thorough introduction to the material should review their lecture notes or read their textbooks.

Every exercise in this study guide workbook applies the same strategy which is solved step-by-step in at least one example within the chapter. Study the examples and then follow them closely in order to complete the exercises. Many of the exercises are broken down into parts to help guide the student through the exercises. Each exercise tabulates the corresponding answers on the same page. Students can find additional help in the hints section at the back of the book, which provides hints, answers to intermediate steps, directions to walk students through every solution, and explanations regarding issues that students commonly ask about.

Every problem in this book can be solved without the aid of a calculator. You may use a calculator if you wish, though it is a valuable skill to be able to perform basic math without relying on a calculator.

The better you truly understand the underlying concepts, the easier it becomes to solve the physics problems.

— Chris McMullen, Ph.D.

1 SINE WAVES

Relevant Terminology

Crest – a point on a wave where there is a **maximum** (the top of a hill).
Trough – a point on a wave where there is a **minimum** (the bottom of a valley).
Equilibrium – a horizontal line about which a sine wave oscillates.
Amplitude – the **vertical** height between a crest and the equilibrium position (or **one-half** of the vertical height between a crest and trough).
Period – the **horizontal** time between two consecutive crests. Period is time it takes for the wave to complete exactly one oscillation.
Frequency – the number of oscillations completed per second.
Angular frequency – the number of radians completed per second. The angular frequency equals 2π times the frequency.
Phase angle – an angle representing how much the wave is shifted horizontally. A **positive** phase angle represents a wave that is shifted to the **left**.

Symbols and SI Units

Symbol	Name	SI Units
x_{max}	the maximum value of x (at a crest)	m
x_{min}	the minimum value of x (at a trough)	m
x_e	the equilibrium position	m
x_0	the initial position	m
x	the instantaneous position	m
A	the amplitude of oscillation	m
t	time	s
T	period	s
δt	the phase shift in seconds	s
φ	phase angle	rad
f	frequency	Hz
ω_0	angular frequency	rad/s

Sine Wave Equations

In physics problems, a sine wave can generally be expressed as

$$x = A\sin(\omega_0 t + \varphi) + x_e$$

where the symbols represent the following:

- x is the position of the oscillating object.
- x_e is the **equilibrium position** (how much the graph is shifted vertically).
- A is the amplitude of the sine wave.
- t is time.
- ω_0 is the **angular frequency** in radians per second.
- φ is the **phase angle** (how much the graph is shifted **horizontally**).

The amplitude (A) is the vertical distance between the maximum value (x_{max}) and the equilibrium position (x_e), or **one-half** of the vertical distance between the maximum value (x_{max}) and the minimum value (x_{min}).

$$A = x_{max} - x_e = \frac{x_{max} - x_{min}}{2}$$

The **angular frequency** (ω_0) equals 2π times the **frequency** (f).

$$\omega_0 = 2\pi f$$

Frequency (f) and **period** (T) share a reciprocal relationship.

$$f = \frac{1}{T} \quad , \quad T = \frac{1}{f}$$

There are three ways to determine the **phase angle** (φ) for a graph of a sine wave:

- Set t equal to zero and algebraically solve for φ. You will get an inverse sine:

  $$\varphi = \sin^{-1}\left(\frac{x_0 - x_e}{A}\right)$$

  It's important to realize that the answer may lie in any of the four Quadrants. (It may help to review the trigonometry essentials from Chapter 9 of Volume 1.) The method of the following bullet point will help you determine the correct Quadrant. Note that x_0 is the **initial position**: It's the value of x for which $t = 0$.
- You can estimate phase angle visually by drawing 0°, 90°, 180°, 270°, and 360° in the usual positions and extrapolating to $t = 0$, as discussed on pages 10-11 and as shown in one of the examples. It's wise to do this in addition to the inverse sine from the first bullet point.
- When there is a graph, you could measure how much the sine wave is shifted horizontally in seconds, δt, and use the following equation.

  $$\frac{\varphi}{2\pi} = \frac{\delta t}{T}$$

  It's important to note that δt may be negative.
  - If you measured a shift to the **right**, δt is **negative**.
  - If you measured a shift to the **left**, δt is **positive**.

Essential Concepts

Consider the general sine wave plotted below. It corresponds to the following equation.

$$x = A\sin(\omega_0 t + \varphi) + x_e$$

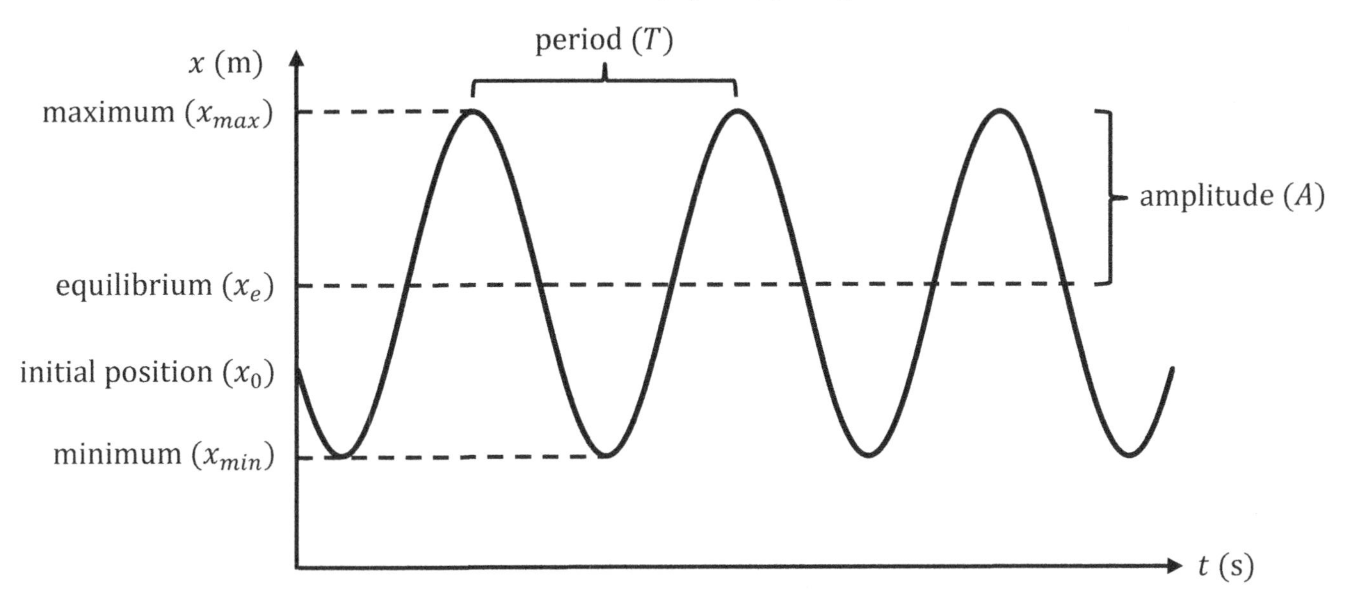

The diagram above visually demonstrates the following quantities:

- The **period** (T) is the **horizontal time** (**not** distance! – note that the horizontal axis has t, not x) between two crests. **Note**: Remember to look at the **crests** (the peaks at the top) and **not** to look along the equilibrium line. (Students who measure points along the equilibrium line instead of between two crests often make a mistake and measure one-half of the period.)
- The **equilibrium** position (x_e) is the vertical value on the graph, which corresponds to a horizontal line that bisects the sine wave between its crests and troughs.
- The crests correspond to x_{max}, which designates the highest points.
- The troughs correspond to x_{min}, which designates the lowest points.
- The **amplitude** (A) is the **vertical** distance between each crest and equilibrium (or **one-half** of the vertical distance from crest to trough).

$$A = x_{max} - x_e = \frac{x_{max} - x_{min}}{2}$$

The following quantities can also be determined from the graph:

- Find the **frequency** (f) from the **period** (T): $f = \frac{1}{T}$.
- Find the **angular frequency** (ω_0) from the **frequency** (f): $\omega_0 = 2\pi f$.
- Find the **phase angle** (φ) from an inverse sine:

$$\varphi = \sin^{-1}\left(\frac{x_0 - x_e}{A}\right)$$

Be sure to pick the correct Quadrant, as discussed on pages 10-11. Note that x_0 is the **initial position**: It's the vertical intercept of the graph.

Phase Angle

The phase angle (φ), represented by a lowercase Greek phi, provides a measure of how much a wave is shifted horizontally compared to a standard sine wave.

- A **positive** phase angle represents a wave that is shifted to the **left**.
- A **negative** phase angle represents a wave that is shifted to the **right**.

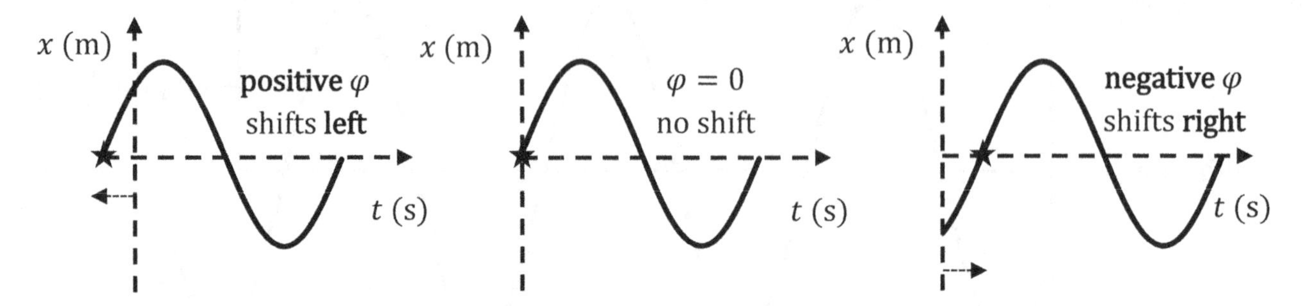

Why does a **positive** phase angle shift the graph to the **left**? You can see this by considering a more familiar function, the basic parabola. The equation $y = x^2$ is an unshifted parabola. If you graph the equation $y = (x + 2)^2$, you will get the same shape, but the **positive** 2 shifts the entire graph to the **left**. Why? Because $y = 0$ when $x = -2$ (since that's the value of x that makes $x + 2$ equal zero). If you graph the equation $y = (x - 2)^2$, you again get the same shape, but the **negative** 2 shifts the entire graph to the **right**. It works the same way with the phase angle and sine waves.

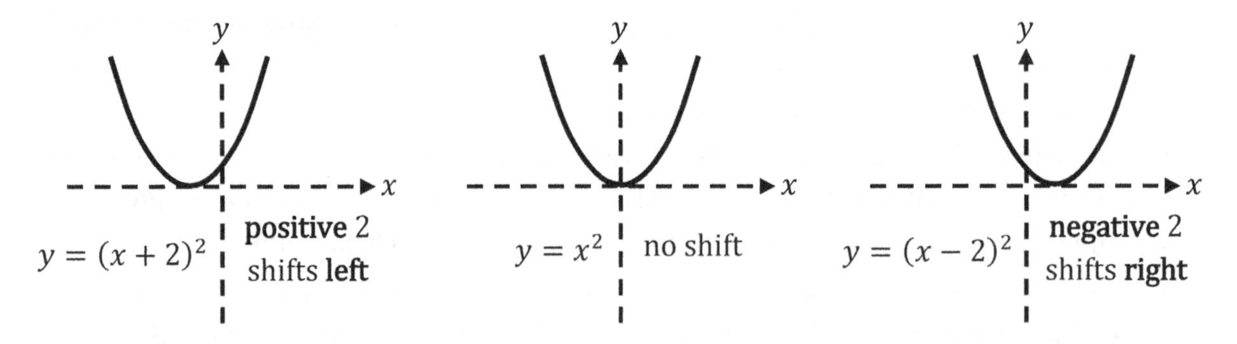

In order to figure out which Quadrant the phase angle lies in, label the angles 0°, 90°, 180°, 270°, and 360° on the sine wave. These are labeled below for a basic sine wave.

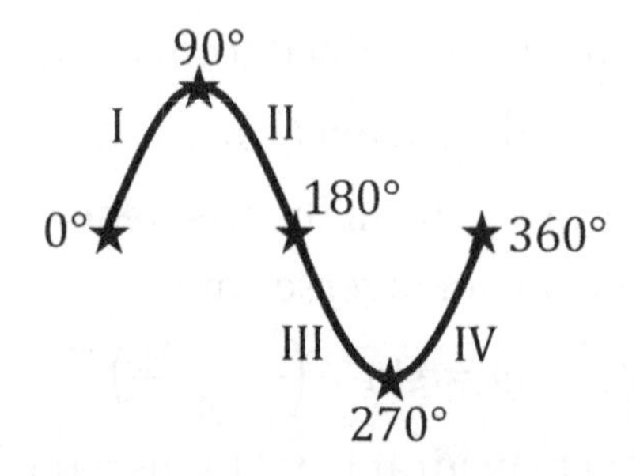

Recall the four **Quadrants**:

- An angle between 0° and 90° lies in Quadrant I.
- An angle between 90° and 180° lies in Quadrant II.
- An angle between 180° and 270° lies in Quadrant III.
- An angle between 270° and 360° lies in Quadrant IV.

Also recall that adding or subtracting 360° to an angle makes an equivalent angle. For example, 0° and 360° are equivalent positions on a sine wave, and −90° and 270° are also equivalent positions.

To determine the correct Quadrant and to also estimate the phase angle, simply label the angles 0°, 90°, 180°, 270°, and 360° on the sine wave, and extrapolate to the vertical axis. Two examples of this are illustrated below.

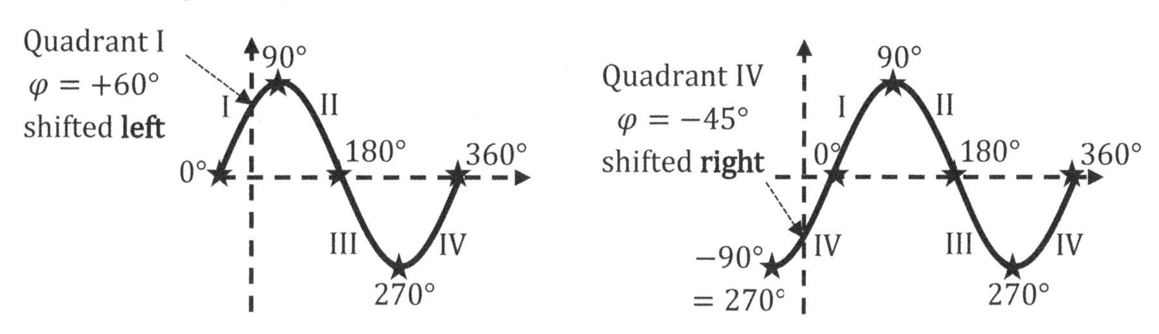

Consider the two graphs shown above.

- In the graph on the left, the vertical axis lies between 0° and 90°. Therefore, the phase angle lies in Quadrant I. The phase angle is **positive** because the graph is shifted to the **left**. In this example, $\varphi = +60°$.
- In the graph on the right, the vertical axis lies between −90° and 0°, which is equivalent to lying between 270° and 360°. (Remember, you can add or subtract 360° to any angle to make an equivalent angle.) Therefore, the phase angle lies in Quadrant IV. The phase angle is **negative** because the graph is shifted to the **right**. In this example, $\varphi = -45°$. **Note**: After labeling 0°, 90°, 180°, 270°, and 360° on the sine wave, we also labeled −90° (which is equivalent to 270°) because the vertical axis is to the left of 0°.

When the graph includes numbers, you can get a more precise answer for the phase angle by reading off the values of the amplitude (A), initial position (x_0), and equilibrium position (x_e), and then using the following equation (which comes from setting $t = 0$ in the equation for the sine wave).

$$\varphi = \sin^{-1}\left(\frac{x_0 - x_e}{A}\right)$$

When taking the inverse sign, be sure to put the answer in the correct Quadrant. This method is illustrated in one of the examples later in this chapter.

Phase Shift

Another way of finding the phase angle is to first read off the **phase shift** (δt) and the **period** (T) in seconds. The phase shift (δt) is the value of the time corresponding to the standard 0° position (see the graphs on the previous pages, and also the graphs below). Note that the phase shift may be negative:

- If you measure a δt for a shift to the **right**, δt is **negative.**
- If you measure a δt for a shift to the **left**, δt is **positive.**

Then use the following ratio to solve for the **phase angle** (φ).

$$\frac{\varphi}{2\pi} = \frac{\delta t}{T}$$

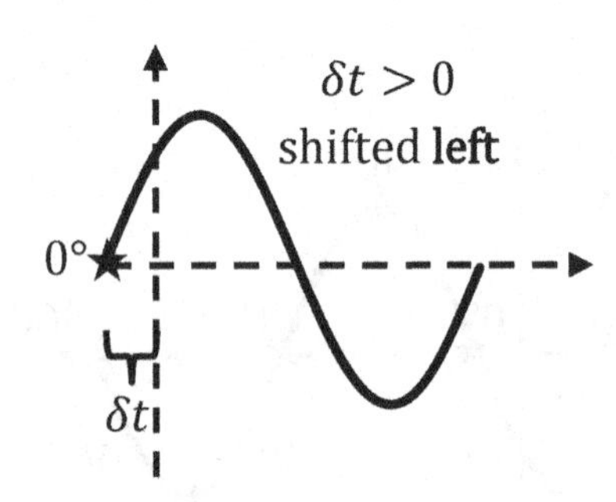

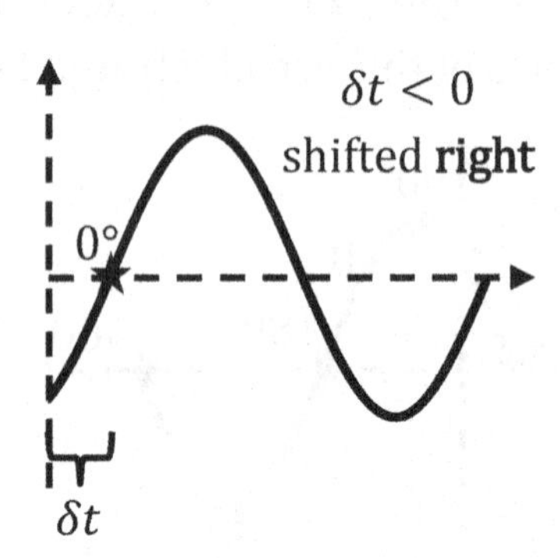

Conversion from Degrees to Radians

In the equation for a sine wave, the **phase angle** (φ) must be in **radians**. The reason is that the angular frequency (ω_0) is in rad/s in the expression $\sin(\omega_0 t + \varphi)$, and the phase angle's units must match the units of $\omega_0 t$ (which has the units of radians, since time is measured in seconds). Use the following conversion factor to convert from degrees to radians.

$$\frac{\pi \text{ rad}}{180°}$$

For example, 120° works out to $\frac{2\pi}{3}$ rad:

$$120° = 120° \frac{\pi \text{ rad}}{180°} = \frac{120}{180}\pi \text{ rad} = \frac{2\pi}{3} \text{ rad}$$

Also, check that your calculator is in **radians** mode (instead of degrees mode) before using the sine function.

Solving for Time

When solving for time in the equation for a sine wave, first isolate the sine function via algebra, and then take an inverse sine of both sides of the equation. There may be multiple solutions, as illustrated in one of the examples.

Important Distinctions

In physics, period and wavelength are two different quantities:

- **Period** is the horizontal **time** between two crests. You determine the period from a graph that has time (t) on the horizontal axis.
- **Wavelength** is the horizontal **distance** between two crests. You determine the wavelength from a graph that has position (x) on the horizontal axis.

It's also important to distinguish between **time** (t) and **period** (T). The period is a specific time: The period is the time it takes to complete exactly one oscillation.

There are two types of frequencies. The frequency (f) is the number of cycles per second, whereas the angular frequency (ω_0) is the number of radians per second.

If you've heard the phrase "peak to peak," you need to interpret this carefully:

- **Period** is a **horizontal** measure between two crests. Period is horizontal because it is a time and time is on the horizontal axis.
- **Amplitude** is a **vertical** measure from a crest to equilibrium. Note that amplitude is **not** measured peak to peak: It's instead measured from peak to equilibrium. If you measure the vertical distance from peak to peak (that is, from crest to trough), you get **twice** the amplitude. Amplitude is a distance because position is on the vertical axis.

Note that x_{max} is **not** the same thing (in general) as the **amplitude**. The amplitude (A) is related to the maximum position (x_{max}) by the equilibrium position (x_e): $A = x_{max} - x_e$. The amplitude only equals x_{max} in the special case that x_e happens to be zero.

Notes Regarding Units

The SI unit of frequency (f) is the Hertz (Hz). Since frequency is the reciprocal of the period, $f = \frac{1}{T}$, one Hertz is equivalent to an inverse second: $\text{Hz} = \frac{1}{\text{s}}$.

The amplitude (A), equilibrium position (x_e), initial position (x_0), maximum value (x_{max}), minimum value (x_{min}), and instantaneous position (x) are only measured in meters (m) when **position** is graphed as a function of time. There are a few cases where these quantities have different units. For example, for a sound wave, we would instead plot pressure as a function of time, for which these quantities would be measured in Pascals (Pa) instead of meters (m), though in that case, we would call the symbols A, P_e, P_{max}, and P_{min} (instead of A, x_e, x_{max}, and x_{min}).

Sine Wave Strategy

How you solve a problem involving a sine wave depends on which kind of problem it is:

- If a problem gives you a **graph** of a sine wave, read the following values directly from the graph.
  - x_{max} is the vertical coordinate of a **crest** (at the top of a peak).
  - x_{min} is the vertical coordinate of a **trough** (at the bottom of a valley).
  - t_{1m} is the horizontal coordinate of the first crest.
  - t_{2m} is the horizontal coordinate of the second crest.
  - x_0 is the **initial position**. It's the value of the sine wave when $t = 0$.

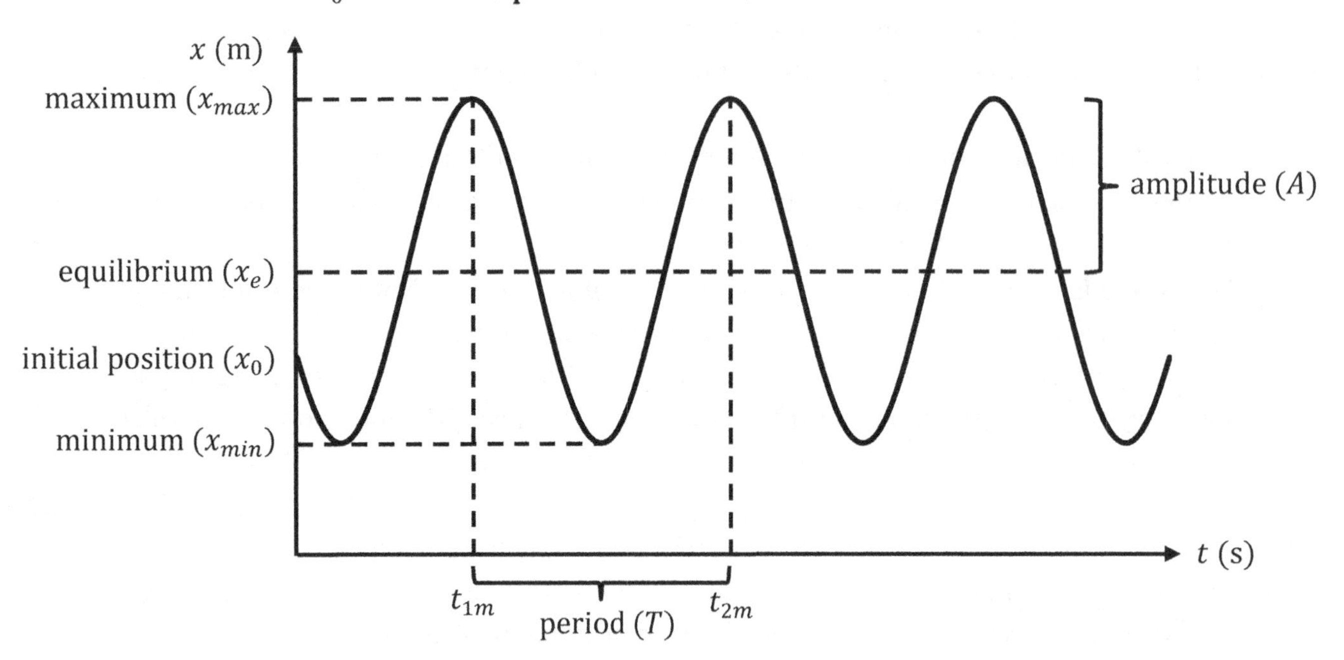

Once you read off the above values, you can calculate the following values.

- The **equilibrium position** is $x_e = \frac{x_{max}+x_{min}}{2}$.
- The **amplitude** is $A = x_{max} - x_e$, which is the same as $A = \frac{x_{max}-x_{min}}{2}$.
- The **period** is $T = t_{2m} - t_{1m}$.
- The **frequency** is $f = \frac{1}{T}$.
- The **angular frequency** is $\omega_0 = 2\pi f$.
- The **phase angle** is $\varphi = \sin^{-1}\left(\frac{x_0 - x_e}{A}\right)$. Be sure to put the phase angle in the correct Quadrant using the technique described on pages 10-11.

Note: If a problem gives you a graph of **velocity** or **acceleration**, or if a problem mentions a **spring** or **pendulum**, also see Chapters 2-4.

- If a problem gives you an **equation** for the position of a sine wave with numbers, compare the equation you are given to the general formula below.[1]

$$x = A\sin(\omega_0 t + \varphi) + x_e$$

You can determine the following quantities directly from the comparison.

  - The **amplitude** (A) is the coefficient of the sine function.
  - The **angular frequency** (ω_0) is the coefficient of time (t).
  - The **phase angle** (φ) in **radians** is added to (or subtracted from, if it is negative) the term $\omega_0 t$ in the argument of the sine function.
  - The **equilibrium position** (x_e) is added to (or subtracted from, if it is negative) the term $A\sin(\omega_0 t + \varphi)$.

Once you read off the above values, you can calculate the following values.

  - The **frequency** is $f = \frac{\omega_0}{2\pi}$.
  - The **period** is $T = \frac{1}{f}$.
  - To find the value of **position** at a given time (t), plug the time into the equation and calculate x. Be sure that your calculator is in **radians** mode.
  - To solve for **time** given x, plug the specified value of x into the equation and solve for t. First isolate the sine function with algebra. Next take an inverse sine of both sides of the equation. In general, there are multiple answers to the inverse sine. For example, $\sin^{-1}\left(\frac{1}{2}\right)$ equals both $\frac{\pi}{6}$ radians and $\frac{5\pi}{6}$ radians (corresponding to 30° and 150°). Be sure to put the inverse sine in **radians** (<u>not</u> degrees). For each case, first subtract φ (in radians) and then divide by ω_0. If your time (t) is negative, add a period (T) to make it positive. In general, you may add n periods to any time and obtain an equivalent answer. This technique is illustrated in an example.
  - The maximum position of the sine wave is $x_{max} = x_e + A$.
  - The minimum position of the sine wave is $x_{min} = x_e - A$.

Note: If a problem mentions **velocity** or **acceleration**, or if a problem mentions a **spring** or **pendulum**, also see Chapters 2-4.

---

[1] Some books use a **cosine** function instead of a **sine** function, especially in the context of a simple pendulum. This simply shifts the phase angle by ninety degrees. It also changes the way that you go about finding the phase angle, as we will see in Chapter 2 when we label the standard angles for a velocity graph (though if you use a cosine function for position, then the velocity function has a negative sine function). Chapter 4 will show you how our equations with a sine wave for position can be rewritten with a cosine wave for position specifically for the case of the simple pendulum (the advantage is that the phase angle equals zero for a system that is released from rest when you work with a cosine wave for position).

Example: The motion of a banana is plotted below.

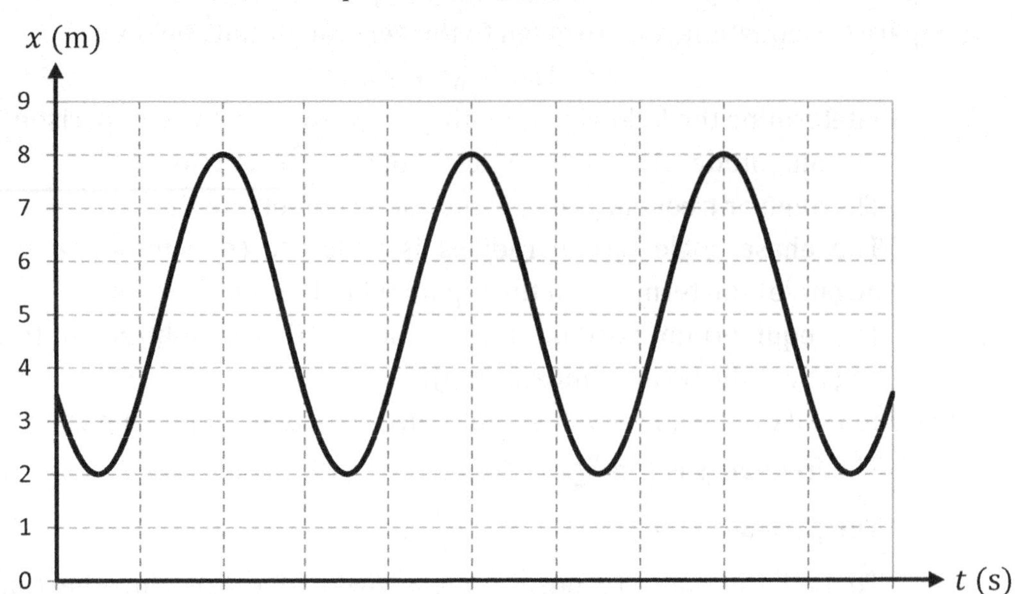

(A) What is the period of oscillation?
The period is the time between two crests. Read off the times for the first two crests.

- The first crest (maximum) occurs at $t_{1m} = 2.0$ s.
- The second crest (maximum) occurs at $t_{2m} = 5.0$ s.

Subtract these two times in order to determine the period.

$$T = t_{2m} - t_{1m} = 5 - 2 = 3.0 \text{ s}$$

The period is $T = 3.0$ s.

Note: In this book, in order to avoid clutter and possible confusion, **during calculations** we will omit units and .0's (for example, we wrote 5 – 2 instead of 5.0 s – 2.0 s) until we reach a final answer. We will include units and significant figures with our **final answers.**

(B) What is the frequency?
Frequency is the reciprocal of the period.

$$f = \frac{1}{T} = \frac{1}{3} \text{ Hz}$$

The frequency is $f = \frac{1}{3}$ Hz. In decimal form, $f = 0.33$ Hz to 2 significant figures.

(C) What is the angular frequency?
Multiply the frequency by 2π radians.

$$\omega_0 = 2\pi f = 2\pi\left(\frac{1}{3}\right) = \frac{2\pi}{3} \text{ rad/s}$$

The angular frequency is $\omega_0 = \frac{2\pi}{3}$ rad/s. If you use a calculator, $\omega_0 = 2.1$ rad/s.

(D) Where is the equilibrium position?
Read off the maximum and minimum positions.

- The maximum position (at a crest) is $x_{max} = 8.0$ m.
- The minimum position (at a trough) is $x_{min} = 2.0$ m.

The equilibrium position is the average of these two values.

$$x_e = \frac{x_{max} + x_{min}}{2} = \frac{8+2}{2} = \frac{10}{2} = 5.0 \text{ m}$$

The equilibrium position is $x_e = 5.0$ m. The sine wave oscillates about this horizontal line.

Note: The graph has x on the vertical axis and t on the horizontal axis. In math class, when a graph has x and y, you instead find x on the horizontal axis. It's different here.

(E) What is the amplitude of oscillation?
Subtract the equilibrium position from the maximum position.

$$A = x_{max} - x_e = 8 - 5 = 3.0 \text{ m}$$

The amplitude is $A = 3.0$ m. Note that you could get the same answer using the equations $A = x_e - x_{min}$ or $A = \frac{x_{max} - x_{min}}{2}$.

(F) What is the phase angle?
Determine the **initial position** (x_0) from the graph. This is the vertical intercept, which is the value of x when $t = 0$. Looking at the graph, the initial position is approximately $x_0 = 3.5$ m. Now apply the equation for phase angle.

$$\varphi = \sin^{-1}\left(\frac{x_0 - x_e}{A}\right) = \sin^{-1}\left(\frac{3.5 - 5}{3}\right) = \sin^{-1}\left(-\frac{1.5}{3}\right) = \sin^{-1}\left(-\frac{1}{2}\right)$$

The reference angle is $\frac{\pi}{6}$ rad (corresponding to 30°), but this is **not** the answer for the phase angle because the sine wave doesn't start in Quadrant I. Label the angles 0°, 90°, 180°, 270°, and 360° on the sine wave, and extrapolate to the vertical axis, as discussed on pages 10-11. See the diagram below.

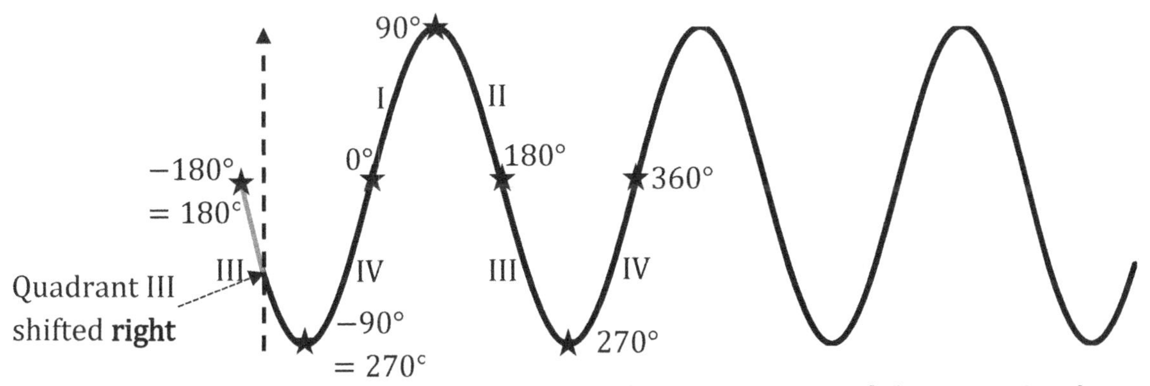

This sine wave begins in Quadrant III, between 180° and 270°. In Quadrant III, add the reference angle to 180° in order to find the phase angle: $\varphi = 180° + 30° = 210°$. Multiply by $\frac{\pi \text{ rad}}{180°}$ to convert to **radians**: $\varphi = 210° \times \frac{\pi \text{ rad}}{180°} = \frac{7\pi}{6}$ rad. The phase angle is $\varphi = \frac{7\pi}{6}$ rad.

There are two ways to think of the phase angle. Note that $\varphi = 210° = \frac{7\pi}{6}$ rad is equivalent to $\varphi = -150° = -\frac{5\pi}{6}$ rad, since we can add or subtract 360° to any angle and obtain an equivalent angle. We can interpret the phase angle for this graph two equivalent ways:

- We can think of the sine wave as being shifted 150° to the **right**, since a **negative** phase angle ($\varphi = -150°$) shifts the graph to the right.
- We can think of the sine wave as being shifted 210° to the **left**, since a **positive** phase angle ($\varphi = 210°$) shifts the graph to the left.

Both answers, $\varphi = 210° = \frac{7\pi}{6}$ rad and $\varphi = -150° = -\frac{5\pi}{6}$ rad, are correct. (However, note that $+150°$ or $-210°$ would be incorrect for this graph.)

There is an alternative way to determine the phase angle for this graph. We could first determine the **phase shift** (δt) by reading off the time corresponding to 0° (the point where a standard sine wave ordinarily begins). If you find this point on the graph, you will see that the phase shift is approximately $\delta t = -1.25$ s. The phase shift is **negative** because the graph is shifted to the **right** (see page 12). We know that it's shifted to the right because we marked 0° to the right of the origin (see page 17). Use the following ratio.

$$\frac{\varphi}{2\pi} = \frac{\delta t}{T}$$

Recall from part (A) that the period is $T = 3.0$ s. Plug numbers into the previous equation.

$$\frac{\varphi}{2\pi} = -\frac{1.25}{3}$$

Multiply both sides of the equation by 2π.

$$\varphi = -\left(\frac{1.25}{3}\right)(2\pi) = -\frac{2.5\pi}{3}$$

Multiply both the numerator and denominator by 2 in order to remove the decimal from the numerator (since it isn't good form to mix fractions and decimals together).

$$\varphi = -\frac{(2.5\pi)(2)}{(3)(2)} = -\frac{5\pi}{6} \text{ rad}$$

Recall that we may add (or subtract) 360° (which is the same as 2π radians) to any angle to obtain an equivalent angle. Since we obtained a negative phase angle, we could get an equivalent positive answer by adding 2π radians.

$$\varphi = -\frac{5\pi}{6} = -\frac{5\pi}{6} + 2\pi = -\frac{5\pi}{6} + \frac{12\pi}{6} = \frac{-5\pi + 12\pi}{6} = \frac{7\pi}{6} \text{ rad}$$

This is the same answer that we obtained on the previous page using a different method. Note that $\varphi = -\frac{5\pi}{6}$ rad and $\varphi = \frac{7\pi}{6}$ rad are **both correct** answers to this question, but that $+\frac{5\pi}{6}$ rad would be **incorrect**.

Example: The position of a banana is given by the following equation, where SI units have been suppressed in order to avoid clutter.

$$x = 7\sin\left(\frac{\pi t}{2} - \frac{\pi}{4}\right) - 1$$

(A) What is the amplitude of oscillation?

Compare the given equation, $x = 7\sin\left(\frac{\pi t}{2} - \frac{\pi}{4}\right) - 1$, to the general equation for a sine wave, $x = A\sin(\omega_0 t + \varphi) + x_e$, in order to see that the amplitude is $A = 7.0$ m.

(B) What is the equilibrium value?

Compare the given equation, $x = 7\sin\left(\frac{\pi t}{2} - \frac{\pi}{4}\right) - 1$, to the general equation for a sine wave, $x = A\sin(\omega_0 t + \varphi) + x_e$, in order to see that the equilibrium value is $x_e = -1.0$ m. Note that the equilibrium value is **negative** in this example.

(C) What is the maximum value of the sine wave?

Add the amplitude to the equilibrium position in order to find the maximum value.

$$x_{max} = x_e + A = -1 + 7 = 6.0 \text{ m}$$

The maximum value of the sine wave is $x_{max} = 4.0$ m.

(D) What is the minimum value of the sine wave?

Subtract the amplitude from the equilibrium position in order to find the minimum value.

$$x_{min} = x_e - A = -1 - 7 = -8.0 \text{ m}$$

The minimum value of the sine wave is $x_{min} = -8.0$ m.

(E) What is the phase angle?

Compare the given equation, $x = 7\sin\left(\frac{\pi t}{2} - \frac{\pi}{4}\right) - 1$, to the general equation for a sine wave, $x = A\sin(\omega_0 t + \varphi) + x_e$, in order to see that the phase angle is $\varphi = -\frac{\pi}{4}$ rad (which equates to $\varphi = -45°$ since π rad $= 180°$). Note that the phase angle is **negative** in this example.

(F) What is the angular frequency?

Compare the given equation, $x = 7\sin\left(\frac{\pi t}{2} - \frac{\pi}{4}\right) - 1$, to the general equation for a sine wave, $x = A\sin(\omega_0 t + \varphi) + x_e$, in order to see that the angular frequency is $\omega_0 = \frac{\pi}{2}$ rad/s.

(G) What is the period of oscillation?

Solve for period in the equation $\omega_0 = \frac{2\pi}{T}$. Multiply by T and divide by ω_0.

$$T = \frac{2\pi}{\omega_0} = \frac{2\pi}{\pi/2} = 2\pi \div \frac{\pi}{2} = 2\pi \times \frac{2}{\pi} = 4.0 \text{ s}$$

The period is $T = 4.0$ s. (To divide by a fraction, multiply by its **reciprocal**.)

(H) Where is the banana at $t = 6.0$ s?
Plug the specified time into the equation for the sine wave.

$$x = 7\sin\left(\frac{\pi t}{2} - \frac{\pi}{4}\right) - 1 = 7\sin\left(\frac{6\pi}{2} - \frac{\pi}{4}\right) - 1$$

Subtract fractions with a **common denominator**. Note that $\frac{6\pi}{2} = \frac{12\pi}{4}$.

$$x = 7\sin\left(\frac{12\pi}{4} - \frac{\pi}{4}\right) - 1 = 7\sin\left(\frac{12\pi - \pi}{4}\right) - 1 = 7\sin\left(\frac{11\pi}{4}\right) - 1$$

If you enter $\sin\left(\frac{11\pi}{4}\right)$ into your calculator and if your calculator is in **radians** mode, you will find that $\sin\left(\frac{11\pi}{4}\right) = \frac{\sqrt{2}}{2}$. Alternatively, you can convert $\frac{11\pi}{4}$ to degrees by multiplying by $\frac{180}{\pi}$ to find that $\frac{11\pi}{4}$ radians equates to 495°. Furthermore, 495° is equivalent to 135° because you can add or subtract 360° to any angle to obtain an equivalent angle. Recall from trig that $\sin 135° = \frac{\sqrt{2}}{2}$ (the reference angle is 45° and sine is positive in Quadrant II).

$$x = 7\sin\left(\frac{11\pi}{4}\right) - 1 = 7\frac{\sqrt{2}}{2} - 1 = \frac{7\sqrt{2}}{2} - \frac{2}{2} = \frac{7\sqrt{2} - 2}{2} \text{ m}$$

The banana is at $x = \frac{7\sqrt{2}-2}{2}$ m when $t = 6.0$ s. Using a calculator, $x = 3.9$ m.

(I) When is the banana at $x = 2.5$ m?
Plug the specified position into the equation for the sine wave.

$$2.5 = 7\sin\left(\frac{\pi t}{2} - \frac{\pi}{4}\right) - 1$$

Add 1 to both sides of the equation.

$$3.5 = 7\sin\left(\frac{\pi t}{2} - \frac{\pi}{4}\right)$$

Divide both sides of the equation by 7.

$$\frac{1}{2} = \sin\left(\frac{\pi t}{2} - \frac{\pi}{4}\right)$$

Take the inverse sine of both sides of the equation.

$$\sin^{-1}\left(\frac{1}{2}\right) = \frac{\pi t}{2} - \frac{\pi}{4}$$

Here is where you can run into trouble using a calculator: There are multiple answers to $\sin^{-1}\left(\frac{1}{2}\right)$. The reference angle is $\frac{\pi}{6}$ rad (corresponding to 30°) because $\sin\frac{\pi}{6} = \frac{1}{2}$. However, sine is positive in Quadrants I and II, so $\frac{5\pi}{6}$ rad (corresponding to 150°) is also a solution. We must plug both angles into the above equation and work out the algebra for each case.

$$\frac{\pi}{6} = \frac{\pi t}{2} - \frac{\pi}{4} \quad , \quad \frac{5\pi}{6} = \frac{\pi t}{2} - \frac{\pi}{4}$$

Add $\frac{\pi}{4}$ to both sides of each equation.

$$\frac{\pi}{6} + \frac{\pi}{4} = \frac{\pi t}{2} \quad , \quad \frac{5\pi}{6} + \frac{\pi}{4} = \frac{\pi t}{2}$$

Add fractions with a **common denominator**.

$$\frac{2\pi}{12} + \frac{3\pi}{12} = \frac{\pi t}{2} \quad , \quad \frac{10\pi}{12} + \frac{3\pi}{12} = \frac{\pi t}{2}$$

$$\frac{5\pi}{12} = \frac{\pi t}{2} \quad , \quad \frac{13\pi}{12} = \frac{\pi t}{2}$$

Multiply both sides of each equation by 2 and divide both sides by π. The π's cancel out.

$$t = \frac{5}{6} \text{ s} \quad \text{or} \quad t = \frac{13}{6} \text{ s}$$

Wait. There are even more solutions! That's because the sine wave repeats itself every period. Recall from part (G) that the period is $T = 4.0$ s. This means that we can add $4n$ seconds (where n represents a nonnegative integer) to each solution above and still obtain a correct answer. Therefore, a better way to express our final answer is:

$$t = \frac{5}{6} \text{ s} + 4n \text{ s} \quad \text{or} \quad t = \frac{13}{6} \text{ s} + 4n \text{ s}$$

The banana is at $x = 2.5$ m when $t = \frac{5}{6}$ s $+ 4n$ s or when $t = \frac{13}{6}$ s $+ 4n$ s, where n is a nonnegative integer (0, 1, 2, 3...). This means that the answers are $t = \frac{5}{6}$ s, $t = \frac{13}{6}$ s, $t = \frac{29}{6}$ s, $t = \frac{37}{6}$ s, $t = \frac{53}{6}$ s, $t = \frac{61}{6}$ s, $t = \frac{77}{6}$ s, $t = \frac{85}{6}$ s, and so on. This is because the sine wave repeats itself every $T = 4.0$ s. We're adding multiples of 24 to the numerators of $\frac{5}{6}$ s and $\frac{13}{6}$ s because we're adding fractions. The way to add fractions is to find a **common denominator**, so we're rewriting the period (which equals 4 s) as $\frac{24}{6}$ s.

1. The motion of a banana is plotted below.

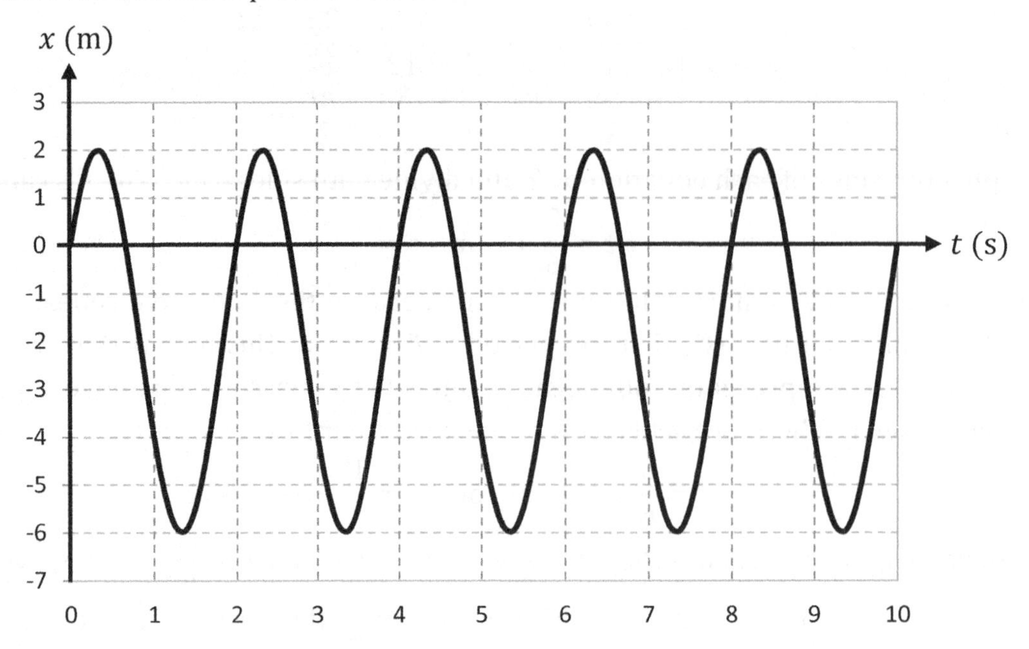

(A) What is the amplitude of oscillation?

(B) What is the period of oscillation?

(C) Where is the equilibrium position?

(D) What is the frequency?

(E) Where is the banana at $t = 5.0$ s?

(F) What is the phase angle?

Want help? Check the hints section at the back of the book.

Answers: (A) 4.0 m (B) 2.0 s (C) -2.0 m (D) 0.50 Hz (E) -4.0 m (F) $\frac{\pi}{6}$ rad

2. The position of a banana is given by the following equation, where SI units have been suppressed in order to avoid clutter.

$$x = 3\sin\left(\pi t - \frac{5\pi}{6}\right) + 2$$

(A) What is the amplitude of oscillation?

(B) Where is the equilibrium position?

(C) What is the angular frequency?

(D) What is the minimum value of x?

(E) What is the period of oscillation?

(F) What is the phase angle in degrees?

(G) Where is the banana at $t = \frac{5}{3}$ s?

(H) When is the banana at $x = 0.5$ m?

Want help? Check the hints section at the back of the book.

Answers: (A) 3.0 m (B) 2.0 m (C) π rad/s (D) -1.0 m (E) 2.0 s (F) $-150°$ (G) 3.5 m (H) $2n$ s, $\frac{2}{3}$ s $+ 2n$ s

3. The motion of a banana is plotted below.

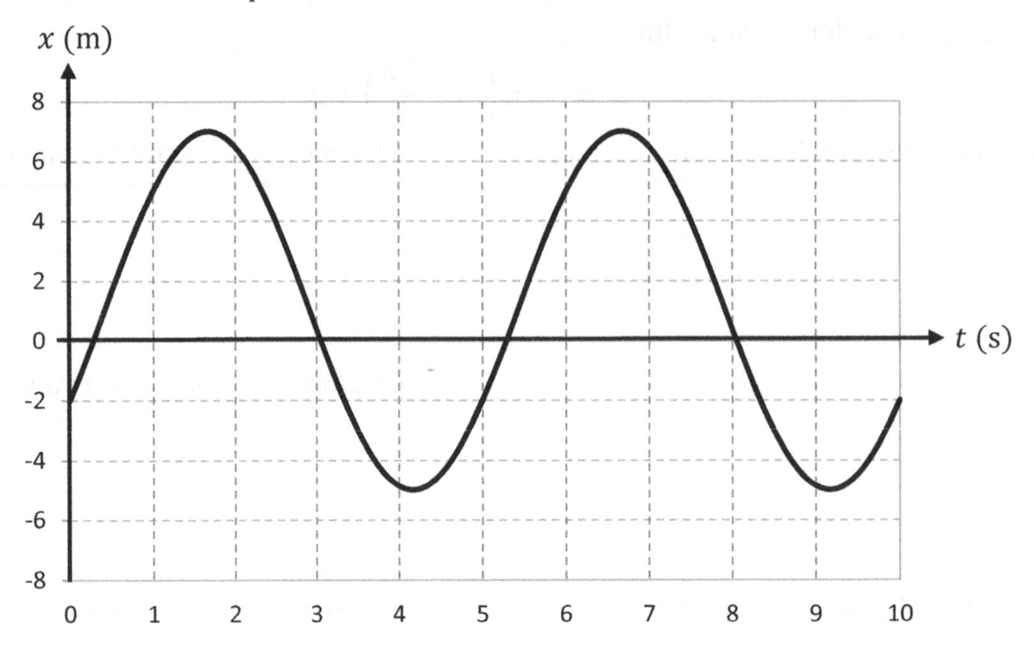

Write down the equation for this sine wave.

Want help? Check the hints section at the back of the book.

Answer: $x = 6\sin\left(\frac{2\pi t}{5} + \frac{11\pi}{6}\right) + 1$

2 SIMPLE HARMONIC MOTION

Relevant Terminology

Velocity – the instantaneous rate at which position changes.
Acceleration – the instantaneous rate at which velocity changes.
Crest – a point on a wave where there is a **maximum** (the top of a hill).
Trough – a point on a wave where there is a **minimum** (the bottom of a valley).
Equilibrium – a horizontal line about which a sine wave oscillates.
Amplitude – the **vertical** height between a crest and the equilibrium position (or **one-half** of the vertical height between a crest and trough).
Period – the **horizontal time** between two consecutive crests. Period is the time it takes for the wave to complete exactly one oscillation.
Frequency – the number of oscillations completed per second.
Angular frequency – the number of radians completed per second. The angular frequency equals 2π times the frequency.
Phase angle – an angle representing how much the wave is shifted horizontally. A **positive** phase angle represents a wave that is shifted to the **left**.

Simple Harmonic Motion

A system oscillates with simple harmonic motion when its **acceleration is proportional to the negative of its displacement from equilibrium**. For one-dimensional motion, we can express this with the following equation.

$$a_x = -\omega_0^2 \Delta x$$

The symbol $\Delta x = x - x_e$ represents the displacement of the system from equilibrium and ω_0 is the angular frequency.

Many oscillating systems undergo simple harmonic motion:

- A simple oscillating spring experiences simple harmonic motion (Chapter 3).
- A simple pendulum approximately undergoes simple harmonic motion (Chapter 4).
- The averaged behavior of the electrons in a wire in a typical AC circuit follows simple harmonic motion.
- Particles may vibrate with simple harmonic motion.
- If you could create a tunnel through the center of the earth, an object dropped through the tunnel would experience simple harmonic motion.

If you want to know whether or not a system experiences simple harmonic motion, apply Newton's second law ($\sum F_x = ma_x$) in order to determine whether or not the acceleration is proportional to $-\Delta x$.

Simple Harmonic Motion Equations

For a system oscillating with simple harmonic motion, the position (x), velocity (v_x), and acceleration (a_x) are given by the following equations.

$$x = A\sin(\omega_0 t + \varphi) + x_e$$
$$v_x = A\omega_0 \cos(\omega_0 t + \varphi)$$
$$a_x = -A\omega_0^2 \sin(\omega_0 t + \varphi)$$

Here, A is the **amplitude** of the position wave (but it's **not** the amplitude of the velocity or acceleration waves), ω_0 is the **angular frequency**, t represents time, φ is the **phase angle**, and x_e is the **equilibrium position** for the position wave. The amplitude of the position wave is related to the peak and equilibrium values by:

$$A = x_{max} - x_e = \frac{x_{max} - x_{min}}{2}$$

The **angular frequency** (ω_0) equals 2π times the **frequency** (f).

$$\omega_0 = 2\pi f$$

Frequency (f) and **period** (T) share a reciprocal relationship.

$$f = \frac{1}{T} \quad , \quad T = \frac{1}{f}$$

For a graph of a sine wave, one way to determine the **phase angle** (φ) is to set t equal to zero and algebraically solve for φ, following the method described in Chapter 1. Note that the equation for φ is slightly different for a velocity or acceleration graph than it is for a position graph.

$$\varphi = \sin^{-1}\left(\frac{x_0 - x_e}{A}\right) \quad , \quad \varphi = \cos^{-1}\left(\frac{v_{x0}}{A\omega_0}\right) \quad , \quad \varphi = \sin^{-1}\left(-\frac{a_{x0}}{A\omega_0^2}\right)$$

The net force ($\sum F_x$) can be found from Newton's second law: Multiply the mass (m) times the acceleration (a_x).

$$\sum F_x = ma_x = -mA\omega_0^2 \sin(\omega_0 t + \varphi)$$

A system that undergoes simple harmonic motion obeys the following equation. The acceleration of the system is proportional to the negative of the displacement ($\Delta x = x - x_e$) from equilibrium.

$$a_x = -\omega_0^2 \Delta x$$

We can verify that the equations for x and a_x listed at the top of the page satisfy the previous equation by plugging them in. First rewrite Δx as $(x - x_e)$.

$$a_x = -\omega_0^2 (x - x_e)$$

Now replace a_x by $-A\omega_0^2 \sin(\omega_0 t + \varphi)$ and $(x - x_e)$ by $A\sin(\omega_0 t + \varphi)$.

$$-A\omega_0^2 \sin(\omega_0 t + \varphi) = -\omega_0^2 A\sin(\omega_0 t + \varphi)$$

We see that both sides are equal, which shows that the equations for x and a_x indeed satisfy the equation for simple harmonic motion ($a_x = -\omega_0^2 \Delta x$).

Symbols and SI Units

Symbol	Name	SI Units
x_{max}	the maximum value of x (at a crest)	m
x_{min}	the minimum value of x (at a trough)	m
x_e	the equilibrium position	m
x_0	the initial position	m
Δx	displacement from equilibrium	m
x	the instantaneous position	m
v_x	the instantaneous velocity	m/s
v_{x0}	the initial velocity	m/s
v_m	the maximum speed	m/s
a_x	the instantaneous acceleration	$\mathrm{m/s^2}$
a_{x0}	the initial acceleration	$\mathrm{m/s^2}$
a_m	the maximum acceleration	$\mathrm{m/s^2}$
A	the amplitude of the position function	m
t	time	s
T	period	s
φ	phase angle	rad
f	frequency	Hz
ω_0	angular frequency	rad/s
$\sum F_x$	net force	N
m	mass	kg

Simple Harmonic Motion Strategy

How you solve a problem involving simple harmonic motion depends on which kind of problem it is:

- If a problem gives you a **graph** of a sine wave, follow the strategy for interpreting the graph of a sine wave from Chapter 1. If the graph or question involves velocity or acceleration, also note the following.
    - If **velocity** (v_x) is on the vertical axis, the **amplitude** of the cosine wave is $v_m = A\omega_0$. If **acceleration** (a_x) is on the vertical axis, and the **amplitude** of the sine wave is $a_m = A\omega_0^2$. The **amplitude** only equals A when **position** (x) is on the vertical axis.
    - For graphs of **velocity** (v_x) and **acceleration** (a_x), the **equilibrium** value is zero: $v_e = 0$ and $a_e = 0$. There is only an **equilibrium** value (x_e) to find for a graph of **position** (x).
    - To solve for the **phase angle**, follow the techniques described in Chapter 1. If you apply the method described on pages 10-11, note that the equation is different for velocity or acceleration than it is for position. In any case, set time (t) equal to zero in the equation for the sine wave and solve for the phase angle (φ). When you do this, you will get one of the equations below (depending on the type of graph).

$$\varphi = \sin^{-1}\left(\frac{x_0 - x_e}{A}\right) \quad , \quad \varphi = \cos^{-1}\left(\frac{v_{x0}}{A\omega_0}\right) \quad , \quad \varphi = \sin^{-1}\left(-\frac{a_{x0}}{A\omega_0^2}\right)$$

    Be sure to put the phase angle in the correct Quadrant using the technique described on pages 10-11, but note that it's a little different for **velocity** and **acceleration** than it is for position (see the examples).
    - Of the three quantities x, v_x, and a_x, if you have a graph of one of these quantities, but a question instead asks you to find one of the other two quantities, there are two ways to solve the problem. One way is to use the three motion equations listed below. (Do **not** use the equations of uniform acceleration, since the acceleration is **not** uniform. Use one of the equations listed below instead.)

$$x = A\sin(\omega_0 t + \varphi) + x_e$$
$$v_x = A\omega_0 \cos(\omega_0 t + \varphi)$$
$$a_x = -A\omega_0^2 \sin(\omega_0 t + \varphi)$$

    Alternatively, you can find the slope of the tangent line or the area under the curve, depending upon which graph you are given and which quantity you are trying to find. This method was taught in Volume 1, Chapter 6, and it applies to graphs of sine waves, too.

Note: If a problem mentions a **spring** or **pendulum**, also see Chapters 3-4.

- If a problem gives you an **equation** for the position (x), velocity (v_x), or acceleration (a_x) of a sine wave with numbers, follow the strategy for interpreting the equation for a sine wave from Chapter 1. If the equation or question involves velocity or acceleration, also note the following.
  - There are three different equations – one for **position** (x), one for **velocity** (v_x), and one for **acceleration** (a_x). In Chapter 1, we only considered the equation for position, but now you must choose the appropriate equation.
  $$x = A\sin(\omega_0 t + \varphi) + x_e$$
  $$v_x = A\omega_0 \cos(\omega_0 t + \varphi)$$
  $$a_x = -A\omega_0^2 \sin(\omega_0 t + \varphi)$$
  Do **not** use the equations of uniform acceleration, since the acceleration is **not** uniform. Use one of the equations listed above instead.
  - Note that the **amplitude** is different in each case: The amplitude of **position** (x) is A, the amplitude of **velocity** (v_x) is $v_m = A\omega_0$, and the amplitude of **acceleration** (a_x) is $a_m = A\omega_0^2$. In each case, the coefficient of the trig function is the amplitude, but the amplitude only equals A for position.
  - For **velocity** (v_x) and **acceleration** (a_x), the **equilibrium** value is zero: $v_e = 0$ and $a_e = 0$. There is only an **equilibrium** value (x_e) for **position** (x).
  - To find the value of **position** (x), **velocity** (v_x), or **acceleration** (a_x) at a given time (t), plug the time into the equation and calculate x, v_x, or a_x. Be sure that your calculator is in **radians** mode.
  - To solve for **time** given x, v_x, or a_x, plug the specified value of x, v_x, or a_x into the equation and solve for t following the strategy discussed in Chapter 1. Note that this will involve an inverse trig function, to which there are generally two answers. Also recall from Chapter 1 that there may be multiple solutions obtained by adding n periods to each time.
  - The **maximum velocity** equals $v_m = A\omega_0$ and the **maximum acceleration** equals $a_m = A\omega_0^2$, whereas the **maximum position** is $x_{max} = x_e + A$.
  - It may be necessary to combine two of the three equations (from the first bullet point above) together in order to solve for an unknown. This happens when you are given one of the three quantities x, v_x, or a_x, and are asked to find one of the other two. One way to do this is to first solve for time, which involves an inverse trig function. A simpler way to do this is to apply one of the two equations below.[1]
  $$a_x = -\omega_0^2(x - x_e) \quad , \quad (x - x_e)^2\omega_0^2 + v_x^2 = A^2\omega_0^2$$
  - It may help to recall the equations $f = \frac{\omega_0}{2\pi}$ and $T = \frac{1}{f}$ from Chapter 1.

Note: If a problem mentions a **spring** or **pendulum**, also see Chapters 3-4.

[1] The first equation is the condition for simple harmonic motion: $a_x = -\omega_0^2 \Delta x^2$. The second equation follows from the trig identity $\sin^2\theta + \cos^2\theta = 1$, or alternatively from the law of conservation of energy.

- If a problem gives you a specific system (like a spring or a pendulum) and asks you to derive an equation for the **period** (or frequency or angular frequency), apply Newton's second law ($\sum F_x = ma_x$) and compare the result with the equation $a_x = -\omega_0^2 \Delta x$. The comparison will give you an equation for the angular frequency (ω_0), from which you can find frequency ($f = \frac{\omega_0}{2\pi}$) or period ($T = \frac{2\pi}{\omega_0}$). This method is illustrated in one of the examples.

Important Distinctions

The words velocity and acceleration are **not** interchangeable: Acceleration describes the rate that velocity changes. In simple harmonic motion, the velocity is zero when the acceleration is maximum, and the acceleration is zero when the velocity is maximum:

- When passing through the **equilibrium** position, the **velocity** is greatest and the **acceleration** is momentarily zero (near equilibrium, velocity changes very little).
- At the **turning points** of the oscillation, the **velocity** is momentarily zero (since the system is changing direction) and the **acceleration** is greatest (velocity is changing the most at this position).

You can see this in the math: The **velocity** equation, $v_x = A\omega_0 \cos(\omega_0 t + \varphi)$, involves a **cosine** function, while the **acceleration** equation, $a_x = -A\omega_0^2 \sin(\omega_0 t + \varphi)$, involves a **sine** function. The cosine function is zero when sine is maximum, and the sine function is zero when the cosine function is maximum.

In general, there may be an **equilibrium** value (x_e) for the position graph and equation, but there are **no** equilibrium values for velocity or acceleration: $v_e = 0$ and $a_e = 0$. The reason has to do with calculus: Velocity is the derivative of position with respect to time, and acceleration is the derivative of velocity with respect to time. When you take a derivative of position with respect to time $\left(\frac{dx}{dt}\right)$, the constant term ($x_e$) disappears: $\frac{dx_e}{dt} = 0$.

The **amplitudes** of position (x), velocity (v_x), and acceleration (a_x) are all different:

- The amplitude of position is A.
- The amplitude of velocity is $v_m = A\omega_0$ (which is the maximum speed).
- The amplitude of acceleration is $a_m = A\omega_0^2$ (which is the maximum acceleration).

In each case, the amplitude is the coefficient of the trig function.

Do **not** use the equations of uniform acceleration, since the acceleration is **not** uniform. Use one of the three equations at the top of page 29 instead.

Example: The **position** of a banana is plotted below.

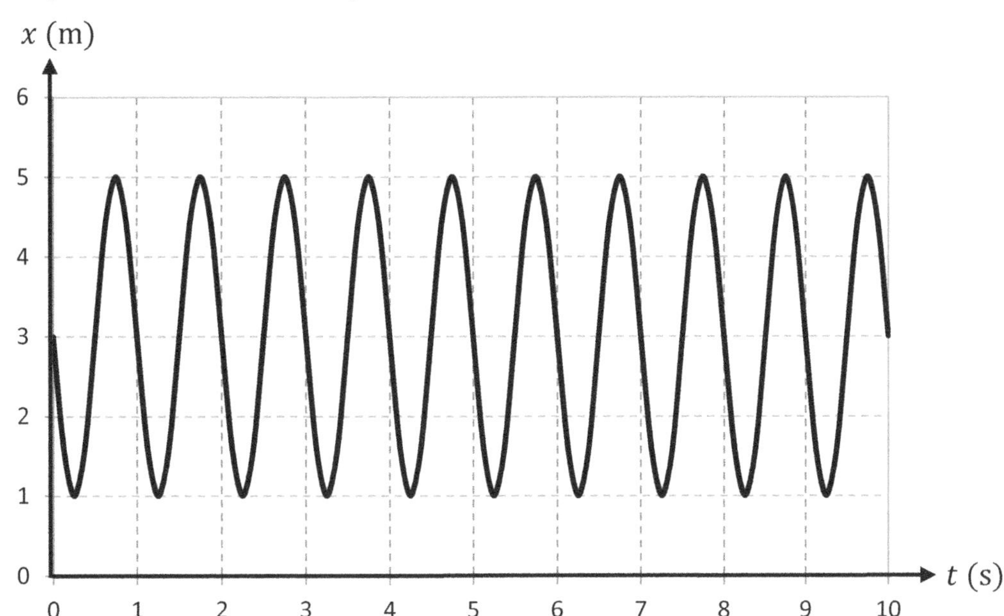

(A) What is the velocity of the banana at $t = 5.0$ s?

We can apply the equation $v_x = A\omega_0 \cos(\omega_0 t + \varphi)$, but first we need to determine the amplitude (A), angular frequency (ω_0), and phase angle (φ). Find these quantities the same way that we did in Chapter 1.

- To find the **amplitude** (A), subtract the minimum position from the maximum position and divide by two. Note that the minimum value is $x_{min} = 1.0$ m, while the maximum value is $x_{max} = 5.0$ m.
$$A = \frac{x_{max} - x_{min}}{2} = \frac{5-1}{2} = \frac{4}{2} = 2.0 \text{ m}$$
The amplitude is $A = 2.0$ m.
- To find the **angular frequency** (ω_0), first find the **period** (T). The period is the time between two crests. Read off the times for the first two crests. The first crest (maximum) occurs at $t_{1m} = 0.75$ s. The second crest (maximum) occurs at $t_{2m} = 1.75$ s. Subtract these two times in order to determine the period.[2]
$$T = t_{2m} - t_{1m} = 1.75 - 0.75 = 1.0 \text{ s}$$
The period is $T = 1.0$ s. To find the angular frequency, divide 2π by the period.
$$\omega_0 = \frac{2\pi}{T} = \frac{2\pi}{1} = 2\pi \text{ rad/s}$$
The angular frequency is $\omega_0 = 2\pi$ rad/s. If you use a calculator, $\omega_0 = 6.3$ rad/s.

[2] Note that there are exactly 10 complete cycles in 10 s. That makes it easy to determine that the period is $T = 1.0$ s. Also, note that it's easy to interpolate the time for this particular graph: Since the sine wave is divided into quarters at 90°, 180°, 270°, and 360°, that's how we know that $t_{1m} = 0.75$ s to two significant figures. In general, it would reduce interpolation error to read off the time for the first crest, the time for the last crest, subtract, and divide by the number of cycles between them: $T = \frac{t_{Lm} - t_{1m}}{N} = \frac{9.75 - 0.75}{9} = \frac{9}{9} = 1.0$ s. (Although there are 10 cycles total on this graph, there are only 9 cycles between the first and last crests.)

- To find the **phase angle** for a **position** graph, first determine the **initial position** (x_0) and **equilibrium position** (x_e) from the graph. The initial position is the vertical intercept, which is the value of x when $t = 0$. Looking at the graph, the initial position is approximately $x_0 = 3.0$ m. To find the equilibrium position, use the following equation.
$$x_e = \frac{x_{max} + x_{min}}{2} = \frac{5+1}{2} = \frac{6}{2} = 3.0 \text{ m}$$
The equilibrium position is $x_e = 3.0$ m. Now apply the equation for the phase angle of a **position** graph.
$$\varphi = \sin^{-1}\left(\frac{x_0 - x_e}{A}\right) = \sin^{-1}\left(\frac{3-3}{3}\right) = \sin^{-1}\left(\frac{0}{3}\right) = \sin^{-1}(0)$$
The reference angle is 0 rad (corresponding to 0°), but this is **not** the answer for the phase angle because the sine wave doesn't start in Quadrant I. Label the angles 0°, 90°, 180°, 270°, and 360° on the sine wave, and extrapolate to the vertical axis, as discussed on pages 10-11. See the diagram below.

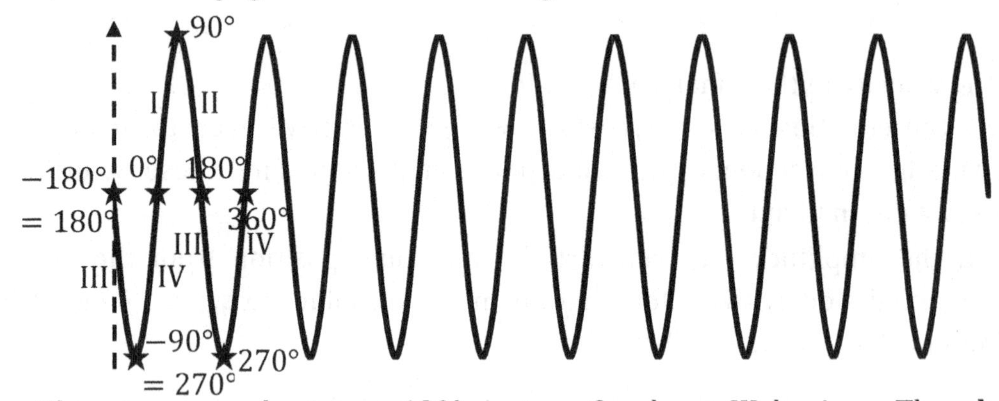

This sine wave begins at 180°, just as Quadrant III begins. The **phase angle** is $\varphi = \pi$ rad (corresponding to 180°).

Now that we know $A = 2.0$ m, $\omega_0 = 2\pi$ rad/s, and $\varphi = \pi$ rad, we can use the equation for the velocity at $t = 5.0$ s.
$$v_x = A\omega_0 \cos(\omega_0 t + \varphi) = (2)(2\pi)\cos[(2\pi)(5) + \pi]$$
$$v_x = 4\pi\cos(10\pi + \pi) = 4\pi\cos(11\pi)$$
Remember that you can add or subtract multiples of 2π radians to any angle and obtain an equivalent angle in radians. Thus, 11π rad is equivalent to π rad (which corresponds to 180°). Recall from trig that $\cos(\pi) = -1$.
$$v_x = 4\pi\cos(11\pi) = 4\pi\cos(11\pi - 10\pi) = 4\pi\cos(\pi) = 4\pi(-1) = -4\pi \text{ m/s}$$
The velocity of the banana at $t = 5.0$ s is $v_x = -4\pi$ m/s. If you use a calculator, this works out to $v_x = -13$ m/s to two significant figures. The significance of the **minus sign** is that the slope of position is **negative** at $t = 5.0$ s (since velocity represents the **slope** of a position graph).

(B) What is the acceleration of the banana at $t = 5.0$ s?
Since we already found $A = 2.0$ m, $\omega_0 = 2\pi$ rad/s, and $\varphi = \pi$ rad in part (A), we can use the equation for the acceleration at $t = 5.0$ s.

$$a_x = -A\omega_0^2 \sin(\omega_0 t + \varphi) = -(2)(2\pi)^2 \sin[(2\pi)(5) + \pi]$$

$$a_x = -(2)(4\pi^2)\sin(10\pi + \pi) = -8\pi^2 \sin(11\pi)$$

Note that $(2\pi)^2 = 2^2\pi^2 = 4\pi^2$ according to the rule $(xy)^n = x^n y^n$. Remember that you can add or subtract multiples of 2π radians to any angle and obtain an equivalent angle in radians. Thus, 11π rad is equivalent to π rad (which corresponds to 180°). Recall from trig that $\sin(\pi) = 0$.

$$a_x = -8\pi^2 \sin(11\pi) = -8\pi^2 \sin(11\pi - 10\pi) = -8\pi^2 \sin(\pi) = -8\pi^2(0) = 0$$

The acceleration of the banana at $t = 5.0$ s is $a_x = 0$. That's because the banana is passing through the equilibrium position at $t = 5.0$ s. The acceleration is momentarily zero at equilibrium, whereas the speed is maximum at equilibrium. (In contrast, the acceleration is maximum at the peaks – the crests and troughs – whereas the speed is momentarily zero at the peaks.)

(C) What is the maximum speed of the banana?
Look at the velocity equation.

$$v_x = A\omega_0 \cos(\omega_0 t + \varphi)$$

The cosine function oscillates between -1 and 1. Therefore, the maximum speed is $A\omega_0$, which is the amplitude of the velocity (not to be confused with the amplitude of position, which is A).

$$v_m = A\omega_0 = (2)(2\pi) = 4\pi \text{ m/s}$$

The maximum speed is $v_m = 4\pi$ m/s. If you use a calculator, the maximum speed is $v_m = 13$ m/s to two significant figures.

(D) What is the maximum acceleration of the banana?
Look at the acceleration equation.

$$a_x = -A\omega_0^2 \sin(\omega_0 t + \varphi)$$

The sine function oscillates between -1 and 1. Therefore, the maximum value of the acceleration is $A\omega_0^2$, which is the amplitude of the acceleration (not to be confused with the amplitude of position, which is A, or the amplitude of velocity, which is $A\omega_0$).

$$a_m = A\omega_0^2 = (2)(2\pi)^2 = (2)(4\pi^2) = 8\pi^2 \text{ m/s}^2$$

Recall that $(2\pi)^2 = 2^2\pi^2 = 4\pi^2$ according to the rule $(xy)^n = x^n y^n$. The maximum acceleration of the banana is $a_m = 8\pi^2$ m/s^2. If you use a calculator, this works out to $a_m = 79$ m/s^2 (which is about 8 gravities[3]).

---

[3] That's a lot of acceleration for a banana oscillating back and forth. We chose numbers that would make the math easy to follow without a calculator, rather than choosing numbers to be realistic.

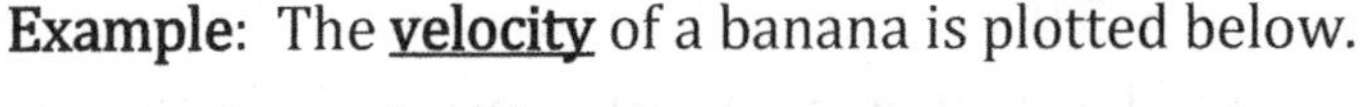

Example: The **velocity** of a banana is plotted below.

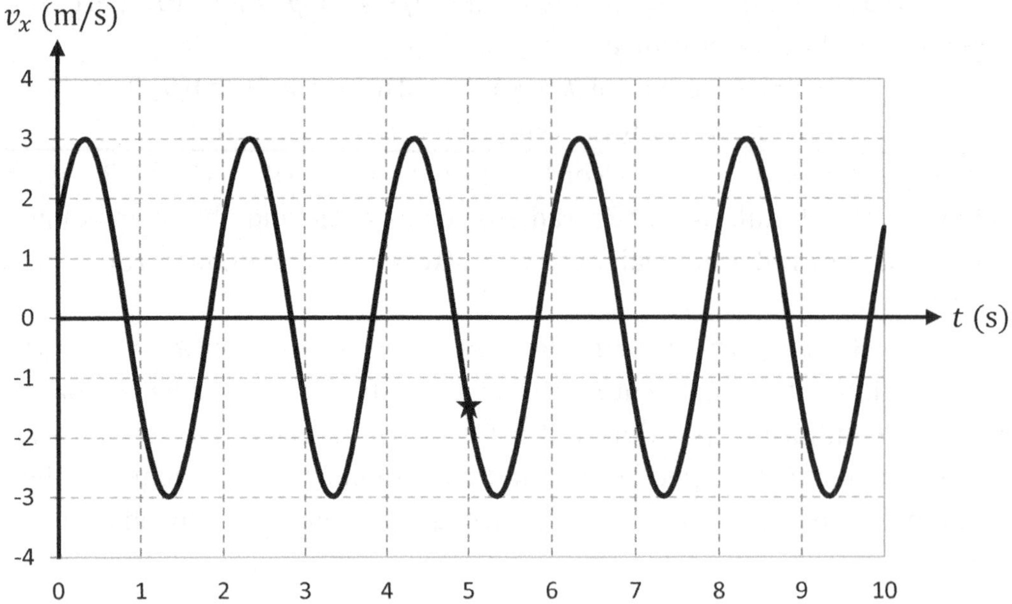

(A) What is the velocity of the banana at $t = 5.0$ s?
This is exactly the same question that we asked in part (A) of the previous example, yet the solution is much different. Why? Because this is a graph of velocity, whereas the previous example had a graph of position.

To find the velocity of the banana at $t = 5.0$ s for a velocity graph, simply read off the vertical value (v_x) when the horizontal value (t) equals 5.0 s. This point is marked with a star (★) on the graph above. The velocity of the banana at $t = 5.0$ s is $v_x = -1.5$ m/s.

(B) What is the maximum speed of the banana?
Since speed is the magnitude of velocity, and since this is a velocity graph, we don't need an equation. We just need to look at the graph. What is the maximum vertical value (v_x) of the sine wave? The peak of the velocity curve is $v_x = 3.0$ m/s, which is the fastest that the banana travels during its oscillation.

(C) What is the acceleration of the banana at $t = 5.0$ s?
Unlike parts (A) and (B), now we need to apply an equation. That's because this question asks about acceleration, which is not shown directly on the graph. We can apply the equation $a_x = -A\omega_0^2 \sin(\omega_0 t + \varphi)$, but first we need to determine the constant A, angular frequency (ω_0), and phase angle (φ). Find these quantities the same way that we did in Chapter 1 and in the previous example. However, as we will see, there will be a significant difference in how we determine the amplitude and the phase angle because the equation for **velocity**, $v_x = A\omega_0 \cos(\omega_0 t + \varphi)$, is significantly different from the equation for **position**, $x = A \sin(\omega_0 t + \varphi) + x_e$ (all of our previous graphs have had position, not velocity).

- To find the **angular frequency** (ω_0), first find the **period** (T). The period is the time between two crests. Read off the times for the first two crests. The first crest (maximum) occurs at about $t_{1m} = 0.3$ s. The second crest (maximum) occurs at about $t_{2m} = 2.3$ s. Subtract these two times in order to determine the period.

$$T = t_{2m} - t_{1m} = 2.3 - 0.3 = 2.0 \text{ s}$$

The period is $T = 2.0$ s. To find the angular frequency, divide 2π by the period.

$$\omega_0 = \frac{2\pi}{T} = \frac{2\pi}{2} = \pi \text{ rad/s}$$

The angular frequency is $\omega_0 = \pi$ rad/s.

- Note that there are three different **amplitudes**: the amplitude of **position** (x) is A, the amplitude of **velocity** (v_x) is $A\omega_0$, and the amplitude of **acceleration** (a_x) is $A\omega_0^2$. Since this example gives a velocity graph, its amplitude is $A\omega_0$. The **amplitude** ($A\omega_0$) of the velocity graph equals the maximum velocity, which is 3.0 m/s (the maximum vertical value on the sinusoidal wave). Set $A\omega_0$ equal to 3.0 m/s.

$$A\omega_0 = 3.0 \text{ m/s}$$

Divide both sides of the equation by ω_0. Recall that $\omega_0 = \pi$ rad/s.

$$A = \frac{3}{\omega_0} = \frac{3}{\pi} \text{ m}$$

The constant A is $A = \frac{3}{\pi}$ m. If you use a calculator, $A = 0.95$ m.

- Finding the **phase angle** for a **velocity** graph is similar to finding the phase angle for a position graph, but there are a couple of significant differences. One difference is that the equation for velocity, $v_x = A\omega_0 \cos(\omega_0 t + \varphi)$, involves a **cosine** function, whereas the equation for position, $x = A\sin(\omega_0 t + \varphi) + x_e$, involves a **sine** function. Therefore, we need to label the angles 0°, 90°, 180°, 270°, and 360° for a cosine graph rather than a sine graph. Another difference is that the equation for phase angle is $\varphi = \cos^{-1}\left(\frac{v_{x0}}{A\omega_0}\right)$ for a velocity graph. It is instructive to compare how we find the phase angle in this example compared to the previous example.

First determine the **initial velocity** (v_{x0}) from the graph. The initial velocity is the vertical intercept, which is the value of v_x when $t = 0$. Looking at the graph, the initial velocity is approximately $v_{x0} = 1.5$ m/s. Now apply the equation for the phase angle of a **velocity** graph. Recall that we already found that $A\omega_0 = 3.0$ m/s.

$$\varphi = \cos^{-1}\left(\frac{v_{x0}}{A\omega_0}\right) = \cos^{-1}\left(\frac{1.5}{3}\right) = \cos^{-1}\left(\frac{1}{2}\right)$$

The reference angle is $\frac{\pi}{3}$ rad (corresponding to 60°) since $\cos\frac{\pi}{3} = \cos 60° = \frac{1}{2}$, but this is **not** the answer for the phase angle because the cosine wave doesn't start in Quadrant I. (If it were a **sine** wave, it would – but this is a **cosine** wave.) Label the angles 0°, 90°, 180°, 270°, and 360° corresponding to a standard **cosine** wave, and extrapolate to the vertical axis, similar to pages 10-11. See the diagram that follows.

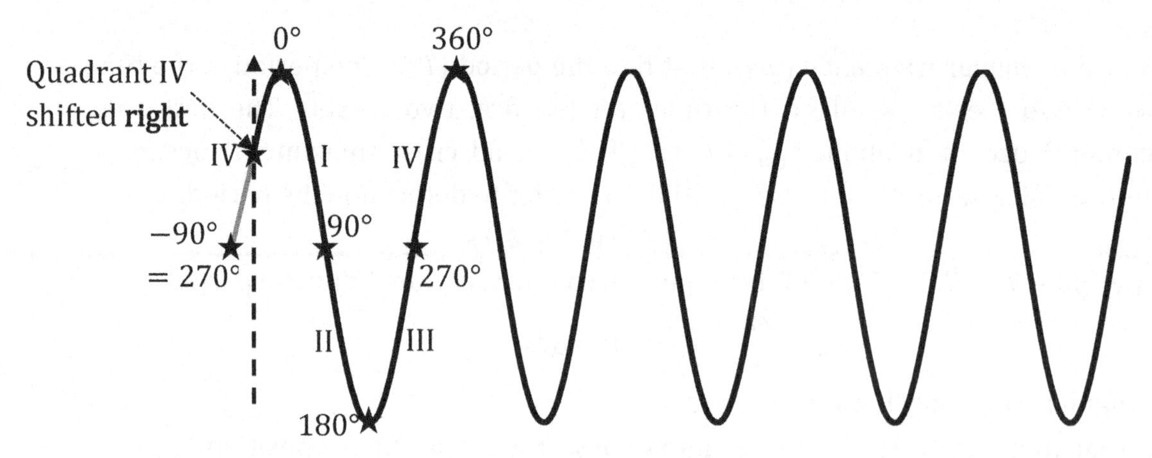

It may help to review how a single cycle of a cosine wave looks. We labeled the angles 0°, 90°, 180°, 270°, and 360° for a standard **cosine** wave below. Compare the shifted graph above to the standard graph below.

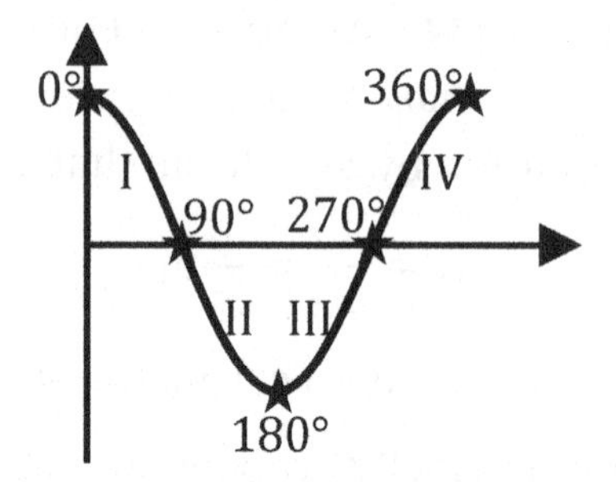

This **cosine** wave begins in Quadrant IV, between 270° and 360°. In Quadrant IV, subtract the reference angle from 360° in order to find the phase angle (recall that the reference angles is 60°): $\varphi = 360° - 60° = 300°$. Multiply by $\frac{\pi \text{ rad}}{180°}$ to convert to **radians:** $\varphi = 300° \times \frac{\pi \text{ rad}}{180°} = \frac{5\pi}{3}$ rad. The phase angle is $\varphi = \frac{5\pi}{3}$ rad.

We can obtain an equivalent angle by subtracting 360°: The phase angle can also be expressed as $\varphi = -\frac{\pi}{3}$ rad (corresponding to −60°). We can think of this sine wave as being shifted $\frac{\pi}{3}$ rad to the **right**, since a **negative** phase angle ($\varphi = -\frac{\pi}{3}$ rad) shifts the graph to the right, or we can think of this sine wave as being shifted $\frac{5\pi}{3}$ rad to the **left**, since a **positive** phase angle ($\varphi = \frac{5\pi}{3}$ rad) shifts the graph to the left.

Now that we know $A = \frac{3}{\pi}$ m, $\omega_0 = \pi$ rad/s, and $\varphi = \frac{5\pi}{3}$ rad (or $\varphi = -\frac{\pi}{3}$ rad), we can use the equation for the acceleration at $t = 5.0$ s.

$$a_x = -A\omega_0^2 \sin(\omega_0 t + \varphi) = -\left(\frac{3}{\pi}\right)(\pi)^2 \sin\left[(\pi)(5) + \frac{5\pi}{3}\right]$$

$$a_x = -3\pi \sin\left(5\pi + \frac{5\pi}{3}\right) = -3\pi \sin\left(\frac{15\pi}{3} + \frac{5\pi}{3}\right) = -3\pi \sin\left(\frac{20\pi}{3}\right)$$

Note that $\frac{\pi^2}{\pi} = \pi$. We rewrote 5π as $\frac{15\pi}{3}$ to make a **common denominator** in order to add the fractions. Remember that you can add or subtract multiples of 2π radians to any angle and obtain an equivalent angle in radians. Thus, $\frac{20\pi}{3}$ rad is equivalent to $\frac{20\pi}{3} - 6\pi = \frac{20\pi}{3} - \frac{18\pi}{3} = \frac{2\pi}{3}$ rad (which corresponds to 120°). Recall from trig that $\sin\left(\frac{2\pi}{3}\right) = \frac{\sqrt{3}}{2}$.

$$a_x = -3\pi \sin\left(\frac{20\pi}{3}\right) = -3\pi \sin\left(\frac{20\pi}{3} - 6\pi\right) = -3\pi \sin\left(\frac{2\pi}{3}\right) = -3\pi\left(\frac{\sqrt{3}}{2}\right) = -\frac{3\pi\sqrt{3}}{2} \text{ m/s}^2$$

The acceleration of the banana at $t = 5.0$ s is $a_x = -\frac{3\pi\sqrt{3}}{2}$ m/s². If you use a calculator, this works out to $a_x = -8.2$ m/s² to two significant figures.

(D) What is the maximum acceleration of the banana?
Look at the acceleration equation.

$$a_x = -A\omega_0^2 \sin(\omega_0 t + \varphi)$$

The sine function oscillates between -1 and 1. Therefore, the maximum value of the acceleration is $A\omega_0^2$, which is the amplitude of the acceleration (not to be confused with the amplitude of velocity, which is $A\omega_0$, or the amplitude of position, which is A). Recall from part (C) that $A = \frac{3}{\pi}$ m and $\omega_0 = \pi$ rad/s.

$$a_m = A\omega_0^2 = \left(\frac{3}{\pi}\right)(\pi)^2 = 3\pi \text{ m/s}^2$$

Note that $\frac{\pi^2}{\pi} = \pi$. The maximum acceleration of the banana is $a_m = 3\pi$ m/s². If you use a calculator, this works out to $a_m = 9.4$ m/s².

Example: The velocity of a banana is given by the following equation, where SI units have been suppressed in order to avoid clutter.

$$v_x = 6\cos\left(\frac{\pi t}{4} + \frac{\pi}{2}\right)$$

(A) What is the amplitude of the velocity cosine wave?

Compare the given equation, $v_x = 6\cos\left(\frac{\pi t}{4} + \frac{\pi}{2}\right)$, to the general equation for the **velocity** of an object oscillating with simple harmonic motion, $v_x = A\omega_0 \cos(\omega_0 t + \varphi)$. The amplitude of the velocity cosine wave is the coefficient of the cosine function: $A\omega_0 = 6.0$ m/s. That's it: The answer is 6.0 m/s. (The amplitude of **velocity** is **not** A. It's $A\omega_0$, which equals 6.0 m/s. Note that A is the amplitude of **position**, as we will see in the next question.)

(B) What is the amplitude of the corresponding position sine wave?

The general equation for the **position** of an object oscillating with simple harmonic motion is $x = A\sin(\omega_0 t + \varphi) + x_e$. The amplitude of the position sine wave is the coefficient of the sine function, which is A.

We learned in part (A) that the amplitude of the **velocity** cosine wave is $A\omega_0 = 6.0$ m/s. Divide both sides of the equation by ω_0.

$$A = \frac{6}{\omega_0}$$

We need to determine the angular frequency (ω_0) in order to find the amplitude of the **position** sine wave. Compare the general equation, $v_x = A\omega_0 \cos(\omega_0 t + \varphi)$, for the velocity of an object oscillating with simple harmonic motion to the equation given in this problem, $v_x = 6\cos\left(\frac{\pi t}{4} + \frac{\pi}{2}\right)$, in order to see that the angular frequency is $\omega_0 = \frac{\pi}{4}$ rad/s. Plug this into the previous equation for the amplitude of the position sine wave.

$$A = \frac{6}{\omega_0} = \frac{6}{\pi/4} = 6 \div \frac{\pi}{4} = 6 \times \frac{4}{\pi} = \frac{24}{\pi} \text{ m}$$

To divide by a fraction, multiply by its **reciprocal**. Note that the reciprocal of $\frac{\pi}{4}$ is $\frac{4}{\pi}$. The amplitude of the **position** sine wave is $A = \frac{24}{\pi}$ m. If you use a calculator, $A = 7.6$ m.

(C) What is the amplitude of the corresponding acceleration sine wave?

The general equation for the **acceleration** of an object oscillating with simple harmonic motion is $a_x = -A\omega_0^2 \sin(\omega_0 t + \varphi)$. The amplitude of the acceleration sine wave is the coefficient of the sine function, which is $A\omega_0^2$ (excluding the minus sign, since the amplitude is positive). Recall that we found $A = \frac{24}{\pi}$ m and $\omega_0 = \frac{\pi}{4}$ rad/s in part (B): $A\omega_0^2 = \left(\frac{24}{\pi}\right)\left(\frac{\pi}{4}\right)^2 = \left(\frac{24}{\pi}\right)\left(\frac{\pi^2}{16}\right) = \frac{24\pi}{16} = \frac{3\pi}{2}$ m/s^2. The amplitude of the acceleration sine wave is $A\omega_0^2 = \frac{3\pi}{2}$ m/s^2. If you use a calculator, this works out to $A\omega_0^2 = 4.7$ m/s^2.

(D) What is the velocity of the banana at $t = 1.0$ s?
Plug $t = 1.0$ s into the given equation for velocity.

$$v_x = 6\cos\left(\frac{\pi t}{4} + \frac{\pi}{2}\right) = 6\cos\left[\frac{\pi(1)}{4} + \frac{\pi}{2}\right] = 6\cos\left(\frac{\pi}{4} + \frac{\pi}{2}\right) = 6\cos\left(\frac{\pi}{4} + \frac{2\pi}{4}\right) = 6\cos\left(\frac{3\pi}{4}\right)$$

Combine fractions by finding a **common denominator**. Note that $\frac{3\pi}{4}$ rad corresponds to 135° (since π rad $= 180°$). Recall from trig that $\cos\left(\frac{3\pi}{4}\right) = -\frac{\sqrt{2}}{2}$.

$$v_x = 6\cos\left(\frac{3\pi}{4}\right) = 6\left(-\frac{\sqrt{2}}{2}\right) = -3\sqrt{2} \text{ m/s}$$

The velocity of the banana at $t = 1.0$ s is $v_x = -3\sqrt{2}$ m/s. If you use a calculator, this works out to $v_x = -4.2$ m/s.

(E) What is the maximum speed of the banana?
This is the same as the amplitude of the velocity, which we found in part (A). The maximum speed of the banana is $v_m = A\omega_0 = 6.0$ m/s.

(F) What is the acceleration of the banana at $t = 1.0$ s?
First, compare the given equation for velocity, $v_x = 6\cos\left(\frac{\pi t}{4} + \frac{\pi}{2}\right)$, to the general equation for velocity, $v_x = A\omega_0\cos(\omega_0 t + \varphi)$, to see that the phase angle is $\varphi = \frac{\pi}{2}$ rad. Plug $A = \frac{24}{\pi}$ m, $\omega_0 = \frac{\pi}{4}$ rad/s, $t = 1.0$ s, and $\varphi = \frac{\pi}{2}$ rad into the general equation for acceleration.

$$a_x = -A\omega_0^2\sin(\omega_0 t + \varphi) = -\left(\frac{24}{\pi}\right)\left(\frac{\pi}{4}\right)^2\sin\left[\left(\frac{\pi}{4}\right)(1) + \frac{\pi}{2}\right] = -\frac{3\pi}{2}\sin\left(\frac{\pi}{4} + \frac{\pi}{2}\right)$$

$$a_x = -\frac{3\pi}{2}\sin\left(\frac{3\pi}{4}\right) = -\frac{3\pi}{2}\left(\frac{\sqrt{2}}{2}\right) = -\frac{3\pi}{4}\sqrt{2} \text{ m/s}^2$$

The acceleration of the banana at $t = 1.0$ s is $a_x = -\frac{3\pi}{4}\sqrt{2}$ m/s^2. If you use a calculator, this works out to $a_x = -3.3$ m/s^2.

(G) What is the maximum acceleration of the banana?
This is the same as the amplitude of the acceleration, which we found in part (C). The maximum acceleration of the banana is $a_m = A\omega_0^2 = \frac{3\pi}{2}$ m/s^2 or $a_m = 4.7$ m/s^2.

Example: An elastic solid obeys the following formula, where F is the stretching force, A is the cross-sectional area, ΔL is the change in length after stretching, L_0 is the original length, and Y is called Young's modulus.

$$\frac{F}{A} = -Y\frac{\Delta L}{L_0}$$

Show that when this force equals the net force exerted on the solid that the solid oscillates with simple harmonic motion. Also, derive an equation for the period of oscillation.

A system undergoes simple harmonic motion when the acceleration is proportional to the negative of the displacement from equilibrium: $a_x = -\omega_0^2 \Delta x^2$. For the first part of our solution, we just need to show that $a_x = -\omega_0^2 \Delta L^2$ (since this example is using the symbol L in place of the symbol x). We can do this by applying Newton's second law (see Volume 1 of this series, Chapter 14).

$$\sum F_x = ma_x$$

Solve for the force in the equation given in this example: Multiply both sides by A.

$$F = -Y\frac{\Delta L}{L_0}A$$

Substitute this for the net force in Newton's second law.

$$-Y\frac{\Delta L}{L_0}A = ma_x$$

Divide both sides of the equation by mass in order to solve for acceleration.

$$a_x = -\frac{YA}{mL_0}\Delta L$$

Since this equation has the same structure as $a_x = -\omega_0^2 \Delta L$, the system undergoes simple harmonic motion. We can determine the angular frequency (ω_0) by comparing these two equations: ω_0^2 is the coefficient of ΔL (excluding the minus sign).

$$\omega_0^2 = \frac{YA}{mL_0}$$

Squareroot both sides of the equation. Apply the equation $\omega_0 = \frac{2\pi}{T}$ in order to determine the period.

$$\omega_0 = \sqrt{\frac{YA}{mL_0}} = \frac{2\pi}{T}$$

Multiply both sides of the equation by T and divide by $\sqrt{\frac{YA}{mL_0}}$. Note that $\frac{1}{\sqrt{\frac{YA}{mL_0}}} = \sqrt{\frac{mL_0}{YA}}$.

$$T = \frac{2\pi}{\sqrt{\frac{YA}{mL_0}}} = 2\pi\sqrt{\frac{mL_0}{YA}}$$

4. The position of a banana is plotted below.

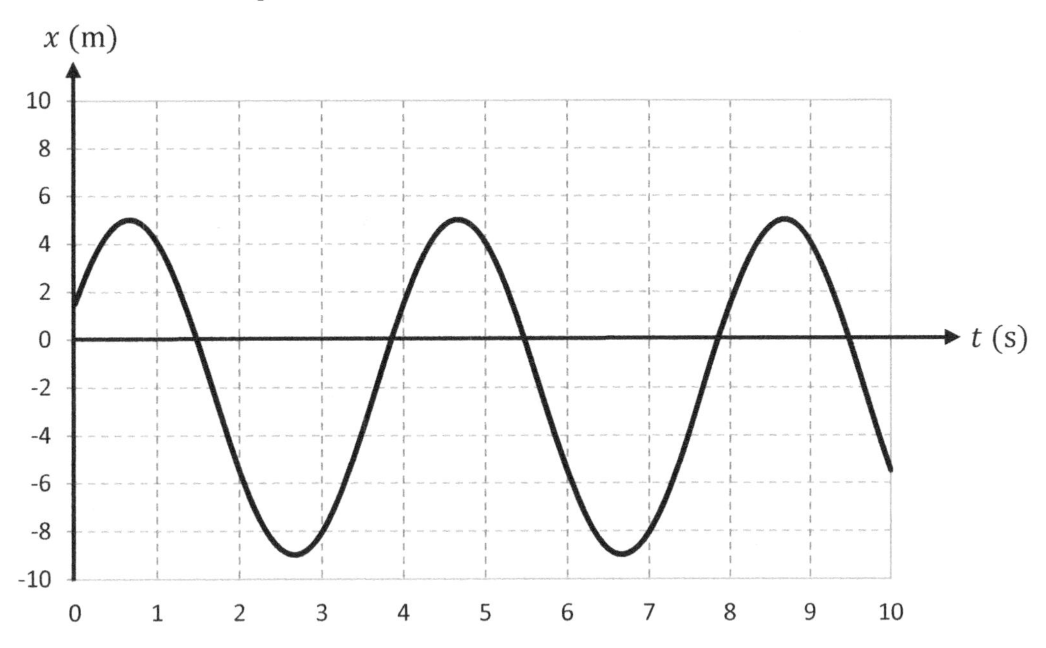

(A) What is the velocity at $t = 6.0$ s?

(B) What is the maximum speed?

(C) What is the acceleration at $t = 6.0$ s?

(D) What is the maximum acceleration?

Want help? Check the hints section at the back of the book.

Answers: (A) $-\frac{7\pi\sqrt{3}}{4}$ m/s $= -9.5$ m/s (B) $\frac{7\pi}{2}$ m/s $= 11$ m/s (C) $\frac{7\pi^2}{8}$ m/s^2 $= 8.6$ m/s^2 (D) $\frac{7\pi^2}{4}$ m/s^2 $= 17$ m/s^2

5. The velocity of a banana is plotted below.

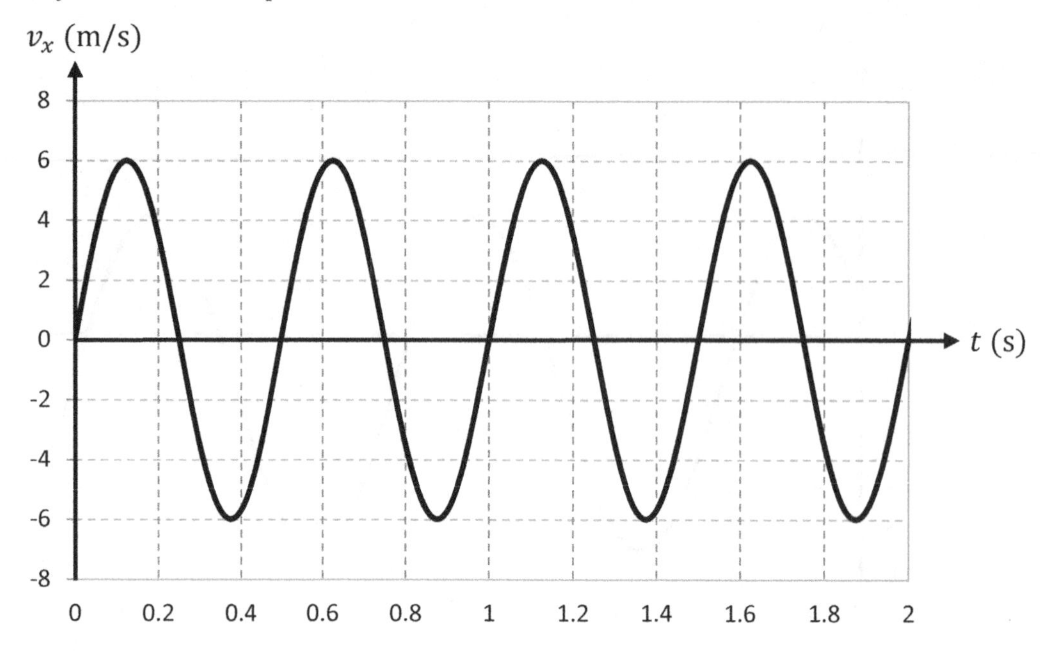

(A) What is the velocity at $t = 1.0$ s?

(B) What is the maximum speed?

(C) What is the acceleration at $t = 1.0$ s?

(D) What is the maximum acceleration?

Want help? Check the hints section at the back of the book.

Answers: (A) 0 (B) 6.0 m/s
(C) 24π m/s^2 = 75 m/s^2 (D) 24π m/s^2 = 75 m/s^2

6. The acceleration of a banana is plotted below.

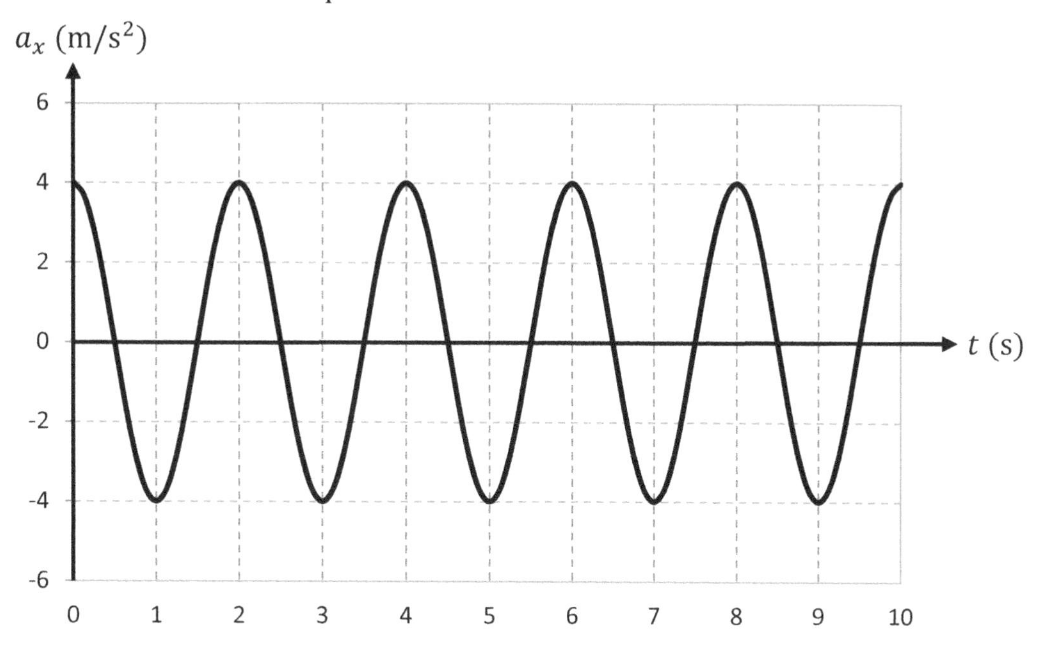

(A) What is the velocity at $t = 3.0$ s?

(B) What is the maximum speed?

(C) What is the acceleration at $t = 3.0$ s?

(D) What is the maximum acceleration?

Want help? Check the hints section at the back of the book.

Answers: (A) 0 (B) $\frac{4}{\pi}$ m/s $= 1.3$ m/s
(C) -4.0 m/s^2 (D) 4.0 m/s^2

7. The position of a banana is given by the following equation, where SI units have been suppressed in order to avoid clutter.

$$x = 3\sin\left(2\pi t - \frac{\pi}{4}\right) - 4$$

(A) What is the velocity of the banana at $t = 0.50$ s?

(B) What is the maximum speed of the banana?

(C) What is the acceleration of the banana at $t = 0.50$ s?

(D) What is the maximum acceleration of the banana?

(E) When does the banana have a speed of 3π m/s?

Want help? Check the hints section at the back of the book.

Answers: (A) $-3\pi\sqrt{2}$ m/s $= -13$ m/s (B) 6π m/s $= 19$ m/s
(C) $-6\pi^2\sqrt{2}$ m/s^2 $= -84$ m/s^2 (D) $12\pi^2$ m/s^2 $= 118$ m/s^2
(E) $\frac{7}{24}$ s $+ n$ s, $\frac{11}{24}$ s $+ n$ s, $\frac{19}{24}$ s $+ n$ s, $\frac{23}{24}$ s $+ n$ s

8. The velocity of a banana is given by the following equation, where SI units have been suppressed in order to avoid clutter.

$$v_x = 2\cos\left(\frac{2\pi t}{3}\right)$$

(A) What is the velocity of the banana at $t = 6.0$ s?

(B) What is the maximum speed of the banana?

(C) What is the acceleration of the banana at $t = 6.0$ s?

(D) What is the maximum acceleration of the banana?

(E) When does the banana have a speed of $\sqrt{3}$ m/s?

Want help? Check the hints section at the back of the book.

Answers: (A) 2.0 m/s (B) 2.0 m/s (C) 0 (D) $\frac{4\pi}{3}$ m/s^2 = 4.2 m/s^2

(E) $\frac{1}{4}$ s + $3n$ s, $\frac{5}{4}$ s + $3n$ s, $\frac{7}{4}$ s + $3n$ s, $\frac{11}{4}$ s + $3n$ s

9. The acceleration of a banana is given by the following equation, where SI units have been suppressed in order to avoid clutter.

$$a_x = -8\sin\left(\frac{\pi t}{2} + \frac{\pi}{2}\right)$$

(A) What is the velocity of the banana at $t = 4.0$ s?

(B) What is the maximum speed of the banana?

(C) What is the acceleration of the banana at $t = 4.0$ s?

(D) What is the maximum acceleration of the banana?

(E) When does the banana have a speed of $\frac{8}{\pi}$ m/s?

Want help? Check the hints section at the back of the book.

Answers: (A) 0 (B) $\frac{16}{\pi}$ m/s = 5.1 m/s (C) -8.0 m/s^2 (D) 8.0 m/s^2

(E) $\frac{1}{3}$ s $+ 4n$ s, $\frac{5}{3}$ s $+ 4n$ s, $\frac{7}{3}$ s $+ 4n$ s, $\frac{11}{3}$ s $+ 4n$ s

10. A fluid oscillates in a cylindrical tube according to the following formula, where F is the force exerted on a portion of the fluid of mass m that is a height Δh above the equilibrium level, g is gravitational acceleration, ρ is the density of the fluid, and A is the cross-sectional area of the tube.

$$\frac{F}{A} = -2\rho g \Delta h$$

Show that the fluid oscillates with simple harmonic motion. Also, derive an equation for the angular frequency with which it oscillates. **Note**: Density equals mass over volume.

Want help? Check the hints section at the back of the book.

Answer: $\omega_0 = \sqrt{\frac{2g}{h}}$

11. A team of supergenius chimpanzees creates a tunnel that passes through the center of the earth, connecting opposite ends of the earth. One of the chimpanzees drops a banana down the tunnel. Show that the banana oscillates with simple harmonic motion.[4] Also, derive an equation for the period of oscillation in terms of the radius of the earth (R_e), the mass of the earth (m_e), and the mass of the banana (m_b).

Want help? Check the hints section at the back of the book.

Answer: $T = 2\pi\sqrt{\frac{R_e^3}{Gm_e}}$

[4] Let's assume that the supergenius chimpanzees first stopped the earth from rotating, removed the earth to the far corners of the universe, insulated the tunnel from earth's core, and any other fanciful achievements which may be needed to render this problem "practical." We'll just focus on the "easy" problem of carrying out the theory, and leave the experimental challenges to the supergenius chimpanzees.

The Chain Rule

Suppose that $f(u)$ is a function of the variable u, and that $u(x)$ itself is a function of another variable x. If you need to take a derivative of f with respect to x, it can be done by applying the chain rule from calculus. The equation for the chain rule is shown below.

$$\frac{df}{dx} = \frac{df}{du}\frac{du}{dx}$$

For example, suppose that $y = (3x - 2)^8$ and you wish to determine $\frac{dy}{dx}$. You wouldn't want to multiply $(3x - 2)$ by itself 8 times to do this. It would be simpler to define $u = 3x - 2$ such that $y = u^8$ and then apply the chain rule to find $\frac{dy}{dx}$ as shown below.

$$\frac{dy}{dx} = \frac{dy}{du}\frac{du}{dx}$$

First plug $y = u^8$ into $\frac{dy}{du}$ to write $\frac{d}{du}(u^8)$, and plug $u = 3x - 2$ into $\frac{du}{dx}$ to write $\frac{d}{dx}(3x - 2)$.

$$\frac{dy}{dx} = \frac{d}{du}(u^8)\frac{d}{dx}(3x - 2)$$

Recall from calculus that $\frac{d}{du}(u^8) = 8u^7$ and that $\frac{d}{dx}(3x - 2) = 3$.

$$\frac{dy}{dx} = (8u^7)(3) = 24u^7$$

Now we plug in $u = 3x - 2$.

$$\frac{dy}{dx} = 24(3x - 2)^7$$

We have thus applied the chain rule to show that $\frac{d}{dx}(3x - 2)^8 = 24(3x - 2)^7$.

Derivatives of Sine and Cosine

Recall from calculus the derivatives of the sine and cosine functions.

$$\frac{d}{d\theta}\sin\theta = \cos\theta \quad , \quad \frac{d}{d\theta}\cos\theta = -\sin\theta$$

In simple harmonic motion, the argument isn't simply θ, but instead involves time. The equations of simple harmonic motion involve $\sin(\omega_0 t + \varphi)$ and $\cos(\omega_0 t + \varphi)$. Recall that velocity and acceleration involve derivatives with respect to time.

$$v_x = \frac{dx}{dt} \quad , \quad a_x = \frac{dv_x}{dt} = \frac{d^2x}{dt^2}$$

When we apply these derivatives to simple harmonic motion, we must apply the chain rule. For example, we can evaluate $\frac{d}{dt}\sin(\omega_0 t + \varphi)$ if we let $f = \sin(\omega_0 t + \varphi)$ and $\theta = \omega_0 t + \varphi$.

$$\frac{df}{dt} = \frac{df}{d\theta}\frac{d\theta}{dt} = \frac{d}{d\theta}[\sin\theta]\frac{d}{dt}(\omega_0 t + \varphi) = (\cos\theta)(\omega_0) = \omega_0\cos\theta$$

$$\frac{d}{dt}\sin(\omega_0 t + \varphi) = \omega_0\cos\theta$$

Simple Harmonic Motion with Calculus

With calculus, we can see how the equations of simple harmonic motion are related. When an object oscillates back and forth with simple harmonic motion, its position is described by a sine wave.

$$x = A\sin(\omega_0 t + \varphi) + x_e$$

The velocity of the object equals a derivative of position with respect to time.

$$v_x = \frac{dx}{dt} = \frac{d}{dt}[A\sin(\omega_0 t + \varphi) + x_e]$$

Distribute the derivative to both terms.

$$v_x = \frac{d}{dt}[A\sin(\omega_0 t + \varphi)] + \frac{d}{dt}(x_e)$$

The derivative of a constant is zero: $\frac{d}{dt}(x_e) = 0$. (Contrast $\frac{dx}{dt}$, which equals v_x, with $\frac{dx_e}{dt}$, which equals 0. The distinction is that the symbol x represents the position of the object and is a **variable**, whereas the symbol x_e is the equilibrium position and is a **constant**.)

$$v_x = \frac{d}{dt}[A\sin(\omega_0 t + \varphi)]$$

Apply the chain rule with $f = A\sin(\omega_0 t + \varphi)$ and $\theta = \omega_0 t + \varphi$.

$$v_x = \frac{df}{dt} = \frac{df}{d\theta}\frac{d\theta}{dt} = \frac{d}{d\theta}[A\sin\theta]\frac{d}{dt}(\omega_0 t + \varphi) = (A\cos\theta)(\omega_0) = A\omega_0\cos\theta$$

$$v_x = A\omega_0\cos(\omega_0 t + \varphi)$$

The acceleration of the object equals a derivative of velocity with respect to time.

$$a_x = \frac{dv_x}{dt} = \frac{d}{dt}[A\omega_0\cos(\omega_0 t + \varphi)]$$

Apply the chain rule with $g = A\omega_0\cos(\omega_0 t + \varphi)$ and $\theta = \omega_0 t + \varphi$.

$$a_x = \frac{dg}{dt} = \frac{dg}{d\theta}\frac{d\theta}{dt} = \frac{d}{d\theta}[A\omega_0\cos\theta]\frac{d}{dt}(\omega_0 t + \varphi) = (-A\omega_0\sin\theta)(\omega_0) = -A\omega_0^2\sin\theta$$

$$a_x = -A\omega_0^2\sin(\omega_0 t + \varphi)$$

Let's put the three equations together:

$$x = A\sin(\omega_0 t + \varphi) + x_e$$
$$v_x = A\omega_0\cos(\omega_0 t + \varphi)$$
$$a_x = -A\omega_0^2\sin(\omega_0 t + \varphi)$$

These are the same three equations from the top of page 26. Calculus shows how the equations for velocity and acceleration follow from the equation for position. At the bottom of page 26, we applied Newton's second law to show that the equations for position and acceleration satisfy the criteria for simple harmonic motion.

3 OSCILLATING SPRING

Relevant Terminology

Spring constant – a measure of the stiffness of a spring. A stiffer spring has a higher value for its spring constant.
Equilibrium – the natural position of a spring. A spring displaced from its equilibrium position tends to oscillate about its equilibrium position.
Displacement from equilibrium – the net displacement of a spring from its equilibrium position. It is a directed distance, telling how far from equilibrium a spring is.
Amplitude – the maximum displacement from equilibrium.
Restoring force – a force that a spring exerts to return toward its equilibrium position.
Velocity – the instantaneous rate at which position changes.
Acceleration – the instantaneous rate at which velocity changes.
Period – the time it takes to complete exactly one oscillation.
Frequency – the number of oscillations completed per second.
Angular frequency – the number of radians completed per second. The angular frequency equals 2π times the frequency.

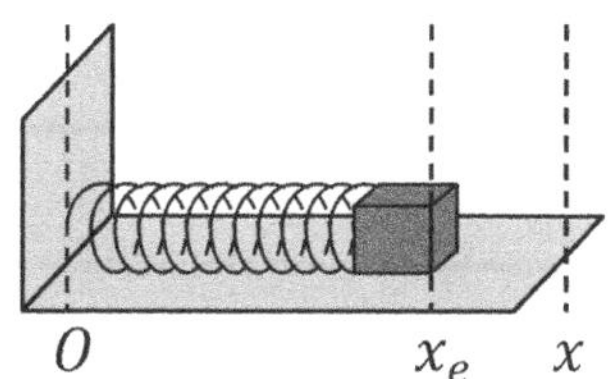

Spring Equations

When one end of a spring is fixed and a mass (m) is connected to the free end of the spring, the mass oscillates with a **period** (T) that depends on the attached mass (m) and the **spring constant** (k) according to the following equation.

$$T = 2\pi\sqrt{\frac{m}{k}}$$

Since angular frequency (ω_0) is related to period (T) by $\omega_0 = \frac{2\pi}{T}$, the equation for the angular frequency of an oscillating spring is:

$$\omega_0 = \sqrt{\frac{k}{m}}$$

A spring obeys **Hooke's law**, which means that it exerts a **restoring force** (F_r) that is proportional to the displacement from equilibrium (Δx) according to $F_r = -k\Delta x$ (see Volume 1, Chapter 15). The potential energy storied in a spring is $PE_s = \frac{1}{2}k\Delta x^2$ (see Volume 1, Chapter 22).

Symbols and SI Units

Symbol	Name	SI Units
x_e	the equilibrium position	m
Δx	displacement from equilibrium	m
x	the instantaneous position	m
v_x	the instantaneous velocity	m/s
a_x	the instantaneous acceleration	m/s^2
A	the amplitude of the spring's oscillation	m
t	time	s
T	period	s
φ	phase angle	rad
f	frequency	Hz
ω_0	angular frequency	rad/s
F_r	restoring force	N
$\sum F_x$	net force	N
m	mass	kg
k	spring constant	N/m or kg/s^2
U_s	the potential energy stored in a spring	J

Important Distinction

The formula for period (T) has $\frac{m}{k}$ in it $\left(T = 2\pi\sqrt{\frac{m}{k}}\right)$, whereas the formula for angular frequency (ω_0) has $\frac{k}{m}$ in it $\left(\omega_0 = \sqrt{\frac{k}{m}}\right)$. You can remember whether it's $\frac{m}{k}$ or $\frac{k}{m}$ by looking at the units. The SI units of spring constant (k) are $\frac{\text{kg}}{\text{s}^2}$, so $\sqrt{\frac{k}{m}}$ puts the second downstairs for ω_0 (in rad/s), whereas $\sqrt{\frac{m}{k}}$ brings the second upstairs for T (in sec). Note that $\frac{1}{1/\text{s}^2} = \text{s}^2$.

Notes Regarding Units

The SI units for the spring constant follow from Hooke's law: $F_r = -k\Delta x$. If you solve for the spring constant, you get $k = -\frac{F_r}{\Delta x}$. Thus, spring constant must have SI units of a Newton (the SI unit of force) per meter (the SI unit of displacement). Recall from Volume 1 how a Newton relates to a kilogram, meter, and second: $1 \text{ N} = 1 \text{ kg} \cdot \text{m/s}^2$. Using this relationship, we see that a Newton per meter equals:

$$1\frac{\text{N}}{\text{m}} = 1\frac{\text{kg}}{\text{s}^2}$$

Therefore, the SI units of a spring constant can alternatively be expressed as a Newton per meter (N/m) or as a kilogram per second squared (kg/s^2).

Strategy for an Oscillating Spring

How you solve a problem involving a spring depends on which kind of problem it is:

- If you need to relate the **spring constant** (k) and **suspended mass** (m) to the **period** (T), **frequency** (f), or **angular frequency** (ω_0), use the following equations.

$$T = 2\pi\sqrt{\frac{m}{k}} \quad , \quad \omega_0 = \sqrt{\frac{k}{m}} \quad , \quad \omega_0 = 2\pi f \quad , \quad \omega_0 = \frac{2\pi}{T} \quad , \quad f = \frac{1}{T}$$

- If a problem tells you the **number of oscillations** completed in a specified amount of time, you can determine the **period** with the following formula.

$$T = \frac{\text{total time}}{\text{number of oscillations}}$$

- For a comparison problem, set up a ratio, as shown in one of the examples. Following are two examples of comparison problems.
  - If you double the suspended mass, what will happen to the period?
  - If you double the spring constant, what will happen to the frequency?

  It may help to review the comparison problems from Volume 1, Chapter 18, pages 168 and 172.
- If a problem gives you a **graph** showing the motion of the spring, follow the strategy for interpreting a simple harmonic motion graph from Chapter 2.
- If a problem gives you an **equation** for the position (x), velocity (v_x), or acceleration (a_x) of a sine wave with numbers, follow the strategy for interpreting the equation for simple harmonic motion from Chapter 2.
- If a problem involves restoring force (F_r), apply Hooke's law: $F_r = -k\Delta x$. If you need to apply Newton's second law to a spring, see Volume 1, Chapter 15.
- If you need to find the potential energy stored in a spring, $PE_s = \frac{1}{2}k\Delta x^2$. If you need to apply conservation of energy to a spring system, see Volume 1, Chapter 22.

Where Does the Formula $T = 2\pi\sqrt{\frac{m}{k}}$ Come From?

The equation $T = 2\pi\sqrt{\frac{m}{k}}$ comes from the strategy that we applied in Chapter 2 on page 40 and in Problems 10-11 (pages 47-48).

First, we apply Newton's second law to the suspended mass.

$$\sum F_x = ma_x$$

If the net force acting on the suspended mass is the restoring force exerted by the spring, Newton's second law becomes:

$$F_r = ma_x$$

According to Hooke's law, the restoring force is:

$$F_r = -k\Delta x$$

Substitute this equation for restoring force into the previous equation for Newton's second law.

$$-k\Delta x = ma_x$$

Divide both sides of the equation by the mass in order to solve for the acceleration.

$$a_x = -\frac{k\Delta x}{m}$$

Since this equation has the same structure as $a_x = -\omega_0^2 \Delta x$, the system undergoes simple harmonic motion. We can determine the angular frequency (ω_0) by comparing these two equations: ω_0^2 is the coefficient of Δx (excluding the minus sign).

$$\omega_0^2 = \frac{k}{m}$$

Squareroot both sides of the equation.

$$\omega_0 = \sqrt{\frac{k}{m}}$$

Apply the equation $\omega_0 = \frac{2\pi}{T}$ in order to determine the period.

$$\frac{2\pi}{T} = \sqrt{\frac{k}{m}}$$

Multiply both sides of the equation by T and divide by $\sqrt{\frac{k}{m}}$. Note that $\frac{1}{\sqrt{\frac{k}{m}}} = \sqrt{\frac{m}{k}}$.

$$T = \frac{2\pi}{\sqrt{\frac{k}{m}}} = 2\pi\sqrt{\frac{m}{k}}$$

Example: On a horizontal frictionless surface, near the surface of the earth, a monkey connects one end of a horizontal spring to a 400-g banana and the other end to a wall. The spring constant is 10 N/m.

(A) What is the period of oscillation?
First identify the known symbols in SI units.

- The suspended mass is $m = 0.40$ kg, where 1 kg $=$ 1000 g.
- The spring constant is $k = 10$ N/m.

Use the equation for the period of a spring.

$$T = 2\pi\sqrt{\frac{m}{k}} = 2\pi\sqrt{\frac{0.40}{10}} = 2\pi\sqrt{0.04} = 2\pi(0.2) = 0.4\pi = \frac{2\pi}{5}\text{ s}$$

The period is $T = \frac{2\pi}{5}$ s. Note that $0.4 = \frac{2}{5}$ and that $\sqrt{0.04} = 0.2$. If you use a calculator, the period works out to $T = 1.3$ s.

(B) What is the frequency?
Frequency is the reciprocal of the period.

$$f = \frac{1}{T} = \frac{1}{2\pi/5}$$

Note that the reciprocal of $\frac{2\pi}{5}$ is $\frac{5}{2\pi}$.

$$f = \frac{5}{2\pi}\text{ Hz}$$

The frequency is $f = \frac{5}{2\pi}$ Hz. If you use a calculator, this works out to $f = 0.80$ Hz.

(C) How many oscillations will the banana make in one minute?
Divide the total time by the period. In SI units, the total time is 60 seconds (since there are 60 seconds in 1 minute).

$$\text{number of oscillations} = \frac{\text{total time}}{T} = \frac{60}{2\pi/5}$$

To divide by a fraction, multiply by its **reciprocal**. Note that the reciprocal of $\frac{2\pi}{5}$ is $\frac{5}{2\pi}$.

$$\text{number of oscillations} = 60 \div \frac{2\pi}{5} = 60 \times \frac{5}{2\pi} = \frac{150}{\pi}$$

The banana will make $\frac{150}{\pi}$ oscillations in one minute. If you use a calculator, this works out to just under 48 oscillations (so the banana will make 47 complete oscillations and part of the 48th).

Example: On a horizontal frictionless surface, near the surface of the earth, a monkey connects one end of a horizontal spring to a banana and the other end to a wall. When the monkey sets the spring into motion, the spring oscillates with a period of 0.50 s. The monkey then replaces the banana with another banana that is four times as heavy as the first banana. What is the period of the new system?

Let's first express what we know with math, and identify the desired unknown.

- The period for the first mass is $T_1 = 0.50$ s.
- The second banana has four times as much weight as the first banana. Since weight equals mass times gravity, it also has four times as much mass. It will be convenient to express this as a ratio: $\frac{m_2}{m_1} = 4$.
- We are looking for the period with the new mass. We will call this T_2.

The phrase "is four times as heavy" classifies this problem as a comparison problem. One way to solve a comparison problem is to write down one equation for each banana, using subscripts 1 and 2 for any quantities that may be different, and then to take a ratio, as we will show below.

Write down an equation for the period of each banana. Since the bananas clearly have different mass, we will use subscripts, m_1 and m_2, to tell the masses apart. Since the question is asking about the period of the second banana, we must allow for the possibility that it is different from the period of the first banana. Therefore, we will also use subscripts, T_1 and T_2, to tell the periods apart. However, the same spring is used in both cases. (If a different spring were used, the problem would need to make this point clear.) Therefore, we will use the same symbol, k, without subscripts for both spring constants. The two equations for period are:

$$T_1 = 2\pi\sqrt{\frac{m_1}{k}} \quad , \quad T_2 = 2\pi\sqrt{\frac{m_2}{k}}$$

Divide the equation for T_2 by the equation for T_1.

$$\frac{T_2}{T_1} = \frac{2\pi\sqrt{\frac{m_2}{k}}}{2\pi\sqrt{\frac{m_1}{k}}}$$

The 2π's cancel out. To divide by a fraction, multiply by its **reciprocal**.

$$\frac{T_2}{T_1} = \sqrt{\frac{m_2}{k}}\sqrt{\frac{k}{m_1}} = \sqrt{\frac{m_2 k}{k m_1}} = \sqrt{\frac{m_2}{m_1}} = \sqrt{4} = 2$$

The spring constant cancels out. Recall that $\frac{m_2}{m_1} = 4$ and that $T_1 = 0.50$ s. Multiply both sides of the equation by T_1.

$$T_2 = 2T_1 = 2(0.5) = 1.0 \text{ s}$$

The period of the new system is $T_2 = 1.0$ s.

12. On a horizontal frictionless surface, near the surface of the earth, a monkey connects one end of a horizontal spring to a 250-g banana and the other end to a wall. The spring constant is 9.00 N/m. What is the period of oscillation?

13. On a horizontal frictionless surface, near the surface of the earth, a monkey connects one end of a horizontal spring to a 500-g banana and the other end to a wall. The spring constant is 8.0 N/m. How many oscillations does the banana complete in 20π seconds?

14. On a horizontal frictionless surface, near the surface of the earth, a monkey connects one end of a horizontal spring to a banana and the other end to a wall. The spring constant is 2.00 N/m and the period is $\frac{\pi}{2}$ seconds. What is the mass of the banana?

Want help? Check the hints section at the back of the book.

Answers: $\frac{\pi}{3}$ s = 1.05 s, 40, 125 g

15. On a horizontal frictionless surface, near the surface of the earth, a monkey connects one end of a horizontal spring to a banana and the other end to a wall. When the monkey sets the spring into motion, the spring oscillates with a period of 2.0 s. The monkey then replaces the banana with another banana that is twice as heavy as the first banana. What is the period of the new system?

16. On a horizontal frictionless surface, near the surface of the earth, a monkey connects one end of a horizontal spring to a block of wood and the other end to a wall. When the monkey sets the spring into motion, the spring oscillates with a period of $\frac{5}{6}$ s. The monkey then transports the system to the moon, where gravity is reduced by a factor of 6. What is the period of the system on the moon?

17. On a horizontal frictionless surface, near the surface of the earth, a monkey connects one end of a horizontal spring to a banana and the other end to a wall. When the monkey sets the spring into motion, the spring oscillates with a period of 1.5 s. The monkey then replaces the spring with another spring that has nine times the spring constant of the first spring. What is the period of the new system?

Want help? Check the hints section at the back of the book.

Answers: $2\sqrt{2}$ s $= 2.8$ s, $\frac{5}{6}$ s $= 0.83$ s, 0.50 s

4 OSCILLATING PENDULUM

Relevant Terminology

Equilibrium – the natural position of a pendulum (when it hangs straight down). A pendulum displaced from its equilibrium position tends to oscillate about equilibrium.
Amplitude – the maximum displacement from equilibrium.
Period – the time it takes to complete exactly one oscillation.
Frequency – the number of oscillations completed per second.
Angular frequency – the number of radians completed per second. The angular frequency equals 2π times the frequency.
Angular velocity – the instantaneous rate at which angular position changes.
Angular acceleration – the instantaneous rate at which angular velocity changes.
Velocity – the instantaneous rate at which position changes.
Acceleration – the instantaneous rate at which velocity changes.

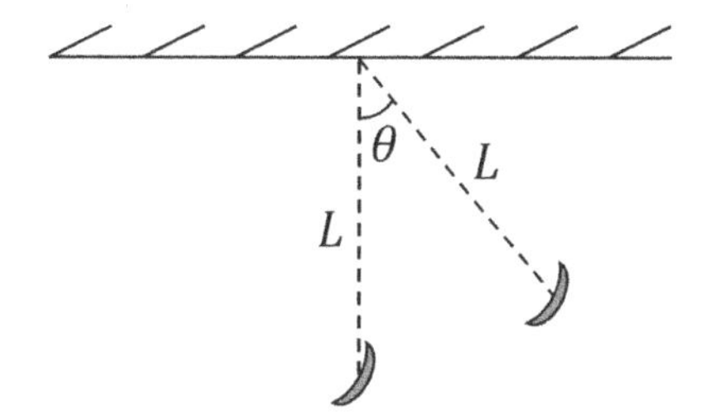

Pendulum Equations

If the pendulum bob is small in size compared to the length of the pendulum, if the cord is approximately inextensible, and if the mass of the cord is negligible compared to the mass of the pendulum bob, then the pendulum is considered to be a **simple pendulum**. The period (T) and angular frequency (ω_0) of a simple pendulum depend on the length (L) of the pendulum and gravitational acceleration (g) according to the following equation.

$$T \approx 2\pi\sqrt{\frac{L}{g}} \quad , \quad \omega_0 \approx \sqrt{\frac{g}{L}}$$

Otherwise, the pendulum is a **physical pendulum**. The period (T) and angular frequency (ω_0) of a physical pendulum depend on the moment of inertia (I) of the pendulum, the mass (m) of the pendulum, gravitational acceleration (g), and the distance (d) from the axis of rotation to the center of mass of the pendulum.

$$T \approx 2\pi\sqrt{\frac{I}{mgd}} \quad , \quad \omega_0 \approx \sqrt{\frac{mgd}{I}}$$

If the axis of rotation doesn't pass through the center of mass of the object, you will need to use the **parallel-axis theorem** (review Volume 1, Chapter 32), where I_{cm} is the moment of inertia about an axis through the center of mass of the object, I is the moment of inertia about an axis parallel to the axis used for I_{cm}, and h is the distance between the two axes.

$$I = I_{cm} + mh^2$$

You can look up the formula for the moment of inertia of many standard geometries about an axis through the center of mass in a table in Volume 1, Chapter 32.

The equations of simple harmonic motion for a pendulum involve angular displacement (θ), angular velocity (ω), and angular acceleration (α).

$$\theta \approx \theta_m \sin(\omega_0 t + \varphi)$$
$$\omega \approx A\omega_0 \cos(\omega_0 t + \varphi)$$
$$\alpha \approx -A\omega_0^2 \sin(\omega_0 t + \varphi)$$

Here, θ_m is the **amplitude** of the angular position (but it's **not** the amplitude of the angular velocity or angular acceleration equations), ω_0 is the **angular frequency** (whereas ω is the angular velocity), t represents time, and φ is the **phase angle**.

If the pendulum is released from rest from angle θ_m at $t = 0$, which is commonly the case, we can rewrite the above equations as follows.

$$\theta \approx \theta_m \cos(\omega_0 t)$$
$$\omega \approx -A\omega_0 \sin(\omega_0 t)$$
$$\alpha \approx -A\omega_0^2 \cos(\omega_0 t)$$

The reason for this is that the phase angle is $\varphi = \sin^{-1}\left(\frac{\theta_m}{\theta_m}\right) = \sin^{-1}(1) = \frac{\pi}{2}$ rad (which corresponds to 90°). If you shift a sine wave 90° to the right, it becomes a cosine wave. That is, $\sin\left(\omega_0 t + \frac{\pi}{2}\right) = \cos(\omega_0 t)$. Another way to see this is to use the trigonometric identity $\sin(x + y) = \sin x \cos y + \sin y \cos x$. Plugging in $x = \omega_0 t$ and $y = \frac{\pi}{2}$, we get $\sin\left(\omega_0 t + \frac{\pi}{2}\right) = \sin \omega_0 t \cos\frac{\pi}{2} + \sin\frac{\pi}{2} \cos \omega_0 t$. If you recall from trig that $\cos\frac{\pi}{2} = 0$ and $\sin\frac{\pi}{2} = 1$, you will see that $\sin \omega_0 t \cos\frac{\pi}{2} + \sin\frac{\pi}{2} \cos \omega_0 t = 0 + \cos \omega_0 t = \cos \omega_0 t$.

You can get the tangential displacement (s_T) from the angular displacement (θ) using the arc length formula ($s = R\theta$), realizing that the length (L) of the pendulum is the radius (R). You can similarly find tangential velocity (v_T) and tangential acceleration (a_T). It may help to review the tangential and angular quantities in Volume 1, Chapter 28.

$$s_T = L\theta \quad , \quad v_T = L\omega \quad , \quad a_T = L\alpha$$

Tip: In physics lab, the proper way to measure the effective length of a simple pendulum is from the axis of rotation to the **center of mass** of the pendulum bob.

Symbols and SI Units

Symbol	Name	SI Units
θ	the instantaneous angular position	rad
θ_m	the amplitude of the pendulum's oscillation	rad
s_T	the instantaneous tangential displacement	m
t	time	s
T	period	s
φ	phase angle	rad
f	frequency	Hz
ω_0	angular frequency	rad/s
ω	the instantaneous angular velocity	rad/s
v_T	the instantaneous tangential velocity	m/s
α	the instantaneous angular acceleration	rad/s^2
a_T	the instantaneous tangential acceleration	m/s^2
m	mass	kg
g	gravitational acceleration	m/s^2
L	length of a simple pendulum, or length of a rod	m
d	distance from the axis of rotation to the center of mass	m
R	radius	m
τ	torque	Nm
F	force	N
r	distance from the axis to the point where the force is applied	m
I	moment of inertia	kg·m^2
I_{cm}	moment of inertia about an axis through the center of mass	kg·m^2
h	distance from the center of mass to the parallel axis	m

Important Distinctions

Note that the symbols ω_0 and ω mean two different things:

- ω_0 is the **angular frequency**. It is 2π times the frequency, and serves as a measure of the periodicity of the pendulum's oscillation. It is a constant.
- ω is the **angular velocity**. It is a variable. The angular velocity changes as the pendulum oscillates.

The equation for the period of a pendulum depends on whether or not it is a **simple** pendulum or a **physical** pendulum. For a physical pendulum, you need to first determine the moment of inertia, as shown in one of the examples.

For a simple pendulum, you can remember whether the formula has $\frac{L}{g}$ or $\frac{g}{L}$ by looking at the units. The SI units of gravitational acceleration (g) are $\frac{\text{m}}{\text{s}^2}$, so $\sqrt{\frac{g}{L}}$ puts the second downstairs for ω_0 (in rad/s), whereas $\sqrt{\frac{L}{g}}$ brings the second upstairs for T (in sec). Note that $\frac{1}{1/\text{s}^2} = \text{s}^2$.

Strategy for an Oscillating Pendulum

How you solve a problem involving a pendulum depends on which kind of problem it is:

- For a **simple** pendulum (where the pendulum bob is small and carries most of the mass), you can relate the **gravitational acceleration** (g) and **length** (L) of the pendulum to the **period** (T), **frequency** (f), or **angular frequency** (ω_0) via the following equations.

$$T \approx 2\pi\sqrt{\frac{L}{g}} \quad , \quad \omega_0 \approx \sqrt{\frac{g}{L}} \quad , \quad \omega_0 = 2\pi f \quad , \quad \omega_0 = \frac{2\pi}{T} \quad , \quad f = \frac{1}{T}$$

- For a **physical pendulum** (either where the pendulum bob has a significant size or where the bob is connected to the pivot by an object such as a rod that carries significant weight), you can relate the **mass** (m), **gravitational acceleration** (g), **moment of inertia** (I), and distance (d) from the axis of rotation to the center of mass to the **period** (T), **frequency** (f), or **angular frequency** (ω_0) via the following equations.

$$T \approx 2\pi\sqrt{\frac{I}{mgd}} \quad , \quad \omega_0 \approx \sqrt{\frac{mgd}{I}} \quad , \quad \omega_0 = 2\pi f \quad , \quad \omega_0 = \frac{2\pi}{T} \quad , \quad f = \frac{1}{T}$$

  You will probably also need to apply the **parallel-axis theorem** (see Volume 1, Chapter 32).

$$I = I_{cm} + mh^2$$

- If a problem tells you the **number of oscillations** completed in a specified amount of time, you can determine the **period** with the following formula.
$$T = \frac{\text{total time}}{\text{number of oscillations}}$$
- For a comparison problem, set up a ratio, as shown in one of the examples. Following are two examples of comparison problems.
  - If you double the gravitational acceleration, what will happen to the period?
  - If you double the length of the cord, what will happen to the frequency?
- If a problem gives you a **graph** showing the motion of the pendulum, follow the strategy for interpreting a simple harmonic motion graph from Chapter 2. However, for a pendulum, you will probably work with θ, ω, and α instead of x, v_x, and a_x.
- If a problem gives you an **equation** for the angular position (θ), angular velocity (ω), or angular acceleration (α) of a sine wave with numbers, follow the strategy for interpreting the equation for simple harmonic motion from Chapter 2. However, you will want to instead use equations from page 60. If you need the tangential displacement (s_T), tangential velocity (v_T), or tangential acceleration (a_T), see the equations at the bottom of page 60.
- If a problem involves applying Newton's second law to a pendulum (for example, to determine the tension in the cord), see Volume 1, Chapter 14.
- If a problem involves applying conservation of energy to a pendulum (for example, to determine the speed at the bottom of the arc), see Volume 1, Chapter 22.
- If a problem involves a **spring**, see Chapter 3.

Where Does the Formula $T \approx 2\pi\sqrt{\frac{L}{g}}$ Come From?

The equation $T \approx 2\pi\sqrt{\frac{L}{g}}$ comes from the strategy that we applied in Chapter 2 on page 40 and in Problems 10-11 (pages 47-48).

First, we sum the **torques** (τ) for the pendulum. Ordinarily, we would apply Newton's second law, but since a pendulum **rotates**, we instead set the sum of the torques equal to moment of inertia times angular acceleration (we learned this in Volume 1, Chapter 33).
$$\sum \tau = I\alpha$$
Recall the equation for **torque** (from Volume 1, Chapter 30).
$$\tau = rF\sin\theta$$
For the **simple** pendulum, $r = L$ (since the length of the pendulum is also the radius), and the force is weight (mg).
$$\tau = Lmg\sin\theta$$

Substitute this expression for torque into the sum of the torques equation (see the previous page), and add a **minus sign** to represent the fact that the torque pulls the pendulum to the left when it is displaced to the right and vice-versa (it is a **restoring torque**, analogous to the restoring force from Hooke's law).

$$-Lmg\sin\theta = I\alpha$$

For a **simple** pendulum, use the equation for the moment of inertia of a pointlike object (see Volume 1, Chapter 32).

$$I = mr^2 = mL^2$$

Again, for a simple pendulum, the length is also the radius: $r = L$. Substitute this equation into the previous equation.

$$-Lmg\sin\theta = mL^2\alpha$$

Divide both sides of the equation by mL^2 in order to solve for the angular acceleration. The mass cancels. Note that $\frac{L}{L^2} = \frac{1}{L}$.

$$\alpha = -\frac{g}{L}\sin\theta$$

This equation doesn't quite have the same structure as simple harmonic motion (Chapter 2). For the motion to be simple harmonic motion, the equation would need to have the form $\alpha = -\omega_0^2\theta$. It almost does: The problem is that it has $\sin\theta$ instead of θ.

It turns out that $\sin\theta$ is **approximately** equal to θ (in **radians**) for **small angles**. Let's try this with 30° and see how true this is. Convert 30° to radians to get $\frac{\pi}{6}$. Enter this on your calculator to see that $\frac{\pi}{6} = 0.5236$. Compare this to $\sin\frac{\pi}{6} = 0.5$. They are approximately equal: For angles up to $\frac{\pi}{6}$ radians, θ and $\sin\theta$ differ by less than 5%.

Therefore, we may approximate $\sin\theta$ as θ in the previous equation, and apply this to a simple pendulum with the stipulation that the amplitude of oscillation should not exceed more than about $\frac{\pi}{6}$ radians (corresponding to 30°). The smaller the amplitude, the better the approximation. Note: **This is the reason we use an approximately equal symbol (≈) in the equations for the period of a pendulum.**

$$\alpha \approx -\frac{g}{L}\theta$$

Since this approximation has the same structure as $\alpha = -\omega_0^2\theta$, the simple pendulum undergoes simple harmonic motion. We can determine the angular frequency (ω_0) by comparing these two equations: ω_0^2 is the coefficient of θ.

$$\omega_0^2 \approx \frac{g}{L}$$

Squareroot both sides of the equation.

$$\omega_0 \approx \sqrt{\frac{g}{L}}$$

Apply the equation $\omega_0 = \frac{2\pi}{T}$ in order to determine the period.

$$\frac{2\pi}{T} \approx \sqrt{\frac{g}{L}}$$

Multiply both sides of the equation by T and divide by $\sqrt{\frac{g}{L}}$. Note that $\frac{1}{\sqrt{\frac{g}{L}}} = \sqrt{\frac{L}{g}}$.

$$T \approx \frac{2\pi}{\sqrt{\frac{g}{L}}} = 2\pi\sqrt{\frac{L}{g}}$$

Where Does the Formula $T \approx 2\pi\sqrt{\frac{I}{mgd}}$ Come From?

Let us return to the following equation, which we found in the previous section for the **simple** pendulum.

$$-Lmg \sin\theta = I\alpha$$

For a **physical** pendulum, we will use the symbol d in place of the symbol L. Whereas L was the length of the simple pendulum, d is the distance from the axis of rotation to the center of mass of the physical pendulum. (For a simple pendulum, d and L are the same.) For a physical pendulum, d is the effective radius of the rotation.

$$-dmg \sin\theta = I\alpha$$

Divide both sides of the equation by I.

$$\alpha = -\frac{dmg}{I}\sin\theta$$

As we did for the simple pendulum, approximate $\sin\theta$ as θ for small angles.

$$\alpha \approx -\frac{dmg}{I}\theta$$

Compare this approximation to the equation for simple harmonic motion: $\alpha = -\omega_0^2\theta$. As usual (see the previous section and Chapter 2), ω_0^2 is the coefficient of θ.

$$\omega_0^2 \approx \frac{mgd}{I}$$

Squareroot both sides of the equation.

$$\omega_0 \approx \sqrt{\frac{mgd}{I}}$$

Apply the equation $\omega_0 = \frac{2\pi}{T}$ in order to determine the period. The algebra is the same as it was in the previous section (for the simple pendulum).

$$T \approx 2\pi\sqrt{\frac{I}{mgd}}$$

Example: Near the surface of the earth, a monkey connects a 250-g banana to a 2.5-m long cord to make a pendulum. What is the period of oscillation?
First identify the known symbols in SI units.

- The suspended mass is $m = 0.250$ kg, where 1 kg = 1000 g. **Note**: Since the equation for period, $T \approx 2\pi\sqrt{\frac{L}{g}}$, doesn't involve mass, this information is irrelevant to the solution: The suspended mass has **no effect** on the period of a simple pendulum.
- The length of the pendulum is $L = 2.5$ m.
- You should also know that earth's surface gravity is $g = 9.81 \text{ m/s}^2$. In this book, we will round 9.81 to ≈ 10, in order to solve the problem without the aid of a calculator.

Use the equation for the period of a **simple** pendulum (since a banana is small in size compared to the 2.5 m long cord).

$$T \approx 2\pi\sqrt{\frac{L}{g}} = 2\pi\sqrt{\frac{2.5}{9.81}} \approx 2\pi\sqrt{\frac{2.5}{10}} = 2\pi\sqrt{\frac{1}{4}} = 2\pi\left(\frac{1}{2}\right) = \pi \text{ s}$$

The period is $T \approx \pi$ s. If you don't round 9.81 to 10, $T \approx 3.2$ s. Note that $\sqrt{\frac{1}{4}} = \frac{1}{2}$.

Example: Near the surface of the earth, a monkey connects a solid sphere with a mass of 5.0 kg and a radius of 0.50 m to a 0.50-m long rod that has negligible mass compared to the sphere. The rod joins to the edge of the sphere, and the system oscillates about a hinge at the free end of the rod. What is the period of oscillation?
First identify the known symbols in SI units.

- The mass of the sphere is $m = 5.0$ kg and its radius is $R = 0.50$ m.
- The length of the rod is $L = 0.50$ m.
- You should also know that earth's surface gravity is $g = 9.81 \text{ m/s}^2 \approx 10 \text{ m/s}^2$.

The moment of inertia of a solid sphere about an axis through its center is (see Volume 1, Chapter 32):

$$I_{cm} = \frac{2}{5}mR^2 = \frac{2}{5}(5)(0.5)^2 = \frac{2}{5}(5)(0.25) = 0.50 \text{ kg·m}^2$$

Apply the **parallel-axis theorem** (Volume 1, Chapter 32) with $h = L + R = 0.5 + 0.5 = 1.0$ m (the distance from the pivot to the center of the solid sphere).

$$I = I_{cm} + mh^2 = 0.50 + (5)(1)^2 = 0.50 + 5 = 5.5 \text{ kg·m}^2$$

Use the equation for the period of a **physical** pendulum with $d = h = 1.0$ m. In this case, d and h happen to be the same because the rod has negligible mass. Otherwise, for d, we would first need to locate the center of mass of the system (see Volume 1, Chapter 27).

$$T \approx 2\pi\sqrt{\frac{I}{mgd}} = 2\pi\sqrt{\frac{5.5}{(5)(9.81)(1)}} \approx 2\pi\sqrt{\frac{5.5}{50}} = 2\pi\sqrt{\frac{11}{100}} \text{ s}$$

The period is $T \approx 2\pi\sqrt{\frac{11}{100}}$ s. If you use a calculator, $T \approx 2.1$ s. Note that $\frac{5.5}{50} = \frac{11}{100}$.

Example: Near the surface of the earth, a monkey connects a banana to cord to make a pendulum. The pendulum oscillates with a period of 1.6 s. The monkey then replaces the cord with another cord, making the length of the pendulum four times as long. What is the period of the new pendulum?

Let's first express what we know with math, and identify the desired unknown.

- The period for the first cord is $T_1 = 1.6$ s.
- The second cord is four times as long as the first cord. It will be convenient to express this as a ratio: $\frac{L_2}{L_1} = 4$.
- We are looking for the period with the new cord. We will call this T_2.

The phrase "is four times as long" classifies this problem as a comparison problem. One way to solve a comparison problem is to write down one equation for each case, using subscripts 1 and 2 for any quantities that may be different, and then to take a ratio, as we will show below.

Write down an equation for the period for each case. Since the cords clearly have different length, we will use subscripts, L_1 and L_2, to tell the lengths apart. Since the question is asking about the period of the second banana, we must allow for the possibility that it is different from the period of the first banana. Therefore, we will also use subscripts, T_1 and T_2, to tell the periods apart. However, the earth's surface gravity is the same in both cases. Therefore, we will use the same symbol, g, without subscripts for the gravitational acceleration for each pendulum. The two equations for period are:

$$T_1 \approx 2\pi\sqrt{\frac{L_1}{g}} \quad , \quad T_2 \approx 2\pi\sqrt{\frac{L_2}{g}}$$

Divide the equation for T_2 by the equation for T_1.

$$\frac{T_2}{T_1} \approx \frac{2\pi\sqrt{\frac{L_2}{g}}}{2\pi\sqrt{\frac{L_1}{g}}}$$

The 2π's cancel out. To divide by a fraction, multiply by its **reciprocal**. The reciprocal of $\sqrt{\frac{L_1}{g}}$ is $\sqrt{\frac{g}{L_1}}$.

$$\frac{T_2}{T_1} \approx \sqrt{\frac{L_2}{g}}\sqrt{\frac{g}{L_1}} = \sqrt{\frac{L_2 g}{g L_1}} = \sqrt{\frac{L_2}{L_1}} = \sqrt{4} = 2$$

Gravitational acceleration cancels out. Recall that $\frac{L_2}{L_1} = 4$ and that $T_1 = 1.60$ s. Multiply both sides of the equation by T_1.

$$T_2 \approx 2T_1 = 2(1.6) = 3.2 \text{ s}$$

The period of the new system is $T_2 \approx 3.2$ s.

18. Near the surface of the earth, a monkey connects a 500-g banana to a 40-m long cord from the roof of a tall building to make a pendulum. What is the period of oscillation?

19. Near the surface of the earth, a monkey connects a 3.0-kg banana to a spring with a spring constant of 6.0 N/m, and separately connects a 1.5-kg banana to a cord to make a pendulum. The spring and pendulum have the same period. How long is the cord?

20. Near the surface of the earth, a monkey fastens a small hinge to one end of a rod that has a length of 0.60 m and a mass of 4.0 kg. The hinge is connected to the ceiling. What is the period of oscillation?

Want help? Check the hints section at the back of the book.

Answers: 4π s $= 13$ s, 5.0 m, $\frac{2\pi}{5}$ s $= 1.3$ s

21. When an alien on planet Monk attaches a rock to the free end of a horizontally oscillating spring (on a frictionless surface) with a spring constant of 20 N/m, the rock oscillates with a period of 3.0 s. When the alien swings the same rock from the end of a 5.0-m cord (still on planet Monk, but no longer connected to a spring), the rock oscillates with a period of 2.0 seconds. How much does the rock weigh on planet Monk?

Want help? Check the hints section at the back of the book.

Answer: 225 N (if you get 4.6, you're only **halfway** through the solution)

22. Near the surface of the earth, a monkey connects a banana to cord to make a pendulum. The pendulum oscillates with a period of $\sqrt{2}$ s. The monkey then replaces the cord with another cord, making the length of the pendulum twice as long. What is the period of the new pendulum?

23. Near the surface of the earth, a monkey connects a banana to cord to make a pendulum. The pendulum oscillates with a period of 1.6 s. The monkey then replaces the banana with another banana that is twice as heavy as the first banana. What is the period of the new pendulum?

24. Near the surface of the earth, a monkey connects a banana to cord to make a pendulum. The pendulum oscillates with a period of $\sqrt{6}$ s. The monkey then transports the pendulum to the moon, where gravity is reduced by a factor of 6. What is the period of the pendulum on the moon?

Want help? Check the hints section at the back of the book.

Answers: 2.0 s, 1.6 s, 6.0 s

5 WAVE MOTION

Relevant Terminology

Crest – a point on a wave where there is a **maximum** (the top of a hill).
Trough – a point on a wave where there is a **minimum** (the bottom of a valley).
Equilibrium – a horizontal line about which a sine wave oscillates.
Amplitude – the **vertical** height between a crest and the equilibrium position (or **one-half** of the vertical height between a crest and trough).
Wavelength – the **horizontal** distance between two consecutive crests.
Period – the time it takes for the wave to complete exactly one oscillation.
Frequency – the number of oscillations completed per second.
Angular frequency – the number of radians completed per second. The angular frequency equals 2π times the frequency, or 2π divided by the period.
Angular wave number – the number of radians traveled per meter. The angular wave number equals 2π divided by the wavelength.
Phase angle – an angle representing how much the wave is shifted horizontally. A **positive** phase angle represents a wave that is shifted to the **left**.
Wave speed – how fast the wave travels.
Propagation – the transmission of a wave through a medium (often, with the sense of spreading out – like the ripples of a water wave – but not necessarily).
Longitudinal – a wave for which the amplitude of oscillation is parallel to the direction that the wave propagates.
Transverse – a wave for which the amplitude of oscillation is perpendicular to the direction that the wave propagates.
Work – work is done when there is not only a force acting on an object, but when the force also contributes toward the displacement of the object.
Energy – the ability to do work, meaning that a force is available to contribute towards the displacement of an object.
Power – the instantaneous rate at which work is done.
Linear mass density – mass per unit length.
Density – mass per unit volume.
Pressure – force per unit area.
Compression – a region where the pressure and density are high.
Rarefaction – a region where the pressure and density are low.
Pitch – the property of a sound wave that corresponds to frequency.
Temperature – a measure of the average kinetic energy of the molecules of a substance.
Bulk modulus – a measure of a liquid's or a solid's resistance to a change in its volume.
Young's modulus – a measure of a solid's resistance to a change in its length.
Intensity – power per unit area.

Essential Wave Concepts

An **oscillation** is a back-and-forth motion. We saw examples of oscillations with the spring and the pendulum in Chapters 3-4. Although a **wave** generally involves oscillations, a wave is a different phenomenon. A **wave** is a **disturbance** that carries energy a large distance compared to the amplitude of the oscillations. For example, consider a large pond that measures 500 m across, where a monkey drops a banana into one end of the pond, and another monkey watches a leaf floating at the edge of the opposite end of the pond. When the banana splashes into the pond, it creates a series of ripples. Each ripple starts out as a small circle from the banana's location and spreads outward. As a ripple reaches a given part of the pond, it disturbs the water level. (This lets you visualize how a wave can be considered a disturbance.) The water level rises upward and downward as each ripple passes through. The amplitude of the water level's oscillation is a few centimeters. When the ripple finally reaches the other end of the lake, the energy of the ripple transports the leaf up and down as it passes underneath it. Thus, we see how a water wave with an amplitude of a few centimeters carried energy a distance of 500 m in this example.

A wave can be characterized as transverse, longitudinal, or a combination of the two.

- A wave is **transverse** if the oscillation is **perpendicular** to the direction that the wave propagates. The water wave that we just considered is primarily transverse: The water level rose up and down while the ripple traveled horizontally. Light (an electromagnetic wave) is also transverse (see Chapter 20).
- A wave is **longitudinal** if the oscillation is parallel and anti-parallel to the direction that the wave propagates. A sound wave is longitudinal: The molecules in the medium vibrate back and forth along the direction of the sound wave. You can find a diagram of this on the following page.

When studying wave motion, it's important not to confuse the following terms:

- **Amplitude** (A) is a **vertical** distance from crest to equilibrium.
- **Wavelength** (λ) is a **horizontal distance** from crest to crest. You could determine wavelength from a graph of y as a function of x.
- **Period** (T) is the **time** it takes to complete one oscillation. You could determine period from a graph of y as a function of t. Period and wavelength are measured the same way on a graph, but one is a graph of time and the other is not.
- **Frequency** (f) is the **number of cycles per second**.
- **Angular frequency** (ω_0) equals 2π divided by the **period**.
- **Angular wave number** (k_0) equals 2π divided by the **wavelength**.
- **Wave speed** (v) tells **how fast** the wave is traveling.

The **units** can help you distinguish among these quantities (see pages 76-77).

Sound Waves

A sound wave involves changes in **pressure** (P) and **density** (ρ). Pressure equals force per unit area (Chapter 9): $P = \frac{F}{A}$. Density equals mass per unit volume (Chapter 8): $\rho = \frac{m}{V}$. As a sound wave travels through a medium, the molecules in the medium vibrate back and forth, creating alternate regions of **compression** (pressure and density higher than equilibrium) and **rarefaction** (pressure and density lower than equilibrium).

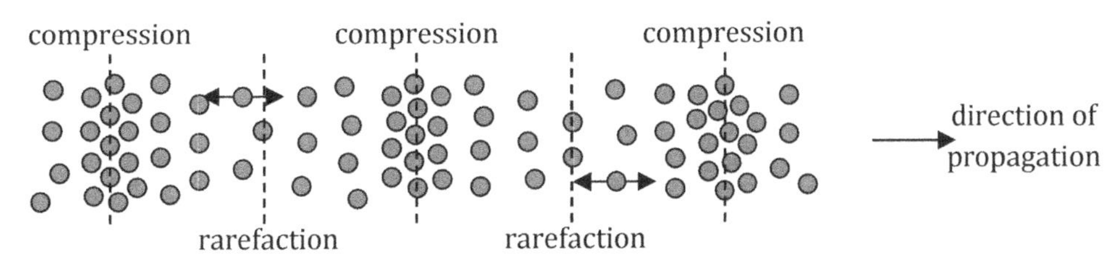

The property of a sound wave that you perceive as **pitch** corresponds to **frequency**: A sound wave with a high pitch has high frequency, while a sound wave with low pitch has low frequency. The property of a sound wave that you perceive as **loudness** relates to intensity. The **intensity** (I) of a wave equals power per unit area (Chapter 20): $I = \frac{P}{A}$. Specifically, the sound level (L) in **decibels** (dB) is related to the intensity by a base-10 logarithm. (For a review of logarithms, see Volume 2, Chapter 17.)

$$L = 10 \log\left(\frac{I}{I_0}\right)$$

The symbol I_0 represents a **reference intensity**, and has the value $I_0 = 1.00 \frac{\text{W}}{\text{m}^2}$. Because of the nature of base-10 logarithms, if the sound level is increased by 10 dB, the sound level is 10 times louder. For example, a 50-dB sound is 1000 times louder than a 20-dB sound. We will elaborate on this in the problem-solving strategy and one of the examples.

Traveling Waves

The position of a sinusoidal **traveling wave** is given by the following equation.

$$y = A \sin(k_0 x - \omega_0 t - \varphi) + y_e$$

Here, A is the **amplitude** of the wave, k_0 is the **angular wave number**, ω_0 is the **angular frequency**, x represents position along the wave's propagation, y represents relative position along the wave's oscillation, t represents time, φ is the **phase angle**, and y_e is the relative **equilibrium position** for the position wave. The **angular wave number** (k_0) and **angular frequency** (ω_0) are related to the wavelength (λ) and period (T) by:

$$k_0 = \frac{2\pi}{\lambda} \quad , \quad \omega_0 = \frac{2\pi}{T}$$

Frequency (f) and **period** (T) share a reciprocal relationship.

$$f = \frac{1}{T} \quad , \quad T = \frac{1}{f}$$

In general, wave speed (v) can be found from any of the formulas on the line below. (In the sections that follow, we will learn additional equations that apply to specific kinds of waves. The equations on the line below also apply to those specific types of waves.)

$$v = \frac{\lambda}{T} = \lambda f = \frac{\omega_0}{k_0}$$

The symbol φ represents **phase angle** (φ). It can be found by a method similar to that discussed in Chapters 1-2, except that the equation for the sine wave is different (see the equation for y on the previous page).

The signs of $k_0 x$ and $\omega_0 t$ tell you which way the wave is traveling.

- When $k_0 x$ and $\omega_0 t$ have **opposite** signs (as in $|k_0|x - |\omega_0|t$ or $-|k_0|x + |\omega_0|t$), the wave is traveling in the **positive** ($+x$) direction. Such a wave has the form $f(x - vt)$.
- When $k_0 x$ and $\omega_0 t$ have the **same** sign (as in $|k_0|x + |\omega_0|t$ or $-|k_0|x - |\omega_0|t$), the wave is traveling in the **negative** ($-x$) direction. Such a wave has the form $f(x + vt)$.

If you factor the k_0 out of $k_0 x + \omega_0 t$, you get $k_0\left(x + \frac{\omega_0}{k_0}t\right) = k_0(x + vt)$. This is how a wave with $|k_0|x - |\omega_0|t$ is a function of $(x - vt)$, for example.

When an object travels with simple harmonic motion (Chapter 2), its associated energy is:

$$E = \frac{1}{2}kA^2 = \frac{1}{2}m\omega_0^2 A^2$$

This k is the **spring constant** (Chapter 3), **not** to be confused with **angular wave number** (k_0). For a sinusoidal wave on a string, the associated power is:

$$P = \frac{1}{2}\mu\omega_0^2 A^2 v$$

Here, μ is the **linear mass density** of the string (see the following section).

Wave Speed

The following equations for **wave speed** (v) apply in general.

$$v = \frac{\lambda}{T} = \lambda f = \frac{\omega_0}{k_0}$$

For a wave traveling along a **string**, wave speed depends on **tension** (F) and **linear mass density** (μ), which is the mass per unit length.[1] If you know the total mass (m_s) of the string and the total length (L_s) of the string, $\mu = \frac{m_s}{L_s}$.

$$v = \sqrt{\frac{F}{\mu}}$$

[1] We're using F for tension instead of T, in order to avoid confusion with the period. You might recall that we used λ for linear mass density for center of mass and moment of inertia integrals in Volume 1: If so, we're using μ now in order to avoid confusion with wavelength.

For a **sound** wave, the formula for wave speed depends on the medium. The speed of sound in a **solid** depends on **Young's modulus** (Y) and **density** (ρ).

$$v = \sqrt{\frac{Y}{\rho}} \quad \text{(solid)}$$

The speed of sound in a **liquid** depends on **bulk modulus** (B) and **density** (ρ).

$$v = \sqrt{\frac{B}{\rho}} \quad \text{(liquid)}$$

The speed of sound in an ideal gas depends on the adiabatic index (γ), the universal gas constant (R), the **absolute temperature** (T) in **Kelvin** (K), and the **molar mass** (M) of the gas in kilograms per mole (kg/mol).[2] Using the ideal gas law (Chapter 16), $PV = nRT$, the formula can alternatively be expressed in terms of pressure (P) and density (ρ). In physics units (SI units), the universal gas constant equals $R = 8.314 \frac{\text{J}}{\text{mol·K}}$.

$$v = \sqrt{\frac{\gamma RT}{M}} = \sqrt{\frac{\gamma P}{\rho}} \quad \text{(ideal gas)}$$

The adiabatic index (γ) equals $\frac{5}{3}$ for a m**onatomic** ideal gas (such as the Noble gases – He, Ne, Ar, Kr, Xe, and Rn) and $\frac{7}{5}$ for a **diatomic** ideal gas (H_2, N_2, O_2, F_2, Cl_2, Br_2, and I_2).[3]

The **speed of sound in air** is about 340 m/s. The exact value depends on factors such as the temperature.

All of these equations for the speed of sound, and even the equation for the speed of a wave along a string $\left(v = \sqrt{\frac{F}{\mu}}\right)$, have the following structure.

$$v = \sqrt{\frac{\text{elastic property (like bulk modulus)}}{\text{inertial property (like density)}}}$$

[2] If you consult a periodic table, beware that it will probably list the molar mass in grams per mole (g/mol) instead of kilograms per mole (kg/mol). The SI units are kg/mol. In Chemistry, it's often more convenient to work with grams, but in physics, it's conventional to use SI units in the mks system (meters, kilograms, seconds), **not** cgs (centimeters, grams, seconds). So you want to convert to kg/mol for physics.

[3] It's really easy to memorize the list of diatomic gases, if you can just remember the number 7. Look at a periodic table: You'll see that N, O, F, Cl, Br, and I form the shape of a "7" on the periodic table. The "7" starts with nitrogen. That's really easy to remember, since nitrogen is atomic number 7. The right side of the "7" corresponds to the halogens. That's also easy to remember, since the halogens are in Group 7 or Group 17 (depending upon which convention for numbering groups your periodic table follows). Then you just need to remember hydrogen. If you know that neutral hydrogen has one valence electron, but that it's shell would prefer to hold two, then it will be easy to remember that hydrogen atoms like to share their electrons to form the covalent bond H_2. It may also be worth noting that neither bromine nor iodine is a gas at STP (standard temperature and pressure). At STP, bromine is a liquid and iodine is a solid.

Symbols and SI Units

Symbol	Name	SI Units
y	the position of a traveling wave relative to equilibrium	m
x	a position coordinate along the direction of propagation	m
y_e	the equilibrium position	m
v	wave speed	m/s
A	the amplitude of the position function	m
t	time	s
T	period	s
λ	wavelength	m
φ	phase angle	rad
f	frequency	Hz
ω_0	angular frequency	rad/s
k_0	angular wave number	rad/m
m	mass	kg
M	molar mass (the mass per mole)	kg/mol
F	tension	N
μ	linear mass density	kg/m
ρ	density	$\frac{\text{kg}}{\text{m}^3}$
P	pressure in Pa (or power in W)	see text to left
B	bulk modulus	$\frac{\text{N}}{\text{m}^2}$
Y	Young's modulus	$\frac{\text{N}}{\text{m}^2}$
γ	adiabatic index	unitless

V	volume	m^3
A	area	m^2
n	number of moles	mol
R	universal gas constant	$\frac{\text{J}}{\text{mol}\cdot\text{K}}$
T	absolute temperature (in Kelvin)	K
E	energy	J
I	intensity	$\frac{\text{W}}{\text{m}^2}$
I_0	reference intensity	$\frac{\text{W}}{\text{m}^2}$
L	sound level	dB

Important Distinction

In physics, period and wavelength are two different quantities:

- **Period** is the horizontal **time** between two crests. You determine the period from a graph that has time (t) on the horizontal axis.
- **Wavelength** is the horizontal **distance** between two crests. You determine the wavelength from a graph that has position (x) on the horizontal axis.

Wave Motion Strategy

How you solve a problem involving wave motion depends on which kind of problem it is:

- If the problem involves **wave speed**, the equations on the line below apply in general. (For a wave on a string or a sound wave, additional equations can be found further below.) The symbols are wavelength (λ), period (T), frequency (f), angular frequency (ω_0), and angular wave number (k_0).

$$v = \frac{\lambda}{T} = \lambda f = \frac{\omega_0}{k_0}$$

For a wave on a **string**, wave speed involves the tension (F) in the string and the linear mass density (μ) of the string. The linear mass density equals the total mass (m_s) of the string divided by the total length (L_s) of the string: $\mu = \frac{m_s}{L_s}$.

$$v = \sqrt{\frac{F}{\mu}}$$

For a **sound wave**, the wave speed depends on the medium. The symbols are: wave speed (v), Young's modulus (Y), density (ρ), bulk modulus (B), adiabatic constant (γ), universal gas constant $\left(R = 8.314 \frac{\text{J}}{\text{mol}\cdot\text{K}}\right)$, absolute temperature ($T$) in Kelvin (K), molar mass (M) in kg/mol, and pressure (P). If the temperature is given in Celsius (°C), convert to Kelvin: $T_K = T_C + 273.15°$. If it is given in Fahrenheit (°F), see Chapter 13. If you're not using a calculator, approximate 8.314 as 25/3.

$$v = \sqrt{\frac{Y}{\rho}} \quad \text{(solid)} \quad , \quad v = \sqrt{\frac{B}{\rho}} \quad \text{(liquid)} \quad , \quad v = \sqrt{\frac{\gamma RT}{M}} = \sqrt{\frac{\gamma P}{\rho}} \quad \text{(ideal gas)}$$

For a sound wave in **air**, the speed of sound is about 340 m/s.

- If a problem tells you the **number of oscillations** completed in a specified amount of time, you can determine the **period** with the following formula.

$$T = \frac{\text{total time}}{\text{number of oscillations}}$$

- The following equation relates **sound level** (L) to **intensity** (I). The reference intensity equals $I_0 = 1.00 \frac{\text{W}}{\text{m}^2}$. Logarithms are reviewed in Volume 2, Chapter 17.

$$L = 10 \log\left(\frac{I}{I_0}\right)$$

To compare the loudness of two sounds in decibels (dB), such as 90 dB and 50 dB:

  - Subtract the two values in dB. Example: 90 dB − 50 dB = 40 dB.
  - Divide by 10 dB. Example continued: $\frac{40 \text{ dB}}{10 \text{ dB}} = 4$.
  - Raise 10 to that power. Example continued: $10^4 = 10{,}000$.

In this example, 90 dB is $10^4 = 10{,}000$ times louder than 50 dB. Also see the examples on the bottom of page 80.

- If a problem gives you an **equation** for a wave, follow the strategy from Chapters 1-2, except for the following differences. The equation for a traveling wave is:
$$y = A\sin(k_0 x - \omega_0 t - \varphi) + y_e$$
Here, y is the relative position of the wave, x a position coordinate along the direction of motion, y_e is the equilibrium position, and k_0 is the angular wave number. For the speed of the wave, see the first bullet point. The **wavelength** (λ) can be found from the angular wave number: $\lambda = \frac{2\pi}{k_0}$. The remaining symbols are the same as they were in Chapters 1-2.
- If a problem gives you a **graph** of a wave, follow the strategy from Chapters 1-2. The difference is how you interpret the horizontal measure between two crests. If the graph has t on the horizontal axis, this measure is the **period** (T), but if the graph instead has x on the horizontal axis, this measure is the **wavelength** (λ).
- If a problem involves the Doppler effect or shock waves, see Chapter 6.
- If a problem involves standing waves (or resonance), see Chapter 7.

Example: A wave travels with a wavelength of 25 cm and a frequency of 200 Hz. What is the wave speed?
First convert the wavelength to SI units: $\lambda = 25 \text{ cm} = 0.25 \text{ m}$. Apply the wave speed formula that involves wavelength and frequency.
$$v = \lambda f = (0.25)(200) = 50 \text{ m/s}$$
The wave speed is $v = 50$ m/s.

Example: A wave travels with a wavelength of 8.0 m and a period of 4.0 s. What is the wave speed?
Apply the wave speed formula that involves wavelength and period.
$$v = \frac{\lambda}{T} = \frac{8}{4} = 2.0 \text{ m/s}$$
The wave speed is $v = 2.0$ m/s.

Example: A 3.0-m long string with a mass of 75 g is clamped at both ends. The tension in the string is 40 N. What would be the speed of a wave traveling along the string?
First convert the mass of the string to SI units: $m_s = 75 \text{ g} = 0.075 \text{ kg}$. Divide the mass of the string by the length of the string in order to determine the linear mass density.
$$\mu = \frac{m_s}{L_s} = \frac{0.075}{3} = 0.025 \text{ kg/m}$$
Apply the equation for wave speed that involves tension and linear mass density.
$$v = \sqrt{\frac{F}{\mu}} = \sqrt{\frac{40}{0.025}} = \sqrt{1600} = 40 \text{ m/s}$$
The wave speed is $v = 40$ m/s.

Example: What is the speed of sound in helium gas at 27°C?

We will treat the helium gas as an ideal gas. Helium (He) is a monatomic gas, so $\gamma = \frac{5}{3}$. In order to estimate the answer without the aid of a calculator, we will approximate the universal gas constant as $R = 8.314 \frac{\text{J}}{\text{mol}\cdot\text{K}} \approx \frac{25}{3} \frac{\text{J}}{\text{mol}\cdot\text{K}}$. (Note that $\frac{25}{3} = 8.333...$) Convert the temperature from Celsius to Kelvin. We will round 273.15 to 273 (which is good to three significant figures – the given temperature was no more precise than that).

$$T_K = T_C + 273.15 = 27 + 273 = 300 \text{ K}$$

Use the equation for the wave speed of an ideal gas. The molar mass of He is 0.004 kg/mol.

$$v = \sqrt{\frac{\gamma RT}{M}} \approx \sqrt{\frac{\left(\frac{5}{3}\right)\left(\frac{25}{3}\right)(300)}{0.004}} = \sqrt{\left(\frac{5}{3}\right)\left(\frac{25}{3}\right)(300)(250)} = \sqrt{\left(\frac{5}{3}\right)(25)\left(\frac{300}{3}\right)(250)}$$

Note that $\frac{1}{0.004} = 250$ and that $\left(\frac{25}{3}\right)(300) = (25)\left(\frac{300}{3}\right)$. This is convenient since 25 and 100 are perfect squares. In the next line, we will write $250 = (25)(10)$ for the same reason.

$$v \approx \sqrt{\left(\frac{5}{3}\right)(25)(100)(250)} = (5)(10)\sqrt{\left(\frac{5}{3}\right)(25)(10)} = (5)(10)(5)\sqrt{\frac{50}{3}} \text{ m/s}$$

Note that $\sqrt{(25)(100)(25)} = \sqrt{25}\sqrt{100}\sqrt{25} = (5)(10)(5) = 250$ and that $\left(\frac{5}{3}\right)(10) = \frac{50}{3}$.

$$v \approx 250\frac{\sqrt{50}}{\sqrt{3}} = 250\frac{\sqrt{50}}{\sqrt{3}}\frac{\sqrt{3}}{\sqrt{3}} = \frac{250}{3}\sqrt{150} = \frac{250}{3}\sqrt{(25)(6)} = \frac{250}{3}(5)\sqrt{6} = \frac{1250\sqrt{6}}{3} \text{ m/s}$$

The speed of sound in helium gas at 27°C is $v \approx \frac{1250\sqrt{6}}{3}$ m/s. If you use a calculator, this works out to $v = 1020$ m/s. (We multiplied by $\frac{\sqrt{3}}{\sqrt{3}}$ to **rationalize** the denominator.)

Example: How much louder is an 80-dB sound compared to a 60-dB sound?

Follow these steps:

- First subtract the two values in decibels: 80 dB − 60 dB = 20 dB.
- Divide 20 dB by 10 decibels: $\frac{20 \text{ dB}}{10 \text{ dB}} = 2$.
- Raise 10 to the power of 2: $10^2 = 100$.

An 80-dB sound is 100 times louder than a 60-dB sound.

Example: How much louder is a 70-dB sound compared to a 20-dB sound?

Follow these steps:

- First subtract the two values in decibels: 70 dB − 20 dB = 50 dB.
- Divide 50 dB by 10 decibels: $\frac{50 \text{ dB}}{10 \text{ dB}} = 5$.
- Raise 10 to the power of 5: $10^5 = 100{,}000$.

A 70-dB sound is 100,000 times louder than a 20-dB sound.

25. A monkey sitting in a tree branch above a pond drops pebbles into the pond with a steady frequency to create a series of concentric circular ripples. A chimpanzee with a meterstick and stopwatch makes the following measurements:

- 30 ripples splash against the edge of the pond every minute.
- The horizontal distance between consecutive crests is 6.0 m.
- The crests are 50 cm higher than the troughs.

(A) Determine the wavelength.

(B) Determine the period.

(C) Determine the frequency.

(D) Determine the amplitude.

(E) Determine the angular wave number.

(F) Determine the wave speed.

26. A cord with a linear mass density of 25 g/cm is clamped at both ends. The tension in the cord is 90 N. What would be the speed of a wave traveling along the cord?

Want help? Check the hints section at the back of the book.

Answers: #25. (A) 6.0 m (B) 2.0 s (C) 0.50 Hz (D) 0.25 m (E) $\frac{\pi}{3}\frac{\text{rad}}{\text{m}}$ (F) 3.0 m/s. #26. 6.0 m/s

27. What is the speed of sound in oxygen gas at 27°C?

Chemistry note: The molar mass of O_2 is twice the molar mass of O.

28. How much louder is a 100-dB sound compared to a 40-dB sound?

Want help? Check the hints section at the back of the book.

Answers: $125\sqrt{7}$ m/s = 331 m/s, 1,000,000×

Partial Derivatives

If f is a function of two independent variables, x and y, when taking a **partial derivative** of f with respect to x, treat y as a constant, and when taking a **partial derivative** of f with respect to y, treat x as a constant. We use the symbol ∂, which is a rounded version of the letter d, to distinguish partial derivatives from total derivatives.

Example: Given $f = 3x^2y$, find a partial derivative of f with respect to x, and also find a partial derivative of f with respect to y.

- To find a partial derivative of f with respect to x, treat y as a constant.
$$\frac{\partial f}{\partial x} = \frac{\partial}{\partial x}(3x^2y) = 3y\frac{\partial}{\partial x}(x^2) = 3y(2x) = 6xy$$
- To find a partial derivative of f with respect to y, treat x as a constant.
$$\frac{\partial f}{\partial y} = \frac{\partial}{\partial y}(3x^2y) = 3x^2\frac{\partial}{\partial y}(y) = 3x^2(1) = 3x^2$$

The Linear Wave Equation

A one-dimensional traveling wave of the form $f(x \pm vt)$ satisfies the linear wave equation:
$$\frac{\partial^2 y}{\partial x^2} = \frac{1}{v^2}\frac{\partial^2 y}{\partial t^2}$$
For example, consider the equation from page 73, which we have repeated below.
$$y = A\sin(k_0x - \omega_0t - \varphi) + y_e$$
When taking a partial derivative of y with respect to x, we treat the independent variable t as a constant. Since y_e is a constant, $\frac{dy_e}{dx} = 0$. Apply the chain rule like we did in Chapter 2 (it may help to review pages 49-50).
$$\frac{\partial y}{\partial x} = \frac{\partial}{\partial x}[A\sin(k_0x - \omega_0t - \varphi) + y_e] = Ak_0\cos(k_0x - \omega_0t - \varphi)$$
Now take a second partial derivative with respect to x.
$$\frac{\partial^2 y}{\partial x^2} = \frac{\partial}{\partial x}[Ak_0\cos(k_0x - \omega_0t - \varphi)] = -Ak_0^2\sin(k_0x - \omega_0t - \varphi)$$
When taking a partial derivative of y with respect to t, we treat the independent variable x as a constant. Since y_e is a constant, $\frac{dy_e}{dt} = 0$.
$$\frac{\partial y}{\partial t} = \frac{\partial}{\partial t}[A\sin(k_0x - \omega_0t - \varphi) + y_e] = -A\omega_0\cos(k_0x - \omega_0t - \varphi)$$
Now take a second partial derivative with respect to t.
$$\frac{\partial^2 y}{\partial t^2} = \frac{\partial}{\partial t}[-A\omega_0\cos(k_0x - \omega_0t - \varphi)] = -A\omega_0^2\sin(k_0x - \omega_0t - \varphi)$$
Plug the expressions for $\frac{\partial^2 y}{\partial x^2}$ and $\frac{\partial^2 y}{\partial t^2}$ into the linear wave equation.

$$\frac{\partial^2 y}{\partial x^2} = \frac{1}{v^2}\frac{\partial^2 y}{\partial t^2}$$

$$-Ak_0^2 \sin(k_0 x - \omega_0 t - \varphi) = \frac{1}{v^2}[-A\omega_0^2 \sin(k_0 x - \omega_0 t - \varphi)]$$

Note that $-A\sin(k_0 x - \omega_0 t - \varphi)$ cancels out.

$$k_0^2 = \frac{\omega_0^2}{v^2}$$

Since $v = \frac{\omega_0}{k}$, we have shown that the equation for a traveling wave from page 73 satisfies the linear wave equation.

Sound waves similarly satisfy the linear wave equation, where we use pressure (P) in place of y. For electromagnetic fields, we use components of the electric field ($\vec{\mathbf{E}}$) or magnetic field ($\vec{\mathbf{B}}$) in place of y.

6 DOPPLER EFFECT AND SHOCK WAVES

Relevant Terminology

Frequency – the number of oscillations completed per second.
Pitch – the property of a sound wave that corresponds to frequency.
Mach number – how much faster a supersonic jet travels compared to the speed of sound.

The Doppler Effect

When there is relative motion between the **source** of a wave (whatever is making the waves) and the **observer** of the wave (the person who is measuring the frequency of a wave, or the person who hears the pitch of a sound wave), the observer measures a shifted frequency (f) compared to the unshifted frequency (f_0) of the source. This phenomenon is known as the Doppler effect. The equation for the **Doppler effect** is:

$$f = f_0 \frac{v \pm v_o}{v \mp v_s}$$

It's important to memorize what the symbols represent. In our notation:

- f is the **shifted** frequency measured by the observer.
- f_0 is the **unshifted** frequency of the source of the wave.
- v (without subscripts) is the **wave speed**. For a **sound wave** in air near normal atmospheric conditions, $v \approx 340$ m/s.
- v_o is the speed of the **observer**.
- v_s is the speed of the **source**. (This "s" stands for "source" – <u>**not**</u> for "sound.")

It's also important to memorize the **sign** conventions. The simple way to understand the sign conventions is to simply **use the upper sign for approaching and the lower sign for receding**. (Note that the plus sign is upper in the numerator for the observer, whereas the minus sign is upper in the denominator for the source. The reason for this is that addition makes a greater frequency in the numerator when the observer is approaching the source, whereas subtraction makes a greater frequency in the denominator when the source is approaching the observer. That is, in the denominator, subtracting the speeds has the effect of making the fraction larger.)

- If the **observer** is heading **toward** the source, use $+$ in the numerator.
- If the **observer** is heading **away from** the source, use $-$ in the numerator.
- If the **source** is heading **toward** the observer, use $-$ in the denominator.
- If the **source** is heading **away from** the observer, use $+$ in the denominator.

Note that the motion of the observer and the source are independent: When you're deciding on the sign in the numerator, only worry about whether the observer is heading toward or away from the source (the source's motion only matters for the sign in the denominator – and in the denominator, it doesn't matter what the observer is doing).

You can see the Doppler effect visually in the diagram below. The left diagram has concentric circles created by a stationary source. In the right diagram, the source is traveling to the right, which shifts the centers of the circular wave fronts. In front of the arrow (the right side of the right diagram), the frequency is higher (there are more ripples per second passing by a given point), whereas behind the arrow (the left side of the right diagram), the frequency is lower (there are fewer ripples per second passing by a given point). Note that higher frequency corresponds to shorter wavelength and vice-versa, in accordance with the equation for wave speed: $v = \lambda f$. That is, wavelength and frequency share a **reciprocal** relationship: $\lambda = \frac{v}{f}$ and $f = \frac{v}{\lambda}$.

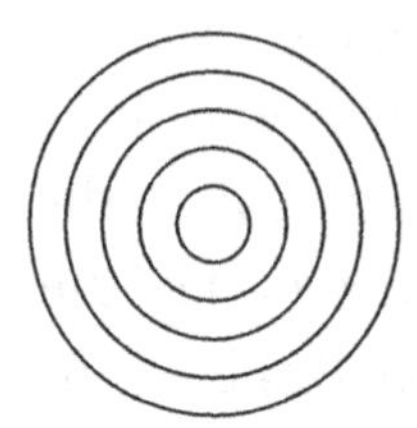 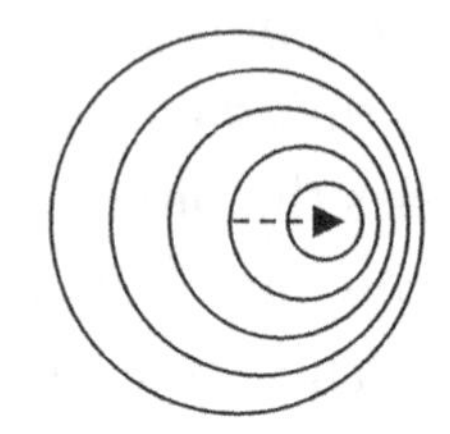

Shock Waves

When the source of a wave travels faster than the wave speed, a **shock wave** is created in the shape of a cone. When an airplane travels faster than the speed of sound, the shock wave is heard as a **sonic boom**. Such an airplane is called **supersonic** (faster than sound).

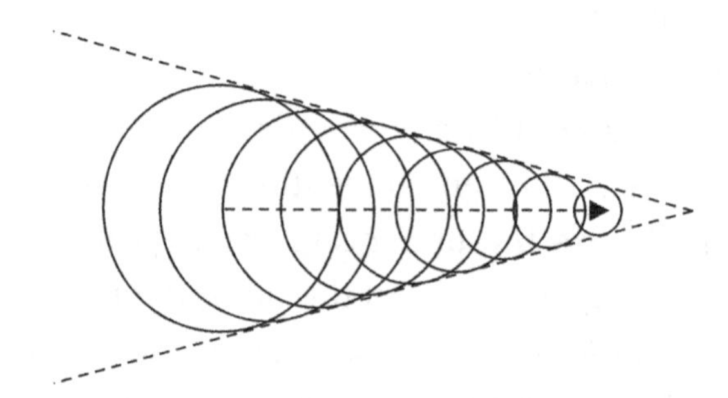

You can see a shock wave visually in the diagram above. As a supersonic aircraft travels horizontally to the right (see the dashed arrow), the sound waves that it creates (these are spherical wave fronts) are all tangent to a cone. As the aircraft travels to the right, it drags the conical shock wave to the right along with it. If and when such a conical shock wave passes through your body, you hear a very loud **sonic boom**. The sonic boom is the result of numerous spherical wave fronts intersecting at the surface of the cone, creating a sound that has a much greater amplitude than the ordinary sound of an aircraft flying.

We use **Mach number** to indicate the speed of a supersonic aircraft.

$$\text{Mach number} = \frac{v_s}{v}$$

It's important to get the notation right.

- v (without subscripts) is the **wave speed**. For a **sound wave** in air near normal atmospheric conditions, $v \approx 340$ m/s.
- v_s is the speed of the source. (This "s" stands for "source" – **not** for "sound.")

For example, an airplane traveling Mach 2 has a speed of $2 \times 340 = 680$ m/s: Mach 2 means twice the speed of sound. As another example, Mach 4 has a speed of $4 \times 340 = 1360$ m/s: Mach 4 means four times the speed of sound.

Consider the diagram below. The symbols represent the following:

- vt is the distance that the sound wave (represented by the dashed circle – which is really a sphere) travels during time t.
- $v_s t$ is the distance that the supersonic jet travels during the same time t. Since the jet is **supersonic**, the jet travels **farther** than the sound wave: $v_s t > vt$.
- h is the altitude of the jet as measured from the ground. Note that $h > vt$, as the line segment BC is a little greater than line segment BD. (It's a common mistake for students to assume that h is the same as vt, but it's **not**.)
- θ is the **half-angle** of the cone. (2θ is the full angle of the cone.)

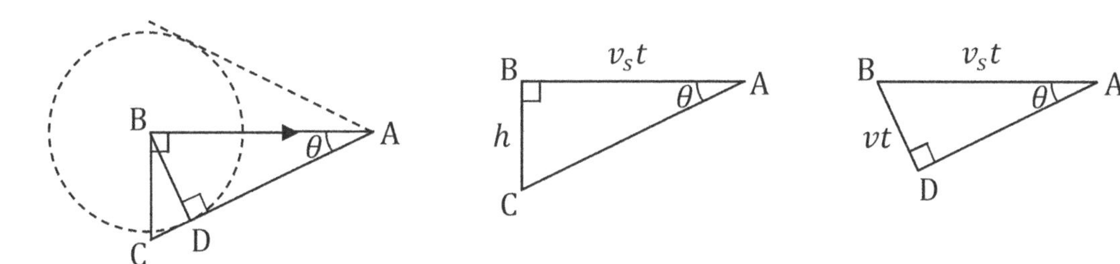

There are two equations to know regarding shock waves. These equations come from two different right triangles in the figure above. The first equation below assumes that the jet is flying **horizontally** (if not, you need to apply trigonometry to find a different equation).

- In right triangle ABCA, h is **opposite** to θ, while $v_s t$ is **adjacent** to θ. For this triangle, the equation involves the **tangent** function.

$$\tan\theta = \frac{h}{v_s t}$$

- In right triangle ABDA, vt is **opposite** to θ, while $v_s t$ is the **hypotenuse**. For this triangle, the equation involves the **sine** function. Observe that time cancels out in the following equation.

$$\sin\theta = \frac{vt}{v_s t} = \frac{v}{v_s}$$

Since $\frac{v_s}{v}$ equals the Mach number, we see that the Mach number also equals $\frac{1}{\sin\theta}$ (which is the same thing as the cosecant of θ).

Symbols and SI Units

Symbol	Name	Units
f	the shifted frequency measured by the observer	Hz
f_0	the unshifted frequency of the source	Hz
v	the speed of the wave (for a sound wave, this is the speed of **sound**)	m/s
v_o	the speed of the **observer**	m/s
v_s	the speed of the **source** ("s" is for "source," **not** "sound") (in a supersonic problem, the speed of the jet)	m/s
h	altitude (height above the ground)	m
θ	the half-angle of the cone-shaped shock wave	° or rad

Important Distinctions

Note that the words "source" and "sound" both start with the letter "s." This makes it easy for students to misinterpret the symbol v_S.

- A v without any subscripts is the speed of the wave. For a sound wave, v is the speed of sound.
- The "s" in v_S refers to the **source**. The symbol v_S represents the speed of the source. (It's **not** the speed of sound.)

The ratio $\frac{v_S}{v}$ is greater than 1 for a **shock wave** problem because the source (supersonic jet) is traveling faster than sound. Similarly, the ratio $\frac{v}{v_S}$ is less than 1 for a shock wave problem. Since the sine function, $\sin\theta$, can't be greater than 1, this makes it easy to reason out that $\frac{v}{v_S} = \sin\theta$. It's also easy to reason out that the bigger speed (the speed of the supersonic jet) goes in the denominator in this equation, whereas the smaller speed (the speed of sound) goes in the numerator. On the other hand, in the equation for Mach number, Mach number $= \frac{v_S}{v}$, the answer is bigger than 1 for a supersonic jet, so in this equation the greater number (the speed of the jet) goes in the numerator, while the smaller number (the speed of sound) goes in the denominator. Students commonly mix these things up in their solutions, but if you take time to reason out what is or isn't less than 1, you can prevent these mistakes.

Doppler Effect Strategy

To solve a problem where there is relative motion between the source of a wave and an observer, follow this strategy:

- Make a list of the known quantities and identify the desired unknown.
  - f is the **shifted** frequency measured by the observer.
  - f_0 is the **unshifted** frequency of the source of the wave.
  - v (without subscripts) is the **wave speed**. For a **sound wave** in air near normal atmospheric conditions, $v \approx 340$ m/s.
  - v_o is the speed of the **observer**.
  - v_s is the speed of the **source**. (This "s" stands for "source" - **not** for "sound.")
- Use the Doppler effect equation.

$$f = f_0 \frac{v \pm v_o}{v \mp v_s}$$

- Reason out the signs correctly.
  - If the **observer** is heading **toward** the source, use $+$ in the numerator.
  - If the **observer** is heading **away from** the source, use $-$ in the numerator.
  - If the **source** is heading **toward** the observer, use $-$ in the denominator.
  - If the **source** is heading **away from** the observer, use $+$ in the denominator.
- Carry out algebra to solve for the desired unknown.

Shock Wave Strategy

To solve a shock wave (or **supersonic** airplane) problem, follow this strategy:

- Make a list of the known quantities and identify the desired unknown.
  - v (without subscripts) is the **wave speed**. For a **sound wave** in air near normal atmospheric conditions, $v \approx 340$ m/s.
  - v_s is the speed of the **source**. (This "s" stands for "source" - **not** for "sound.") For a supersonic jet, v_s is **greater than** 340 m/s. It is **not** 340 m/s.
  - h is the altitude of the jet as measured from the ground.
  - θ is the **half-angle** of the cone. (2θ is the full angle of the cone.)
  - t is the time that the supersonic jet travels horizontally from the moment it passes over an observer's head until the observer hears the **sonic boom**.
- Use the equations for shock waves.

$$\text{Mach number} = \frac{v_s}{v}$$

$$\tan\theta = \frac{h}{v_s t} \quad , \quad \sin\theta = \frac{v}{v_s}$$

- Carry out algebra to solve for the desired unknown.

Example: An orangutan rides a bicycle 20 m/s to the east, while a train travels 60 m/s to the east away from the orangutan. The train blows a whistle with a frequency of 1000 Hz. At what frequency does the orangutan hear the whistle?

Drawing a diagram helps to get the signs right. Both objects travel to the east, with the train heading away from the orangutan. This situation is drawn below.

bicycle   train
20 m/s ⟶   ⟶ 60 m/s

Identify the source and the observer.

- The train is the **source**. The train's whistle is creating the sound waves.
- The orangutan (riding the bicycle) is the **observer**.

List the known quantities and identify the desired unknown.

- The unshifted frequency is $f_0 = 1000$ Hz.
- The speed of sound in air is $v = 340$ m/s.
- The speed of the observer is $v_o = 20$ m/s.
- The speed of the source is $v_s = 60$ m/s.
- We are solving for the shifted frequency (f) heard by the observer (the orangutan).

Write down the equation for the Doppler effect.

$$f = f_0 \frac{v \pm v_o}{v \mp v_s}$$

Reason out the **signs**.

- Since the orangutan (the observer) is heading **towards** the train (the source), we use the **upper** sign in the numerator, which is **positive** $(+)$.
- Since the train (the source) is heading **away from** the orangutan (the observer), we use the **lower** sign in the denominator, which is **positive** $(+)$.

In this example, the equation for the Doppler effect becomes:

$$f = f_0 \frac{v + v_o}{v + v_s}$$

Plug numbers into this equation.

$$f = (1000)\frac{340 + 20}{340 + 60} = (1000)\frac{360}{400} = (1000)\frac{9}{10} = 900 \text{ Hz}$$

The orangutan hears the train whistle at a frequency of $f = 900$ Hz.

Example: A monkey looks straight upward and happens to see a supersonic jet directly overhead that is flying horizontally. The jet is traveling Mach 2. The monkey hears the sonic boom 6.0 s later.

(A) How fast is the jet traveling in SI units?
The equation for Mach number is:

$$\text{Mach number} = \frac{v_s}{v}$$

The speed of sound in air is $v = 340$ m/s.

$$2 = \frac{v_s}{340}$$

Multiply both sides of the equation by 340.

$$v_s = (2)(340) = 680 \text{ m/s}$$

The speed of the jet is $v_s = 680$ m/s. (Mach 2 means twice the speed of sound.)

(B) At what angle above the horizontal will the monkey see the jet when the monkey hears the sonic boom?
Use the shock wave equation that relates v, v_s, and θ.

$$\sin\theta = \frac{v}{v_s} = \frac{340}{680} = \frac{1}{2}$$

Take the inverse sine of both sides of the equation.

$$\theta = \sin^{-1}\left(\frac{1}{2}\right) = 30°$$

The jet will be $\theta = 30°$ above the horizontal (relative to the monkey) when the monkey hears the sonic boom.

(C) What is the altitude of the jet?
Use the shock wave equation that has altitude (h).

$$\tan\theta = \frac{h}{v_s t}$$

Multiply both sides of the equation by $v_s t$. Recall from trig that $\tan 30° = \frac{\sqrt{3}}{3}$.

$$h = v_s t \tan\theta = (680)(6)\tan 30° = (4080)\left(\frac{\sqrt{3}}{3}\right) = 1360\sqrt{3} \text{ m}$$

The altitude of the jet is $h = 1360\sqrt{3}$ m. If you use a calculator, this works out to 2.4 km.

29. A speeding chimpanzee drives a car 65 m/s to the east. A police officer pursues the chimpanzee, driving 90 m/s to the east. The police car blares a siren with a frequency of 800 Hz. At what frequency does the chimpanzee hear the police siren?

30. A gorilla rides a bicycle 20 m/s to the north, while a train travels 40 m/s to the south towards the gorilla. The train blows a whistle with a frequency of 500 Hz. At what frequency does the gorilla hear the whistle?

31. An ambulance travels 60 m/s to the west, while an orangutan drives 40 m/s to the east, heading away from the ambulance. The ambulance blares a siren with a frequency of 1200 Hz. At what frequency does the orangutan hear the siren?

Want help? Check the hints section at the back of the book.

Answers: 880 Hz, 600 Hz, 900 Hz

32. A chimpanzee drives 60 m/s to the east towards a train that also travels to the east. The train blows a whistle with a frequency of 850 Hz. The chimpanzee hears the whistle with a frequency of 800 Hz. What is the speed of the train?

33. A monkey is parked at a railroad crossing. When the train is approaching, the monkey hears the train's whistle at a frequency of 750 Hz. As soon as the train passes the monkey and is getting further away, the monkey hears the train's whistle at a frequency of 500 Hz. What is the speed of the train? **Hint**: <u>Neither</u> frequency given equals f_0.

Want help? Check the hints section at the back of the book.

Answers: 85 m/s, 68 m/s

34. A monkey looks straight upward and happens to see a jet airplane directly overhead that is flying horizontally. Exactly $5\sqrt{2}$ seconds later, the monkey hears the sonic boom from the supersonic jet, when the jet is 45° above the horizon from the monkey.

(A) What is the speed of the jet?

(B) What is the altitude of the jet?

Want help? Check the hints section at the back of the book.

Answers: (A) $340\sqrt{2}$ m/s = 481 m/s (B) 3.4 km

7 STANDING WAVES

Relevant Terminology

Crest – a point on a wave where there is a **maximum** (the top of a hill).
Trough – a point on a wave where there is a **minimum** (the bottom of a valley).
Amplitude – the **vertical** height between a crest and the equilibrium position (or **one-half** of the vertical height between a crest and trough).
Wavelength – the **horizontal** distance between two consecutive crests.
Frequency – the number of oscillations completed per second.
Wave speed – how fast the wave travels.
Linear mass density – mass per unit length.
Temperature – a measure of the average kinetic energy of the molecules of a substance.

Essential Concepts

If you send a train of traveling waves through a medium (like a cord, rod, or pipe), in general the reflected waves will partially cancel one another and nothing special happens. However, if the wavelength of the traveling waves is just right, the reflected waves will reinforce one another, creating a **standing wave** with a much larger amplitude than any of the individual waves. This phenomenon is known as **resonance**.

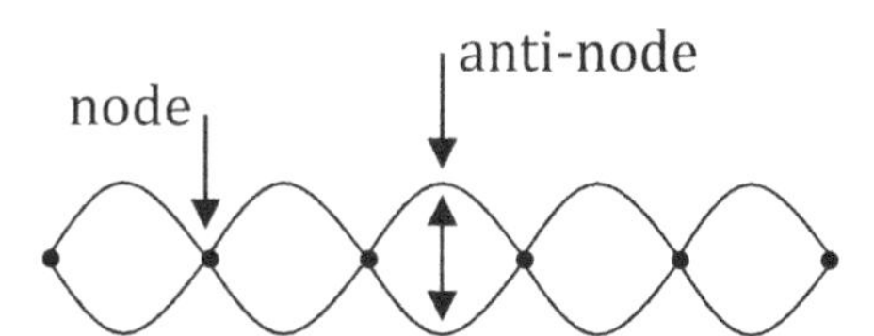

The shapes and wavelengths of the standing waves depend upon the **boundary conditions**.

- A closed or fixed end (boundary) will have a **node**. A node is a point on a standing wave where the amplitude is zero. There is no oscillation at a node.
- A free end will have an **anti-node**. An anti-node is a point on a standing wave where the amplitude is maximum. There is maximum oscillation at an anti-node.

The ends of the standing waves must be nodes or anti-nodes, depending upon whether each end of the medium is fixed or free. However, the points on the medium between the ends do not need to be nodes or anti-nodes (they can be somewhere in between these two extremes). If you produce standing waves on a string such as the diagram of a standing wave shown above, the string oscillates up and down so rapidly that your eye perceives the string to be both at a crest and a trough at the same time at each anti-node (when really the string is oscillating up and down between the crest and trough several times per second).

Standing Wave Equations

The **wavelength** (λ) of a standing wave is related to the **active length** (L) of the medium by:

$$\binom{\text{number of}}{\text{cycles}}\lambda = L$$

Note that **one cycle** looks like this:

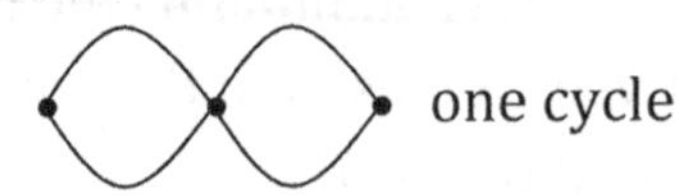

Note that one "football" shape is only <u>**half**</u> of a cycle.

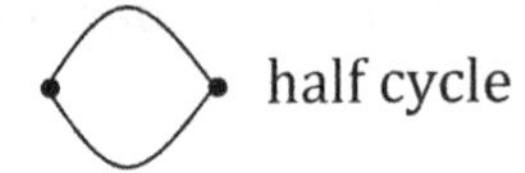

Why? Think about what one full cycle of a sine wave looks like. If we make part of a full cycle look like a dashed curve, you should see that the remaining solid curve looks exactly like a complete cycle of a sine wave.

The **resonance frequency** (f) is related to the **wavelength** (λ) and **wave speed** (v) by:

$$v = \lambda f$$

You may need to use a second equation (from Chapter 5) involving **wave speed** (v), depending upon what you know and what you're solving for. For standing waves on a string, the wave speed depends on the **tension** (F) in the string and the **linear mass density** (μ) of the string. The linear mass density equals the total mass (m_s) of the string divided by the total length[1] (L_s) of the string: $\mu = \frac{m_s}{L_s}$.

$$v = \sqrt{\frac{F}{\mu}} \quad \text{(string)}$$

For a standing wave in a pipe filled with an ideal gas, the wave speed depends on the absolute **temperature** (T) in Kelvin (K), the **molar mass** (M) of the gas (in kg/mol), and whether the gas is **monatomic** ($\gamma = \frac{5}{3}$), like He, Ne, Ar, Kr, Xe, and Rn, or **diatomic** ($\gamma = \frac{7}{5}$), like H_2, N_2, O_2, F_2, Cl_2, Br_2, and I_2. The universal gas constant is $R = 8.314 \frac{\text{J}}{\text{mol·K}} \approx \frac{25}{3} \frac{\text{J}}{\text{mol·K}}$.

$$v = \sqrt{\frac{\gamma R T}{M}} \quad \text{(ideal gas)}$$

For a **sound wave in air** near standard conditions, $v = 340$ m/s.

---

[1] The **active** length (L) of the medium may be different from the **total** length (L_s) of the string. This happens in lab, for example, where some of the string hangs over a pulley to connect to a mass hanger, and some of the string is used to tie knots. The **active** length (L) is just the part where the standing waves form.

Symbols and SI Units

Symbol	Name	Units
λ	wavelength	m
f	frequency	Hz
v	the speed of the wave	m/s
L	the active length of the medium	m
L_s	the total length of a string	m
m_s	the total mass of a string	kg
M	the molar mass of a gas (in kilograms per mole)	kg/mol
F	tension	N
μ	linear mass density (mass per unit length)	kg/m
γ	adiabatic index	unitless
R	universal gas constant	$\frac{\text{J}}{\text{mol}\cdot\text{K}}$
T	absolute temperature	K

Important Distinctions

The active length (L) of a string is just the portion of the string where standing waves form, and may thus be different from the total length (L_s) of the string.

one cycle

It's a **common mistake** for students to associate a full cycle with a "football" shaped standing wave, but one "football" is only **half** of a cycle.

As we will see in the strategy that follows, there are two popular conventions for labeling the standing waves. One convention calls them the fundamental, first overtone, second overtone, etc. Another convention calls them the first harmonic, second harmonic, third harmonic, etc. The numbering is different: For example, the second overtone is the same thing as the third harmonic. (The terms "harmonic" and "overtone" are distinguished in subtle ways in the physics of music. We will not get into such detail in this text.)

Standing Waves Strategy

To solve for the wavelength, speed, or resonance frequency of a standing wave, follow this strategy:[2]

1. Identify the boundary conditions. First, decide whether each end of the medium is fixed or free. See the examples for help determining this.
   - A fixed end will have a **node** (a fixed point).
   - A free end will have an **anti-node** (with maximum oscillation).

   Is there anything restrictive inside the medium (that is, anywhere other than the ends)? For example, if a rod or string is clamped down in its middle, there will also be a node at the position of the **clamp**. The last example features such a clamp.
2. Draw a long train of waves like the one below. This will help you visualize the answers to the next step. The symbol n is a **node**, while a is an **anti-node**.

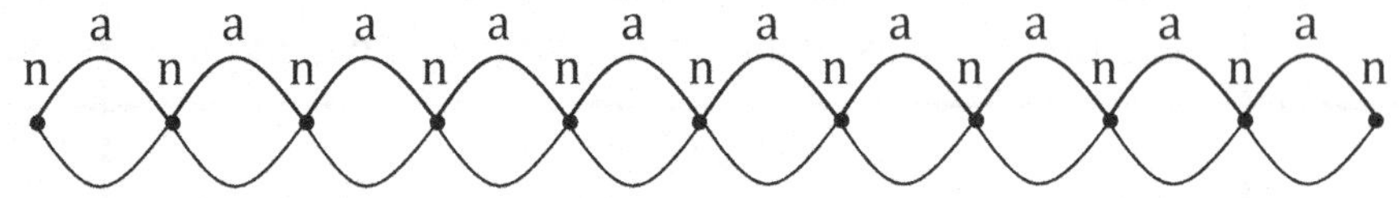

3. Draw the shortest section of the train of waves from Step 2 that matches **<u>all</u>** of the boundary conditions from Step 1. The shortest section is called the **fundamental** or the **first harmonic**. Also draw the second shortest and third shortest sections of the train of waves that match **<u>all</u>** of the boundary conditions from Step 1. These are called the **first overtone, second overtone**, etc., or the **second harmonic, third harmonic**, etc. This technique is illustrated in the examples that follow.
4. Apply the following equation to each standing wave that you drew in Step 3.
$$\begin{pmatrix}\text{number of} \\ \text{cycles}\end{pmatrix} \lambda = L$$
5. Apply the following equation to relate the wavelength (λ), resonance frequency (f), and wave speed (v).
$$v = \lambda f$$
6. You may also need to use equations from Chapter 5. See page 96 for important details regarding these equations. For a wave on a **string**, $v = \sqrt{\frac{F}{\mu}}$ and $\mu = \frac{m_s}{L_s}$. For a sound wave in air near standard conditions, $v = 340$ m/s. For a sound wave in an ideal gas, $v = \sqrt{\frac{\gamma RT}{M}}$, where $R = 8.314 \frac{\text{J}}{\text{mol·K}} \approx \frac{25}{3} \frac{\text{J}}{\text{mol·K}}$.

---

[2] Most textbooks present a formula involving an index n for a few common situations. Following our strategy has two advantages over jumping straight into that formula. First, the strategy breaks the solution down into simple steps in a way that allows the student to understand both the solution and underlying concepts. In addition, this strategy can be applied to a wider variety of standing wave problems. For example, if compressional waves are sent along a rod that has a clamp anywhere other than at its endpoint, the standard textbook formulas involving an index n won't apply, but this strategy will still work. The last example and last problem of this chapter feature a rod with such a clamp.

Example: A monkey sets up standing waves in a cord tied down at two ends. The cord has a mass of 70 g and a total length of 3.5 m, yet the distance between the two knots is 3.0 m. The tension in the cord is 18 N. Determine first three resonance frequencies.
First identify the **boundary conditions**: Since both ends are tied down, there will be a **node** at each end. Look at the train of waves drawn below.

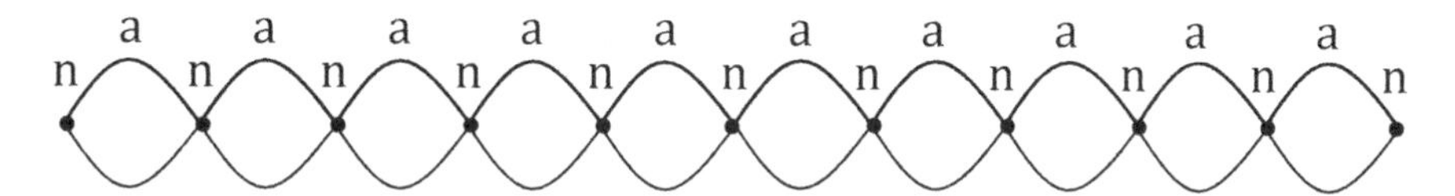

What is the shortest section of this train that has a node (n) at both ends? The answer is a "football" shaped section, like the one below. This is the **fundamental** (or **first harmonic**) for this problem.

The second shortest section that has a node (n) at both ends looks like two "footballs." This is the **first overtone** (or **second harmonic**).

first overtone

The third shortest section that has a node (n) at both ends looks like three "footballs." This is the **second overtone** (or **third harmonic**).

second overtone

Now we apply the following equation to each standing wave that we drew.

$$\begin{pmatrix}\text{number of} \\ \text{cycles}\end{pmatrix} \lambda = L$$

Recall that one full cycle looks like the picture below.

one cycle

Compare the diagram for the **fundamental** to the picture above. In this example, the fundamental is one-half of a cycle. Plug $\frac{1}{2}$ into the above equation for the fundamental. Since each standing wave will have a different wavelength, we will use subscripts (as in λ_0).

$$\frac{1}{2}\lambda_0 = L$$

Compare the diagram for the **first overtone** to the picture above. The first overtone is exactly one cycle. Plug 1 into the above equation for the first overtone.

$$1\lambda_1 = \lambda_1 = L$$

Compare the diagram for the **second overtone** to the picture above. The second overtone is one and one-half cycles: $1 + \frac{1}{2} = \frac{3}{2}$. Plug $\frac{3}{2}$ into the above equation for the second overtone.

$$\frac{3}{2}\lambda_2 = L$$

Use algebra to solve for wavelength in each of the previous equations. Note that the **active** length of the string is the distance between the two knots: $L = 3.0$ m.

$$\frac{1}{2}\lambda_0 = L \quad \to \quad \lambda_0 = 2L = (2)(3) = 6.0 \text{ m}$$

$$\lambda_1 = L \quad \to \quad \lambda_1 = L = 3.0 \text{ m}$$

$$\frac{3}{2}\lambda_2 = L \quad \to \quad \lambda_2 = \frac{2L}{3} = \frac{(2)(3)}{3} = 2.0 \text{ m}$$

The wavelengths of the fundamental and first two overtones are $\lambda_0 = 6.0$ m, $\lambda_1 = 3.0$ m, and $\lambda_2 = 2.0$ m. We're not finished because the problem asked for **frequency**. We could use the equation $v = \lambda f$ to find frequency, but first we need to find the wave speed. Since this is a string, we can use the equation for the speed of a wave along a string, $v = \sqrt{\frac{F}{\mu}}$. We know that the tension is $F = 18$ N, but first we need to determine the linear mass density (μ). To find the linear mass density, divide the total mass of the string ($m_s = 70$ g) by the total length of the string ($L_s = 3.5$ m). Convert the mass from grams (g) to kilograms (kg): $m_s = 70 \text{ g} = 0.070$ kg.

$$\mu = \frac{m_s}{L_s} = \frac{0.07}{3.5} = 0.020 \text{ kg/m}$$

Plug the linear mass density ($\mu = 0.020$ kg/m) and tension ($F = 18$ N) into the equation for the speed of a wave along a string.

$$v = \sqrt{\frac{F}{\mu}} = \sqrt{\frac{18}{0.02}} = \sqrt{(18)(50)} = \sqrt{900} = 30 \text{ m/s}$$

The wave speed is $v = 30$ m/s. Now we can solve for the resonance frequencies using the equation $v = \lambda f$. Divide both sides of the equation by the wavelength: $f = \frac{v}{\lambda}$. We will include subscripts for wavelength and frequency, which are different for each standing wave in this example (whereas the wave speed is the same for each in this example).[3]

$$f_0 = \frac{v}{\lambda_0} = \frac{30}{6} = 5 \text{ Hz}$$

$$f_1 = \frac{v}{\lambda_1} = \frac{30}{3} = 10 \text{ Hz}$$

$$f_2 = \frac{v}{\lambda_2} = \frac{30}{2} = 15 \text{ Hz}$$

The first three resonance frequencies are $f_0 = 5$ Hz, $f_1 = 10$ Hz, and $f_2 = 15$ Hz.

[3] That's because this example involves a single string with a fixed active length. Not every problem is the same way, though. A variation of this problem is to specify a fixed frequency, and ask how long the cord should be to create the first three standing waves. In that variation, the frequency would be the same for each standing wave, but the wavelength and wave speed would be different for each.

Example: A monkey howls near the open end of a 4.25-m long pipe. As shown below, the left end is closed while the right end is open. Determine first three resonance frequencies.

First identify the **boundary conditions**:

- There will be a **node** at the left end because the left end is **closed**. (The air at the very left edge doesn't have the freedom to vibrate horizontally.)
- There will be an **anti-node** at the right end because the right end is **open**.

Look at the train of waves drawn below.

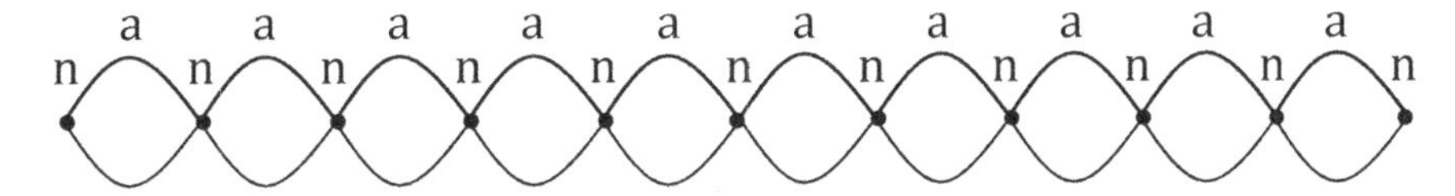

What is the shortest section of this train that has a node (n) at the left end and an anti-node (a) at the right end? The answer is half of a "football" shaped section, like the one below. This is the **fundamental** (or **first harmonic**) for this problem.

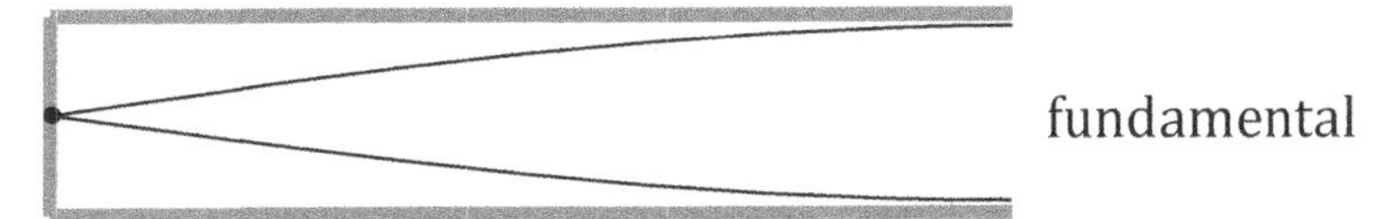
fundamental

The second shortest section that has a node (n) at the left end and an anti-node (a) at the right end looks like one and a half "footballs." This is the **first overtone** (or **second harmonic**).

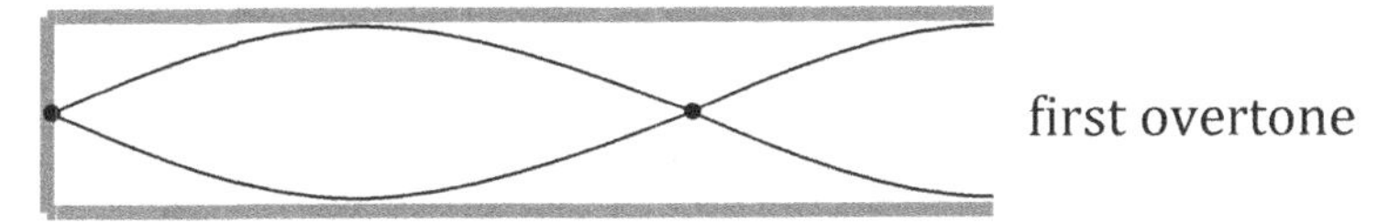
first overtone

The third shortest section that has a node (n) at the left end and an anti-node (a) at the right end looks like two and a half "footballs." This is the **second overtone** (or **third harmonic**).

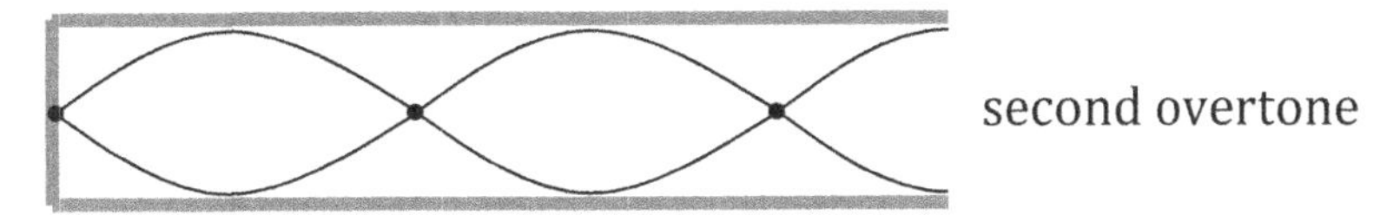
second overtone

Now we apply the following equation to each standing wave that we drew.

$$\begin{pmatrix}\text{number of}\\ \text{cycles}\end{pmatrix}\lambda = L$$

Important note: The number of cycles is **<u>not</u>** equal to the number of "footballs." Recall that one full cycle looks like the picture below. Therefore, one "football" is **<u>half</u>** of a cycle.

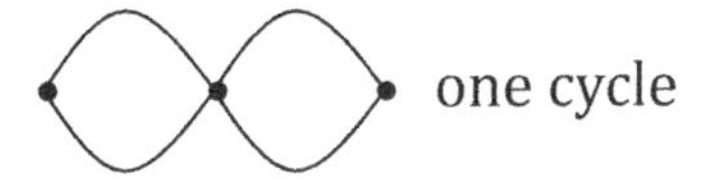
one cycle

Compare the diagram for the **fundamental** to the picture for one cycle. In this example, the fundamental is one-fourth of a cycle. (The fundamental is one-half of a "football," but a "football" is one-half of a cycle, which means that the fundamental is one-fourth of a cycle, since $\frac{1}{2} \times \frac{1}{2} = \frac{1}{4}$. If it helps, you can count the number of "footballs" and **divide by two.**) Plug $\frac{1}{4}$ into the equation involving wavelength. Since each standing wave will have a different wavelength, we will use subscripts (as in λ_0).

$$\frac{1}{4}\lambda_0 = L$$

Compare the diagram for the **first overtone** to the picture for one cycle. The first overtone is three-fourths of a cycle (one and a half "footballs" divided by two: $\frac{1.5}{2} = 0.75 = \frac{3}{4}$). Plug $\frac{3}{4}$ into the equation involving wavelength.

$$\frac{3}{4}\lambda_1 = L$$

Compare the diagram for the **second overtone** to the picture for one cycle. The second overtone is one and one-fourth cycles (two and a half "footballs" divided by two: $\frac{2.5}{2} = 1.25 = \frac{5}{4}$). Plug $\frac{5}{4}$ into the equation involving wavelength.

$$\frac{5}{4}\lambda_2 = L$$

Use algebra to solve for wavelength in each of the previous equations. Recall that the length of the pipe is $L = 4.25$ m.

$$\frac{1}{4}\lambda_0 = L \quad \rightarrow \quad \lambda_0 = 4L = (4)(4.25) = 17 \text{ m}$$

$$\frac{3}{4}\lambda_1 = L \quad \rightarrow \quad \lambda_1 = \frac{4L}{3} = \frac{(4)(4.25)}{3} = \frac{17}{3} \text{ m}$$

$$\frac{5}{4}\lambda_2 = L \quad \rightarrow \quad \lambda_2 = \frac{4L}{5} = \frac{(4)(4.25)}{5} = \frac{17}{5} \text{ m}$$

Use the equation $f = \frac{v}{\lambda}$ to find frequency. Assume that there is air inside the pipe near standard conditions (since the problem didn't specify otherwise). Recall that the speed of sound in air near standard conditions is $v = 340$ m/s. We will include subscripts for wavelength and frequency, which are different for each standing wave in this example (whereas the speed of sound is the same for each).

$$f_0 = \frac{v}{\lambda_0} = \frac{340}{17} = 20 \text{ Hz}$$

$$f_1 = \frac{v}{\lambda_1} = \frac{340}{17/3} = (340)\left(\frac{3}{17}\right) = (20)(3) = 60 \text{ Hz}$$

$$f_2 = \frac{v}{\lambda_2} = \frac{340}{2} = (340)\left(\frac{5}{17}\right) = (20)(5) = 100 \text{ Hz}$$

The first three resonance frequencies are $f_0 = 20$ Hz, $f_1 = 60$ Hz, and $f_2 = 100$ Hz. To divide by a fraction, multiply by its **reciprocal**.

Example: A 6.0-m long metal rod is clamped at its midpoint and compressional waves are sent through the rod with a speed of 4800 m/s. (The ends of the rod are **not** clamped – just the middle.) Find the resonance frequency for the fundamental and first overtone.
First identify the **boundary conditions**:

- Both ends will be **anti-nodes** because the ends are **not** clamped (they are **free**).
- In addition, there will be a **node** at the center of the rod because the midpoint of the rod is clamped down (it is **fixed**).

Look at the train of waves drawn below.

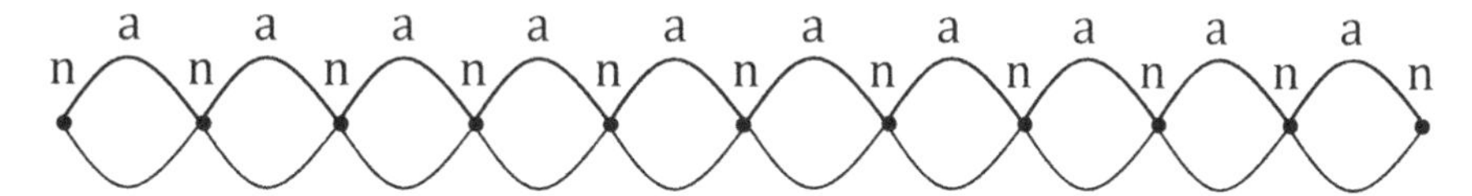

What is the shortest section of this train that has an anti-node (a) at the left end, another anti-node (a) at the right end, and a node (n) at the center? The answer is the standing wave shown below. The wave below is the **fundamental** for this problem.

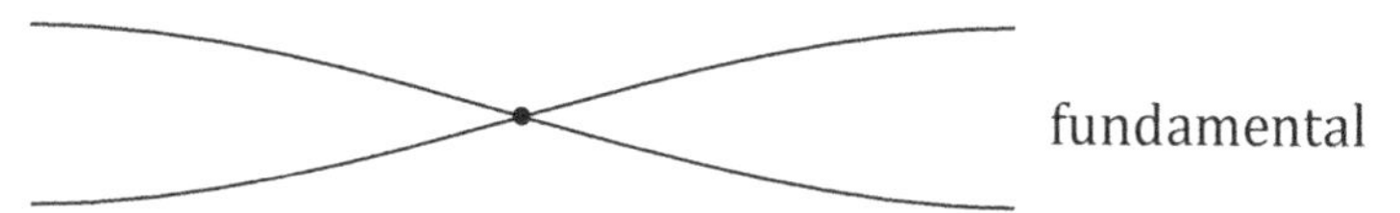
fundamental

The second shortest section that has an anti-node (a) at the left end, another anti-node (a) at the right end, and a node (n) at the center looks like the standing wave shown below. (Anything shorter – other than the fundamental shown above – would either not have the required node at the exact center or wouldn't have full anti-nodes at both ends.) The wave below is the **first overtone** for this problem.

first overtone

Now we apply the following equation to each standing wave that we drew.

$$\begin{pmatrix}\text{number of}\\ \text{cycles}\end{pmatrix}\lambda = L$$

Recall that one full cycle looks like the picture below, such that a "football" is **half** of a cycle.

one cycle

Compare the diagram for the **fundamental** to the picture for one cycle. In this example, the fundamental is one-half of a cycle. (The fundamental consists of two "half-footballs," and two halves make a whole "football." Recall that one "football" is half of a cycle.) Plug $\frac{1}{2}$ into the equation involving wavelength. Since each standing wave will have a different wavelength, we will use subscripts (as in λ_0).

$$\frac{1}{2}\lambda_0 = L$$

Compare the diagram for the **first overtone** to the picture for one cycle. The first overtone is one and one-half cycles. (Add two complete "footballs" to two half "footballs" to get three "footballs," and then divide by two since one "football" is half of a cycle). Plug $\frac{3}{2}$ into the equation involving wavelength.

$$\frac{3}{2}\lambda_1 = L$$

Use algebra to solve for wavelength in each of the previous equations. Recall that the length of the rod is $L = 6.0$ m.

$$\frac{1}{2}\lambda_0 = L \quad \rightarrow \quad \lambda_0 = 2L = (2)(6) = 12 \text{ m}$$

$$\frac{3}{2}\lambda_1 = L \quad \rightarrow \quad \lambda_1 = \frac{2L}{3} = \frac{(2)(6)}{3} = 4.0 \text{ m}$$

Use the equation $f = \frac{v}{\lambda}$ to find frequency. The problem states that the compressional waves have a speed of $v = 4800$ m/s. We will include subscripts for wavelength and frequency, which are different for each standing wave in this example (whereas the speed of the compressional waves is the same for each).

$$f_0 = \frac{v}{\lambda_0} = \frac{4800}{12} = 400 \text{ Hz}$$

$$f_1 = \frac{v}{\lambda_1} = \frac{4800}{4} = 1200 \text{ Hz}$$

The resonance frequencies for the fundamental and first overtone are $f_0 = 400$ Hz and $f_1 = 1200$ Hz.

35. A monkey sets up standing waves in a cord clamped at both ends. The cord has a linear mass density of 0.30 g/cm. The distance between the two clamps is 1.5 m. The tension in the cord is 27 N.

(A) Draw the fundamental, first overtone, and second overtone.

(B) Determine the wavelengths of the fundamental and first two overtones.

(C) Determine first three resonance frequencies.

Want help? Check the hints section at the back of the book.

Answers: (B) 3.0 m, 1.5 m, 1.0 m (C) 10 Hz, 20 Hz, 30 Hz

36. A monkey howls near the open end of a 34-cm long pipe. As shown below, the left end is open while the right end is closed.

(A) Draw the fundamental, first overtone, and second overtone in the space provided.

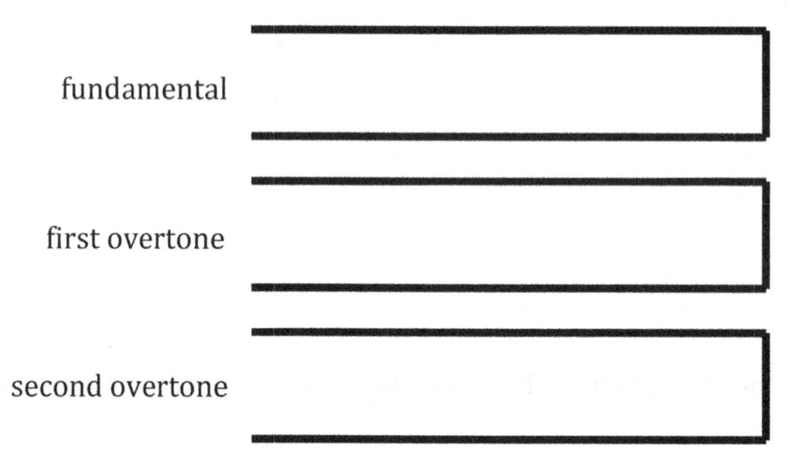

(B) Determine the wavelengths of the fundamental and first two overtones.

(C) Determine first three resonance frequencies.

Want help? Check the hints section at the back of the book.

Answers: (B) $\frac{34}{25}$ m, $\frac{34}{75}$ m, $\frac{34}{125}$ m (C) 250 Hz, 500 Hz, 1250 Hz

37. Both ends of the pipe illustrated below are open. The lab (including the pipe) is filled with banana gas (named for its intoxicating effect on monkeys). The speed of sound in banana gas is 300 m/s. The length of the pipe is 4.5 m.

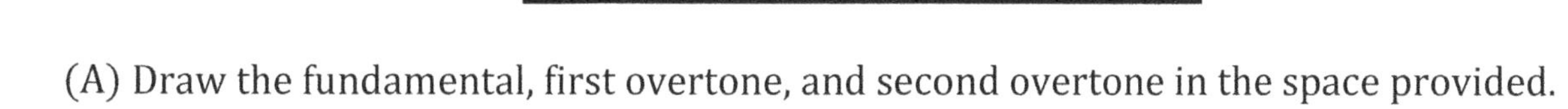

(A) Draw the fundamental, first overtone, and second overtone in the space provided.

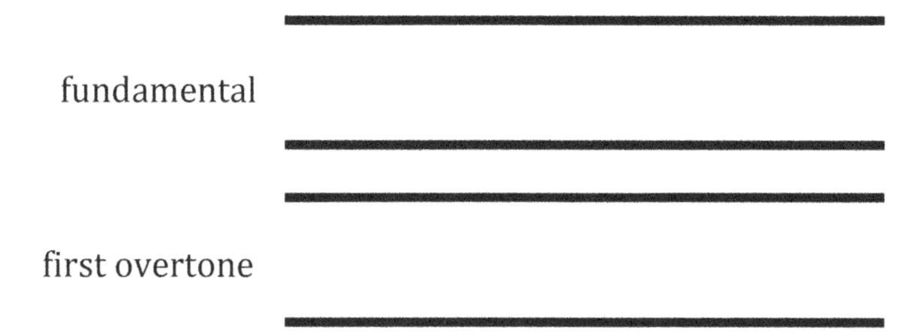

fundamental

first overtone

second overtone

(B) Determine the wavelengths of the fundamental and first two overtones.

(C) Determine first three resonance frequencies.

Want help? Check the hints section at the back of the book.

Answers: (B) 9.0 m, 4.5 m, 3.0 m (C) $\frac{100}{3}$ Hz, $\frac{200}{3}$ Hz, 100 Hz

38. A 6.0-m long horizontal metal rod is clamped at point that is one-fourth the length of the rod from its left end and compressional waves are sent through the rod with a speed of 6000 m/s. (The ends of the rod are **not** clamped – just the point one-fourth the length of the rod from its left end.) Find the resonance frequency for the fundamental and first overtone.

Want help? Check the hints section at the back of the book.

Answers: 1.0 kHz, 3.0 kHz

8 DENSITY

Relevant Terminology

Inertia – the natural tendency of all objects to maintain constant momentum.
Mass – a measure of **inertia**. More massive objects are more difficult to accelerate as they have greater inertia to overcome.
Weight – the **gravitational force** that an object experiences. Weight equals mass times gravitational acceleration.
Volume – the amount of space that an object occupies. Also referred to as **capacity**.
Density – a measure of the compactness of a substance. Density is **mass per unit volume**.

Essential Concepts

The **density** of a substance provides a measure of the compactness of the substance. Since density equals mass divided by volume, an object that puts more mass in less space is more dense. Density is a characteristic property of a substance, which depends on the size and mass of its atoms, along with the way that the atoms are arranged.

Consider the following "trick" question. **Which is heavier – a brick or a pile of feathers?** See if you can figure out the "trick" on your own before you read on.

It's impossible to answer this question correctly without more information. It depends on how large the pile of feathers is. Consider the following extreme case: Suppose you fill an entire football stadium with feathers. You wouldn't be able to lift millions of feathers at once, but you could easily lift a single brick. On the other hand, if the pile of feathers is small, the brick will weigh more.

While we can't conclude anything about the weight (or heaviness) without additional information, we can say that the brick is more **dense** than the feathers. If the pile of feathers has exactly the same volume as the brick, the brick will be heavier because it has greater density.

Another way to compare the densities of objects is to drop them in a fluid. For example, if you put a lead block and aluminum block in a swimming pool, the lead block will fall with greater acceleration because it is more dense than aluminum. Both materials are more dense than water because they sink. If instead you place a block of ice in a cup of water, the ice cube will float because it is less dense than water. We will explore the physics of sinking and floating further in Chapter 10.

The Density Equation

The **density** (ρ) of an object equals the ratio of its **mass** (m) to its **volume** (V).

$$\rho = \frac{m}{V}$$

To solve for mass, multiply both sides of the equation by volume.

$$m = \rho V$$

To solve for volume, divide both sides of the previous equation by mass.

$$V = \frac{m}{\rho}$$

Notice that mass is **never** downstairs in the density equation.

Following are the formulas for the volume of a few standard geometries.

Shape	Dimensions	Volume
Cube	length L	$V = L^3$
Cuboid (a rectangular box)	length L width W height H	$V = LWH$
Sphere	radius R	$V = \frac{4\pi R^3}{3}$
Cylinder (right circular)	radius R height H	$V = \pi R^2 H$
Cone (right, circular base)	radius R height H	$V = \frac{\pi R^2 H}{3}$

Important Distinctions

It's important to distinguish between the following terms.

- **Volume** is the amount of space that an object occupies. It refers to **size**.
- **Weight** is gravitational force. It refers to **heaviness**.
- **Density** is mass per unit volume. It refers to **compactness**. The density of an object depends on both its mass and its volume.

Symbols and SI Units

Symbol	Name	SI Units
ρ	density	$\frac{\text{kg}}{\text{m}^3}$
m	mass	kg
V	volume	m^3
L	length	m
W	width	m
H	height	m
R	radius	m

Notes Regarding Units

The SI units for density (ρ) are $\frac{\text{kg}}{\text{m}^3}$. This follows from the equation $\rho = \frac{m}{V}$, since the SI unit for mass (m) is the kilogram (m) and the SI units of volume (V) are cubic meters (m^3).

In physics, we work with the **mks** system of units: the meter (m), kilogram (kg), and second. This is convenient for mechanics problems. For example, if you throw a baseball, it's usually more convenient to work with the ball's mass in kilograms than grams or the distance traveled in meters than centimeters. However, in other branches of science, the **cgs** system of units (centimeters, grams, seconds) is more common. For example, if you work in a lab where you work with spoons and test tubes, it may be more convenient to work with grams and centimeters than kilograms and meters.

Thus, if you have studied other subjects, you may know that the density of water is about one in grams per cubic centimeter ($\frac{\text{g}}{\text{cm}^3}$). However, in physics, the density of water is about 1000 in kilograms per cubic meter ($\frac{\text{kg}}{\text{m}^3}$). Note that $1 \ \frac{\text{g}}{\text{cm}^3} = 1000 \ \frac{\text{kg}}{\text{m}^3}$ because $1 \text{ kg} = 10^3 \text{ g}$ and $1 \text{ m}^3 = (1 \text{ m})^3 = (100 \text{ cm})^3 = 100^3\text{cm}^3 = 10^6 \text{ cm}^3$.

In physics problems, it's a good habit to convert all distances to **meters** (m) and all masses to **kilograms** (kg).

Density Problem Strategy

How you solve a problem involving density depends on which kind of problem it is:

- For a textbook problem, you may need to look up the value of ρ in a table.
- To relate **density** (ρ), **mass** (m), and **volume** (V), apply the density equation.

$$\rho = \frac{m}{V}$$

The SI unit of mass (m) is the kilogram (kg), and the SI units of volume (V) are m^3. Note the following conversions.[1]

$$1 \text{ kg} = 10^3 \text{ g}$$
$$1 \text{ m}^3 = 10^6 \text{ cm}^3$$
$$1 \frac{\text{g}}{\text{cm}^3} = 1000 \frac{\text{kg}}{\text{m}^3}$$

Note: You may need to look up the **formula** for **volume** in the table on page 110.

- For a conceptual question involving density that compares two objects, first write down the density equation for each object with subscripts for any quantities that may be different for the two objects. Then solve for the quantity that is the same in both equations. This lets you eliminate one quantity from the equations, as shown in one of the examples.
- If a problem may involve **buoyancy**, sinking, or floating, see Chapter 10.
- If a problem involves fluid dynamics, see Chapter 11.

Example: A monkey begins with a 2.0 m × 2.0 m × 2.0 m cube of wet sand with a density of 2000 $\frac{\text{kg}}{\text{m}^3}$. The monkey then sculpts the sand into a giant banana (using 100% of the sand). How much does the giant banana weigh (while the sand is still wet)?

Since the monkey uses 100% of the sand for his sculpture, the mass of the banana is the same as it was when it was the shape of a cube. First find the volume of the cube.

$$V = L^3 = (2)^3 = 8.0 \text{ m}^3$$

Apply the density equation.

$$\rho = \frac{m}{V}$$

Multiply both sides of the equation by volume.

$$m = \rho V = (2000)(8) = 16{,}000 \text{ kg}$$

Weight equals mass times gravity.

$$mg = (16{,}000)(9.81) \approx (16{,}000)(10) = 160{,}000 \text{ N} = 160 \text{ kN}$$

The giant banana sculpture weighs $mg = 160$ kN, which is the same as $mg = 160{,}000$ N.

[1] Note that $1 \frac{\text{g}}{\text{cm}^3} = 1 \frac{\text{g}}{\text{cm}^3} \times \frac{1 \text{ kg}}{10^3 \text{ g}} \times \frac{10^6 \text{ cm}^3}{1 \text{ m}^3} = 10^3 \frac{\text{kg}}{\text{m}^3} = 1000 \frac{\text{kg}}{\text{m}^3}$ and that $1 \text{ m}^3 = (10^2 \text{ cm})^3 = 10^6 \text{ cm}^3$.

Example: A banana made out of platinum is twice as dense as a banana made out of silver. Both bananas have the same weight. Which banana has more volume, and by what factor?

Write down the density equation for each banana. We will use subscripts "p" and "s" (for "platinum" and "silver") for density and volume, which are different. We will not use a subscript for mass, since the bananas have the same mass (and the same weight).

$$\rho_p = \frac{m}{V_p} \quad , \quad \rho_s = \frac{m}{V_s}$$

Solve for mass in both equations since each banana has the same mass. Multiply both sides of each equation by the respective volume.

$$m = \rho_p V_p \quad , \quad m = \rho_s V_s$$

Since $\rho_p V_p$ and $\rho_s V_s$ both equal the same mass, we may set them equal to each other.

$$\rho_p V_p = \rho_s V_s$$

The problem states that the platinum banana is twice as dense as the silver banana. Express this in an equation.

$$\rho_p = 2\rho_s$$

Substitute this expression into the previous equation.

$$(2\rho_s)V_p = \rho_s V_s$$

Divide both sides of the equation by ρ_s.

$$2V_p = V_s$$

The volume of the silver banana, V_s, is **twice** the volume of the platinum banana, V_p. Since the silver banana is **less dense**, and since the two bananas have the same mass, the silver banana occupies **more space.**[2]

[2] In case you're wondering, yes, you could solve this problem with much less work. It's much easier for a student who catches on quickly to ignore extra steps in a long solution than it is for a struggling student to fully understand a very concise solution. Experience shows that many students struggle with the concept of density, ratios, and division by fractions (which we managed to circumvent here – yet division by fractions comes up frequently in physics, as you must have seen by now), so we're trying to break this solution down as much as possible. (Of course, density **should** be simple: Look at the equation. It's one of the simplest equations we use in physics.)

39. A monkey bakes a giant banana cake in the shape of a rectangular box with dimensions of $0.50 \text{ m} \times 0.40 \text{ m} \times 0.20 \text{ m}$. The cake has a mass of 32 kg (remember, it's giant). What is the density of the banana cake?

40. A banana made out of lead is four times as dense as a banana made out of aluminum. Both bananas have the same volume. Which banana weighs more, and by what factor?

41. A banana made out of lead is four times as dense as a banana made out of aluminum. Both bananas have the same mass. Which banana has more volume, and by what factor?

42. A monkey cuts a cylindrical aluminum rod in half and throws the other half in a trash can. Comparing the portion that the monkey keeps to the original aluminum rod, which one has greater density, and by what factor?

Want help? Check the hints section at the back of the book.

Answers: $800 \frac{\text{kg}}{\text{m}^3}$, lead (4×), aluminum (4×), same (1×)

9 PRESSURE

Relevant Terminology

Force – a push or a pull.
Pressure – force per unit area.
Fluid – a liquid or a gas.
Density – a measure of the compactness of a substance. Density is **mass per unit volume.**
Stress – force per unit area that causes an object to **deform.**
Strain – a measure of the **deformation** of an object that is under stress.

Essential Concepts

Pressure equals **force per unit area.** It's important to realize that both force and area factor into pressure. For example, consider a pencil that has a flat eraser on one end and a sharp lead point at the other end. If you push the pencil with the flat eraser pressed against your skin, you can apply quite a bit of force without feeling more than mild discomfort. In contrast, if you push the pencil with the sharp lead point against your skin, a small force can be painful. One distinction is that the sharp lead point has much less surface area compared to the flat eraser. When the force is spread out over less area, there is greater pressure (since area is in the denominator of the pressure equation – see the next section).

As another example, consider a nail. It would be quite painful to step barefoot on a nail with the point of the nail pushing into your skin. That's because the tip of a nail has very little surface area, resulting in high pressure. However, it can be comfortable to lie down on a bed of nails (created by hammering several hundred closely spaced nails through a piece of wood). When the weight of your body is distributed over several hundred nails, there is much more area, significantly lowering the pressure.

An object under stress results in strain.

- **Stress** is the force per unit area that causes the object to deform.
- **Strain** is a measure of the degree to which the object is deformed.

The ratio of the stress to the strain is called the **elastic modulus.**

$$\text{elastic modulus} = \frac{\text{stress}}{\text{strain}}$$

Three common types of deformation include:

- **Young's modulus** (Y) provides a measure of how a solid resists a change in **length.**
- **Shear modulus** (S) provides a measure of how a solid resists a change in **shape.**
- **Bulk modulus** (B) provides a measure of how a material resists a change in **volume.**

Pressure Equations

Pressure (P) equals **force** (F) per unit **area** (A). This equation for pressure applies in general, but it's most practical when applied to a **solid**.

$$P = \frac{F}{A}$$

For a **fluid** (a liquid or a gas), it's usually more helpful to express the **pressure** (P) in terms of the **density** (ρ), the **depth** (h), and gravitational acceleration (g). For a liquid, note that h is **depth**, measured downward from the top of the liquid level (it is **not** height – that is, **don't** measure upward from the bottom of the container) and P_0 is the pressure (usually, it's atmospheric pressure, approximately $P_0 \approx 1.0 \times 10^5$ Pa) at the top of the liquid.

$$P = P_0 + \rho g h$$

Young's modulus (Y) is the ratio of the **tensile** stress to the tensile strain, where the tensile strain equals the ratio of the change in length (ΔL) to the original length (L_0) of the solid.

$$Y = \frac{F/A}{\Delta L / L_0}$$

Shear modulus (S) is the ratio of the **shear** stress to the shear strain for a solid with two parallel flat faces for which two opposing forces pull the faces in opposite directions, distorting the object's shape as shown below. (The shape below is really a rectangular box in three dimensions, and the two parallel faces referred to are the top and bottom faces.)

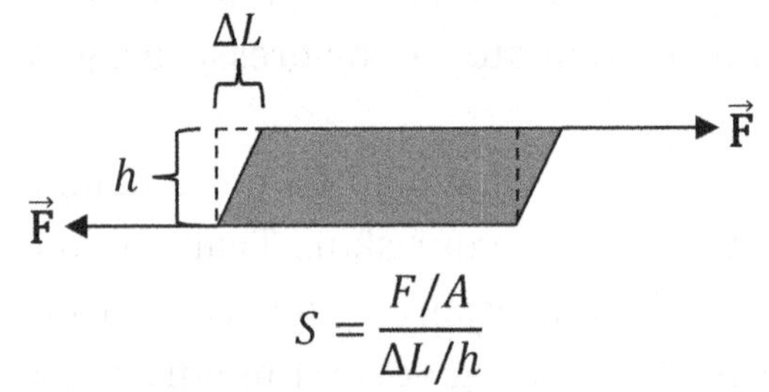

$$S = \frac{F/A}{\Delta L / h}$$

Bulk modulus (B) is the ratio of the **volume** stress (called **pressure**) to the volume strain, where the volume strain equals the ratio of the change in volume (ΔV) to the original volume (V_0) of the substance, when the external forces are perpendicular to the object's surface. This is common when a solid is submerged in a **fluid**. Such an object experiences a change in volume, but not in shape. A minus sign reflects that the pressure causes the volume to decrease (such that ΔV is negative). Since ΔV is negative, the two minus signs make the bulk modulus (B) positive.

$$B = -\frac{F/A}{\Delta V / V_0}$$

Symbols and SI Units

Symbol	Name	SI Units
P	pressure	Pa
F	force	N
A	area	m^2
ρ	density	$\frac{\mathrm{kg}}{\mathrm{m}^3}$
g	gravitational acceleration	$\frac{\mathrm{m}}{\mathrm{s}^2}$
h	depth (measured down from the top of the liquid level)	m
Y	Young's modulus	$\frac{\mathrm{N}}{\mathrm{m}^2}$
S	shear modulus	$\frac{\mathrm{N}}{\mathrm{m}^2}$
B	bulk modulus	$\frac{\mathrm{N}}{\mathrm{m}^2}$
L	length	m
W	width	m
R	radius	m
V	volume	m^3

Notes Regarding Units

The SI unit for pressure (P) is the **Pascal** (Pa). From the equation $P = \frac{F}{A}$, it follows that a Pascal equals $1 \text{ Pa} = 1 \frac{\text{N}}{\text{m}^2}$ since the SI unit for force (F) is the Newton (N) and the SI units of area (A) are square meters (m^2).

There are many other units of pressure, including atmospheres (atm), pounds per square inch (psi), Torricelli (torr), bars (bar), and millimeters of mercury (mm Hg). In physics, if a problem gives you pressure in any of these units, look up the conversion to Pascals (Pa). If you know that standard car tires in the United States are measured in pounds per square inch (psi), this may help you remember that pressure equals force per unit area (but in physics, you want to use Newtons and square meters, not pounds and square inches).

Pressure Problem Strategy

How you solve a problem involving pressure depends on which kind of problem it is:

- To relate **pressure** (P), **force** (F), and **area** (A), apply the following equation. In practice, this equation is usually best suited to a **solid**.
$$P = \frac{F}{A}$$
It may help to recall the formulas for the area of a few common shapes.
$$A = L^2 \text{ (square)} \quad , \quad A = LW \text{ (rectangle)} \quad , \quad A = \pi R^2 \text{ (circle)}$$
For a **fluid** (a liquid or a gas), apply the following equation. Note that ρ is the **density** of the fluid and h is **depth**. For a liquid, note that the depth (h) is measured down from the liquid level (**<u>don't</u>** measure the height up from the bottom) and P_0 is the pressure at the top of the liquid. If the top of a container of liquid is open to the atmosphere near standard conditions, then $P_0 \approx 1.0 \times 10^5$ Pa.
$$P = P_0 + \rho g h$$
- If a problem involves an elastic modulus, use the appropriate equation.
  - **Young's modulus** (Y) involves a change in **length** (ΔL): $Y = \frac{F/A}{\Delta L/L_0}$.
  - **Shear modulus** (S) involves a **shear**, with h and ΔL defined by the figure on page 116: $S = \frac{F/A}{\Delta L/h}$.
  - **Bulk modulus** (B) involves a change in **volume** (ΔV): $B = -\frac{F/A}{\Delta V/V_0}$.
- If a problem may involve Archimedes' principle, see Chapter 10.
- If a problem involves fluid dynamics, see Chapter 11.
- If a problem may involve Pascal's principle, see Chapter 12.

Example: A 40-kg monkey balances himself on top of a light cane. The bottom of the cane is a circle with a radius of 0.010 m. What pressure does the monkey exert on the ground?
Use the formula for pressure that depends on force and area, where the force is weight (mg) with $g = 9.81 \text{ m/s}^2 \approx 10 \text{ m/s}^2$ and the area is the area of a circle ($A = \pi R^2$).
$$P = \frac{F}{A} = \frac{mg}{\pi R^2} = \frac{(40)(9.81)}{\pi(0.01)^2} \approx \frac{(40)(10)}{\pi(0.01)^2} = \frac{400}{\pi(0.0001)} = \frac{4}{\pi} \times 10^6 \text{ Pa}$$
The pressure is $P = \frac{4}{\pi} \times 10^6$ Pa. If you use a calculator, this works out to $P = 1.2 \times 10^6$ Pa.

Example: What is the pressure at a depth of 100 m below the surface of the ocean? The density of water is approximately $1.0 \times 10^3 \frac{\text{kg}}{\text{m}^3}$.
Use the formula for pressure in a fluid with $P_0 \approx 1.0 \times 10^5$ Pa and $g = 9.81 \text{ m/s}^2 \approx 10 \text{ m/s}^2$.
$$P = P_0 + \rho g h = 1 \times 10^5 + (1 \times 10^3)(9.81)(100) \approx 1 \times 10^5 + (1 \times 10^3)(10)(100)$$
$$P = 10^5 + 10^3 10^3 = 10^5 + 10^6 = 0.1 \times 10^6 + 1.0 \times 10^6 = 1.1 \times 10^6 \text{ Pa}$$
The pressure is $P = 1.1 \times 10^6$ Pa at a depth of 100 m below the surface of the ocean.

43. A 20-kg box of bananas with dimensions of 0.10 m × 0.20 m × 4.0 m stands freely on its smallest side. What pressure does the box exert on the ground?

44. The three containers illustrated below are all filled with banana juice. The liquid level is the same height above the bottom of the container in each case. Rank the pressure at the bottom of each container from highest to lowest.

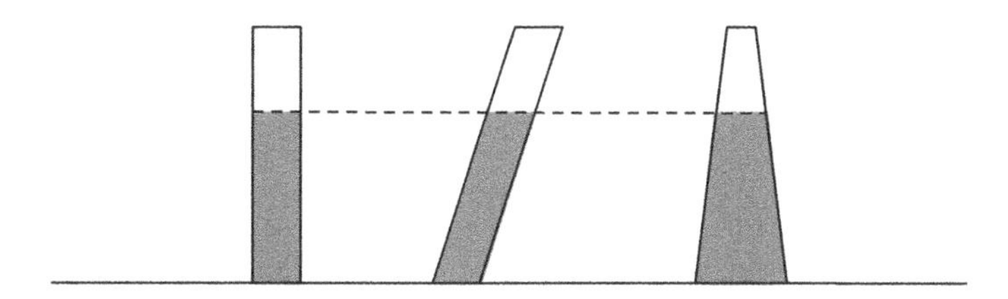

Want help? Check the hints section at the back of the book.

Answers: 10 kPa, the same

45. A monkey goes scuba diving at a depth of 50 m below the surface of the ocean. The density of water is approximately $1.0 \times 10^3 \frac{\text{kg}}{\text{m}^3}$. The monkey's scuba diving suit features a window in front of his face in the shape of a circle with a 10-cm diameter.

(A) What is the pressure at that depth?

(B) What force is exerted on the window?

Want help? Check the hints section at the back of the book.

Answers: 600 kPa, 1500π N = 4.7 kN

10 ARCHIMEDES' PRINCIPLE

Relevant Terminology

Fluid – a liquid or a gas.
Force – a push or a pull.
Mass – a measure of **inertia**. More massive objects are more difficult to accelerate as they have greater inertia to overcome.
Weight – the **gravitational force** that an object experiences. Weight equals mass times gravitational acceleration.
Volume – the amount of space that an object occupies. Also referred to as **capacity**.
Density – a measure of the compactness of a substance. Density is **mass per unit volume**.
Pressure – force per unit area.
Buoyancy – the net upward pressure exerted on an object submerged in a fluid.

Essential Concepts

Recall from Chapter 9 that **pressure** equals force per unit area, $P = \frac{F}{A}$, but that specifically for a **fluid** (a liquid or a gas), pressure depends on **depth** according to $P = P_0 + \rho g h$. There is greater upward pressure at the bottom of the submerged object (since $\rho g h$ is greater at a greater depth) and less downward pressure at the top of the submerged object (since $\rho g h$ is smaller at a smaller depth). This results in a net upward **buoyant force**. According to **Archimedes' principle**, the magnitude of the buoyant force equals the **weight of the displaced fluid**. (When an object is submerged in water, for example, the water level rises. The "displaced fluid" refers to the amount of water that rises due to the submerged object's volume. If you weigh the displaced fluid, it will equal the buoyant force exerted on the submerged object.) Archimedes' principle is the reason that you feel a little lighter when you're taking a bath, or why it is easier to lift somebody inside of a swimming pool.

If an object is fully submerged in a fluid:

- A fully submerged object will **sink** if the object is **more dense** than the fluid.
- A fully submerged object will **rise** upward if the object is **less dense** than the fluid.
- A fully submerged object will **float** (that is, neither sink nor rise) if the object has the **same** density as the fluid.

Ice cubes are less dense than liquid water because ice cubes float at the top of a cup of water. An ice cube is more dense than air because an ice cube falls down when dropped in air, whereas helium is less dense than air since a helium balloon rises upward through air. Steel is much more dense than water, yet a boat made out of steel can float because the large amount of air inside of its hull makes the overall density less dense than water.

Buoyancy Equations

According to Archimedes' principle, the magnitude of the **buoyant force** (F_B) equals the **weight of the displaced fluid** ($m_f g$). (Recall that weight equals mass times gravitational acceleration.)

$$F_B = m_f g$$

Newton's second law states that the net force acting on an object equals the object's mass (m_o) times the object's acceleration (a_y).

$$\sum F_y = m_o a_y$$

When an object is submerged in a fluid, the weight of the object ($m_o g$) pulls downward while the buoyant force pushes upward (F_B). Setting up our coordinate system with $+y$ pointing upward, the net force is thus $\sum F_y = F_B - m_o g$. Substitute this into Newton's second law.

$$F_B - m_o g = m_o a_y$$

Substitute the equation for Archimedes' principle ($F_B = m_f g$) into the previous equation.

$$m_f g - m_o g = m_o a_y$$

Recall from Chapter 8 that density (ρ) equals mass per unit volume: $\rho = \frac{m}{V}$. If we multiply both sides of the density equation by volume, we get $m = \rho V$. The mass of the object (m_o) can thus be expressed in terms of the density of the object (ρ_o) and the volume of the object (V_o) as $m_o = \rho_o V_o$, and the mass of the **displaced fluid** (m_f) can be expressed in terms of the density of the fluid (ρ_f) and the volume of the **displaced fluid** (V_f) as $m_f = \rho_f V_f$. Substitute these two expressions for mass into the previous line.

$$\rho_f V_f g - \rho_o V_o g = \rho_o V_o a_y$$

There are two special cases of the above equation that are common in physics problems:

- For a **fully submerged object**, the volume of the object equals the volume of the displaced fluid. In this case, $V_f = V_o$ and volume cancels out.

$$\rho_f g - \rho_o g = \rho_o a_y$$

- For an object **floating** (that is, neither rising nor sinking) at the top of a liquid, the acceleration is zero ($a_y = 0$) and gravitational acceleration cancels out.

$$\rho_f V_f = \rho_o V_o$$

$$\frac{V_f}{V_o} = \frac{\rho_o}{\rho_f}$$

For a floating cube sticking out of the liquid:

$$\begin{matrix}\text{the fraction of a floating cube} \\ \text{sticking out of the liquid}\end{matrix} = 1 - \frac{V_f}{V_o} = 1 - \frac{\rho_o}{\rho_f}$$

If additional mass (m_a) is placed on the floating cube, use the following equation.

$$\rho_f V_f = m_o + m_a$$

Symbols and SI Units

Symbol	Name	SI Units
ρ_o	density of the submerged object	$\frac{\text{kg}}{\text{m}^3}$
ρ_f	density of the fluid	$\frac{\text{kg}}{\text{m}^3}$
m_o	mass of the submerged object	kg
m_a	added mass	kg
m_f	mass of the displaced fluid	kg
V_o	volume of the submerged object	m^3
V_f	volume of the displaced fluid	m^3
h	depth	m
F	force	N
F_B	buoyant force	N
F_y	y-component of force	N
a_y	y-component of acceleration	m/s^2
g	gravitational acceleration	m/s^2
A	area	m^2
P	pressure	Pa
P_0	pressure at the top of the fluid	Pa

Important Distinctions

Archimedes' principle involves the **weight of the displaced fluid** ($m_f g$). This is **not** the weight of the whole fluid – just the weight of the portion of the fluid that is displaced when the object is submerged in the fluid. In general, the weight of the **object** ($m_o g$) is **not** the same as the weight of the **displaced fluid** ($m_f g$). However, if the object is **floating** (that is, neither rising nor sinking), in that special case the weight of the object and the weight of the **displaced fluid** are equal.

Buoyancy Strategy

How you solve a problem involving buoyancy depends on which kind of problem it is:

- If an object is **fully submerged** in a fluid (liquid or gas), and the only forces acting on the object are the object's weight and the buoyant force, Newton's second law gives:
$$F_B - m_o g = m_o a_y$$
In this case, the volume of the object equals the volume of the displaced fluid ($V_o = V_f$), and the above equation can be rewritten as:
$$\rho_f g - \rho_o g = \rho_o a_y$$
- For a problem with an object **floating** (that is, neither rising nor sinking) at the top of a liquid, the acceleration is zero ($a_y = 0$) and the following equation applies.
$$\rho_f V_f = \rho_o V_o$$
$$\frac{V_f}{V_o} = \frac{\rho_o}{\rho_f}$$
The fraction of a floating cube sticking out of the liquid is:
$$\text{the fraction of a floating cube sticking out of the liquid} = 1 - \frac{V_f}{V_o} = 1 - \frac{\rho_o}{\rho_f}$$
If additional mass (m_a) is placed on the floating cube, use the following equation.
$$\rho_f V_f = m_o + m_a$$
Here, V_f is the volume of the displaced fluid and m_a is the added mass. (This is really the same equation as $m_f = m_o + m_a$ or $\rho_f V_f = \rho_o V_o + m_a$, since $m_f = \rho_f V_f$ and $m_o = \rho_o V_o$. If the added mass is negligible it reduces to the equation at the beginning of this bullet point.)
- One way to calculate the **buoyant force** is to apply Archimedes' principle.
$$F_B = m_f g = \rho_f V_f g$$
If the object is **fully submerged** in the liquid (as in the first bullet point above), then $V_o = V_f$ and $F_B = \rho_f V_o g$. Another way to calculate buoyant force is to solve for F_B in Newton's second law (see the next bullet point and the first bullet point).
- If a submerged object is otherwise constrained – for example, it may be connected to a cord – begin with Newton's second law. It will be similar to page 122, except for the addition of other forces (such as tension in the cord).
$$\sum F_y = m_o a_y$$
Apply Archimedes' principle to the buoyant force: $F_B = m_f g = \rho_f V_f g$.
- The **apparent weight** of the object is, in general, different from the actual weight of the object. A scale reads normal force, tension, or a spring force, which equals the **apparent weight**. To find the apparent weight, you need to solve for the relevant force (such as tension). To do this, apply Newton's second law (as described in the previous bullet point), as shown in the last example of this chapter.

Example: An object with a mass of 6.0 kg and density of 750 $\frac{\text{kg}}{\text{m}^3}$ is placed in the center of a large container of water (where it is fully submerged) and released from rest. The water has a density of 1000 $\frac{\text{kg}}{\text{m}^3}$.

(A) What is the acceleration of the object?
Use the equation for a fully submerged object. See the previous page.

$$\rho_f g - \rho_o g = \rho_o a_y$$

Divide both sides by the density of the object. **Factor** out the gravitational acceleration. We will approximation $g = 9.81 \text{ m/s}^2$ as $g \approx 10 \text{ m/s}^2$.

$$a_y = \frac{\rho_f g - \rho_o g}{\rho_o} = \frac{\rho_f - \rho_o}{\rho_o} g = \frac{1000 - 750}{750}(9.81) \approx \frac{1000 - 750}{750}(10)$$

$$a_y \approx \frac{250}{750}(10) = \frac{10}{3} \frac{\text{m}}{\text{s}^2}$$

The acceleration is $a_y \approx \frac{10}{3} \text{ m/s}^2$. If you use a calculator, $a_y = 3.3 \text{ m/s}^2$. The direction of the acceleration is **upward**. (We know this because the object is less dense than the fluid.)

(B) What buoyant force is exerted on the object?
Apply Newton's second law to the fully submerged object. The buoyant force pushes up, while the weight of the object pulls down.

$$\sum F_y = m_o a_y$$

$$F_B - m_o g = m_o a_y$$

Add the weight of the object to both sides of the equation. **Factor** out the mass.

$$F_B = m_o g + m_o a_y = m(g + a_y)$$

Add fractions with a **common denominator**. Note that $9.81 \approx 10 = \frac{30}{3}$.

$$F_B = 6\left(9.81 + \frac{10}{3}\right) \approx 6\left(10 + \frac{10}{3}\right) = 6\left(\frac{30}{3} + \frac{10}{3}\right) = 6\left(\frac{30 + 10}{3}\right) = 6\left(\frac{40}{3}\right) = 80 \text{ N}$$

The buoyant force is $F_B = 80$ N. The buoyant force ($F_B = 80$ N) is greater than the weight of the object ($m_o g = 60$ N), causing the object to accelerate upward.

Example: Water with a density of 1000 $\frac{\text{kg}}{\text{m}^3}$ is poured into a cup. An ice cube with a density of 900 $\frac{\text{kg}}{\text{m}^3}$ is also placed in the cup. What fraction of the ice cube sticks out of the water?

Apply the handy equation from page 124.

$$\begin{matrix}\text{the fraction of an ice cube} \\ \text{sticking out of the liquid}\end{matrix} = 1 - \frac{\rho_o}{\rho_f} = 1 - \frac{900}{1000} = 1 - 0.9 = 0.1 = \frac{1}{10}$$

The fraction is $\frac{1}{10}$. As a percentage, 10% of the ice cube sticks out of the water.

Example: A monkey suspends a chunk of iron from a vertical thread connected to a spring scale. Initially, the scale reading is 8.0 N. The monkey then fully submerges the chunk of iron in a glass of water, as illustrated below. The density of the iron is 8000 $\frac{\text{kg}}{\text{m}^3}$ and the density of the water is 1000 $\frac{\text{kg}}{\text{m}^3}$.

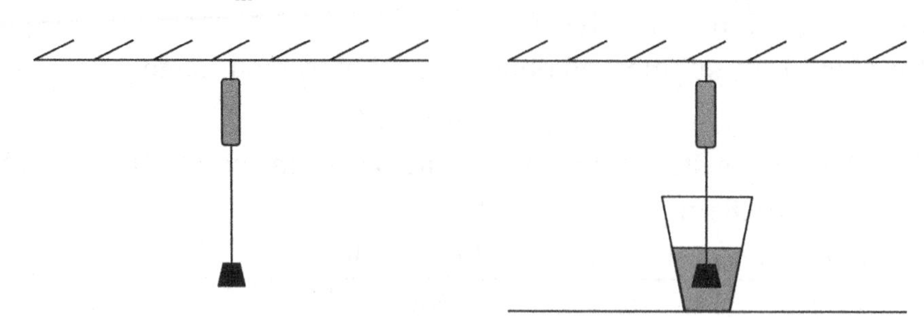

(A) What is the buoyant force after the chunk of iron is fully submerged in the water?

Apply Archimedes' principle: The magnitude of the buoyant force (F_B) equals the weight of the **displaced fluid** ($m_f g$). Note that $m_f = \rho_f V_f$ follows from $\rho_f = \frac{m_f}{V_f}$.

$$F_B = m_f g = \rho_f V_f g$$

Since the chunk of iron is **fully submerged** in the water, the volume of the object equals the volume of the displaced fluid ($V_o = V_f$).

$$F_B = \rho_f V_o g$$

Initially, the tension in the cord equals the weight of the object. That is, $m_o g = 8.0$ N. Divide by gravitational acceleration to find the mass of the object.

$$m_o = \frac{8}{g} = \frac{8}{9.81} \approx \frac{8}{10} = 0.80 \text{ kg}$$

Use the equation $\rho_o = \frac{m_o}{V_o}$ to find the volume of the object. Multiply both sides of the equation by V_o and divide by ρ_o.

$$V_o = \frac{m_o}{\rho_o} = \frac{0.80}{8000} = 0.00010 \text{ m}^3$$

Plug this into the equation for buoyant force that we found previously.

$$F_B = \rho_f V_o g = (1000)(0.0001)(9.81) \approx (1000)(0.0001)(10) = 1.0 \text{ N}$$

The buoyant force is $F_B = 1.0$ N.

(B) What does the scale read after the chunk of iron is fully submerged in the water?

Draw a free-body diagram (FBD) for the chunk of iron. Initially, tension ($\vec{\mathbf{T}}_i$) pulls up while the weight of the object ($m_o\vec{\mathbf{g}}$) pulls down (left diagram). After the iron is fully submerged in water, there is also a buoyant force ($\vec{\mathbf{F}}_B$) pushing up in addition to tension ($\vec{\mathbf{T}}_f$) pulling up and the weight of the object ($m_o\vec{\mathbf{g}}$) pulling down (right diagram). The subscripts "i" and "f" on $\vec{\mathbf{T}}_i$ and $\vec{\mathbf{T}}_f$ represent that the tension changes after submersion in the water.

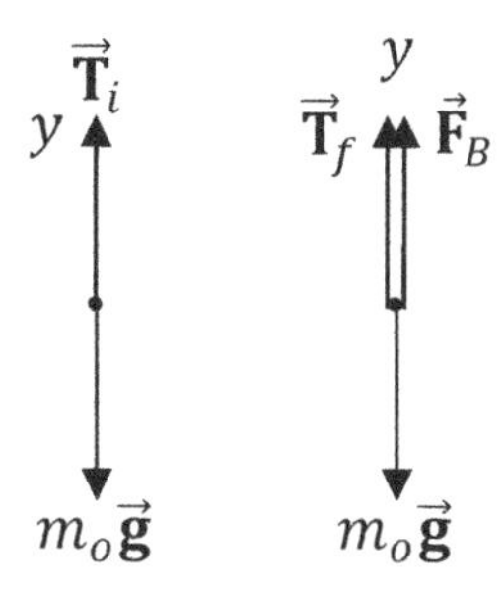

Apply Newton's second law initially and finally. In both cases, the cord prevents the chunk of iron from accelerating, such that $a_y = 0$.

$$\sum F_{iy} = m_o a_y \quad , \quad \sum F_{fy} = m_o a_y$$

$$T_i - m_o g = 0 \quad , \quad T_f + F_B - m_o g = 0$$

Add $m_o g$ to both sides of both equations, and subtract F_B from the right equation.

$$T_i = m_o g \quad , \quad T_f = m_o g - F_B$$

Since $T_i = m_o g$, we may replace $m_o g$ with T_i in the right equation.

$$T_f = T_i - F_B$$

Recall that the problem gives the initial tension ($T_i = 8.0$ N) and that we found the buoyant force ($F_B = 1.0$ N) in part (A).

$$T_f = T_i - F_B = 8 - 1 = 7.0 \text{ N}$$

The tension in the cord is $T_f = 7.0$ N after the chunk of iron is fully submerged in the water. The scale reads the tension in the cord (which is also the apparent weight of the chunk of iron).

Note: Initially, there is also a buoyant force when the chunk of iron is submerged in **air**, but this buoyant force is negligible compared to the effect of submerging the chunk of iron in water.

46. When a 5.0-kg object is fully submerged in a tub of water, the object falls downward with an acceleration of 6.0 m/s^2. The water has a density of $1000 \frac{\text{kg}}{\text{m}^3}$.

(A) What is the buoyant force?

(B) What is the density of the object?

Want help? Check the hints section at the back of the book.

Answers: 20 N, $2500 \frac{\text{kg}}{\text{m}^3}$

47. A monkey creates a drink that he names "banana juice." The monkey pours the banana juice into a cup. When the monkey adds an ice cube to the drink, 25% of the ice cube sticks out of the banana juice. The density of the ice cube is 900 $\frac{\text{kg}}{\text{m}^3}$. What is the density of the banana juice?

Want help? Check the hints section at the back of the book.

Answer: 1200 $\frac{\text{kg}}{\text{m}^3}$

48. A piece of metal carved into the shape of a monkey is fully submerged in a cup of water. The water has a density of 1000 $\frac{\text{kg}}{\text{m}^3}$. The piece of metal has an actual weight of 20 N and an apparent weight of 12 N. What is the density of the metal?

Want help? Check the hints section at the back of the book.

Answer: 2500 $\frac{\text{kg}}{\text{m}^3}$

49. A 0.25 m × 0.25 m × 0.25 m block of wood with a density of 640 $\frac{\text{kg}}{\text{m}^3}$ is placed in a swimming pool where the water has a density of 1024 $\frac{\text{kg}}{\text{m}^3}$.

(A) What percentage of the block of wood sticks out of the water?

(B) When a small monkey climbs onto the top of the block of wood, the top surface of the block of wood just barely drops down to the water level. What is the mass of the monkey?

Want help? Check the hints section at the back of the book.

Answers: 37.5%, 6.0 kg

50. A monkey pumps 2.0 m^3 of helium into a balloon. The helium has a density of 0.18 $\frac{\text{kg}}{\text{m}^3}$. The mass of the balloon plus the load that it carries equals 1.64 kg (this does **not** include the mass of the helium inside the balloon). The air has a density of 1.30 $\frac{\text{kg}}{\text{m}^3}$.

(A) What is the buoyant force?

(B) What is the acceleration?

Want help? Check the hints section at the back of the book.

Answers: 26 N, 3.0 m/s^2

11 FLUID DYNAMICS

Relevant Terminology

Fluid – a liquid or a gas.
Force – a push or a pull.
Mass – a measure of **inertia**. More massive objects are more difficult to accelerate as they have greater inertia to overcome.
Weight – the **gravitational force** that an object experiences. Weight equals mass times gravitational acceleration.
Volume – the amount of space that an object occupies. Also referred to as **capacity**.
Density – a measure of the compactness of a substance. Density is **mass per unit volume.**
Pressure – force per unit area.
Streamline – the path taken by a particle in an ideal fluid under steady flow.

Fluid Dynamics Equations

The **continuity equation** below applies to an ideal fluid under steady flow. In such a case, the fluid particles travel along a **streamline.** A_1 and A_2 are the cross-sectional areas of the streamline or pipe at positions 1 and 2, while v_1 and v_2 are the respective speeds. The continuity equation represents **conservation of mass** for a fluid. If a fluid flows through a pipe, for example, no mass should be gained or lost between positions 1 and 2.

$$A_1 v_1 = A_2 v_2$$

The product Av is the **flow rate** (Q): $Q = Av$. Recall that the area of a circle is $A = \pi R^2$.

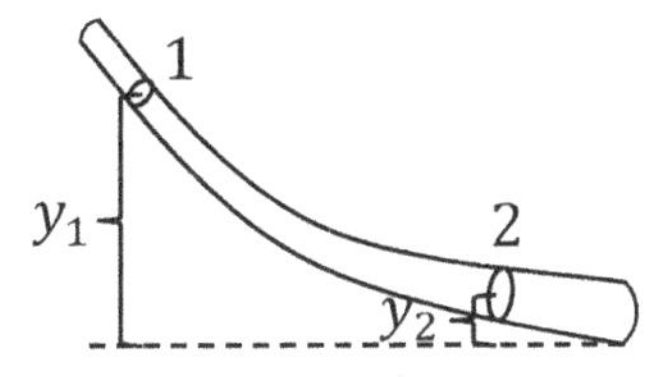

Bernoulli's equation below expresses **conservation of energy** to an ideal fluid for two points along a streamline. (If you divide kinetic energy, $\frac{1}{2}mv^2$, by volume, you get $\frac{1}{2}\rho v^2$, and if you divide gravitational potential[1] energy, mgy, by volume, you get $\rho g y$.)

$$P_1 + \frac{1}{2}\rho v_1^2 + \rho g y_1 = P_2 + \frac{1}{2}\rho v_2^2 + \rho g y_2$$

According to **Torricelli's law,** if a small hole is made in a container of liquid, the **speed of efflux** (v) through the hole is related to the depth (h) as follows:

$$v = \sqrt{2gh}$$

[1] We usually write gravitational potential energy as mgh, but this poses a problem in the context of fluids, where we use h to represent **depth** rather than height. In fluid dynamics, we use y for height and h for depth.

Symbols and SI Units

Symbol	Name	SI Units
A_1	cross-sectional area at position 1	m^2
A_2	cross-sectional area at position 2	m^2
P_1	pressure at position 1	Pa
P_2	pressure at position 2	Pa
ρ	density	$\frac{\text{kg}}{\text{m}^3}$
y_1	height at position 1	m
y_2	height at position 2	m
v_1	speed at position 1	m/s
v_2	speed at position 2	m/s
Q	flow rate	$\frac{\text{m}^3}{s}$
g	gravitational acceleration	m/s^2
V	volume	m^3
h	depth	m
F	force	N
R	radius	m
D	diameter	m

Important Distinctions

Don't confuse the lowercase Greek symbol rho (ρ), which represents **density**, with the uppercase letter P, which represents **pressure**.

Note that y represents the **height** above a common reference point, whereas h represents the **depth** below a surface.

Fluid Dynamics Strategy

How you solve a problem involving fluid dynamics depends on which kind of problem it is:

- If a problem tells you the **flow rate** (Q) in $\frac{\text{m}^3}{\text{s}}$ (or equivalent units), this is the same as cross-sectional area times speed: $Q = Av$.
- You can relate the cross-sectional areas at two points along a streamline to the speeds using the **continuity equation**.

$$A_1 v_1 = A_2 v_2$$

- You can relate the pressure at two points along a streamline to the speeds and heights using **Bernoulli's equation**.

$$P_1 + \frac{1}{2}\rho v_1^2 + \rho g y_1 = P_2 + \frac{1}{2}\rho v_2^2 + \rho g y_2$$

- Some fluid dynamics problems require combining **both** equations – the flow rate and Bernoulli's equation. We will see this in the example featuring the Venturi tube.
- If a hole is made in a container of liquid, the **speed of efflux** (v) is related to the **depth** (h) of the hole below the liquid level according to **Torricelli's law**. If the liquid is open to the atmosphere at the top and originally stationary:

$$v = \sqrt{2gh}$$

Example: Water flows through a pipe with circular cross section. The speed of flow is 3.0 m/s in a section of the pipe that has a diameter of 4.0 cm. What is the speed of flow in a section of the pipe that has a diameter of 2.0 cm?

Identify the given information in SI units.

- The diameter is $D_1 = 0.040$ m at the first position.
- The speed of flow is $v_1 = 3.0$ m/s at the first position.
- The diameter is $D_2 = 0.020$ m at the second position.

Divide each diameter by 2 in order to find the corresponding radius.

$$R_1 = \frac{D_1}{2} = \frac{0.04}{2} = 0.020 \text{ m}$$

$$R_2 = \frac{D_2}{2} = \frac{0.02}{2} = 0.010 \text{ m}$$

Use the formula for the area of a circle to find the cross-sectional areas of the pipe.

$$A_1 = \pi R_1^2 = \pi(0.02)^2 = 0.0004\pi \text{ m}^2$$

$$A_2 = \pi R_2^2 = \pi(0.01)^2 = 0.0001\pi \text{ m}^2$$

Apply the continuity equation.

$$A_1 v_1 = A_2 v_2$$

$$v_2 = \frac{A_1}{A_2} v_1 = \frac{0.0004\pi}{0.0001\pi}(3.0) = 12 \text{ m/s}$$

The speed of flow is $v_2 = 12$ m/s in the section of the pipe with a diameter of 2.0 cm.

Example: Water flows through a pipe with circular cross section. The flow rate is $108 \frac{\text{m}^3}{\text{min.}}$. What is the speed of flow in a section of the pipe that has a diameter of 6.0 cm?

Identify the given information.

- The diameter is $D = 0.060$ m.
- The flow rate is $Q = 108 \frac{\text{m}^3}{\text{min.}}$.

Divide the diameter by 2 in order to find the radius.

$$R = \frac{D}{2} = \frac{0.06}{2} = 0.030 \text{ m}$$

Use the formula for the area of a circle to find the cross-sectional area of the pipe.

$$A = \pi R^2 = \pi(0.03)^2 = 0.0009\pi \text{ m}^2$$

Convert the flow rate to SI units.

$$Q = 108 \frac{\text{m}^3}{\text{min.}} \times \frac{1 \text{ min.}}{60 \text{ s}} = \frac{108}{60} \frac{\text{m}^3}{\text{s}} = \frac{9}{5} \frac{\text{m}^3}{\text{s}} = 1.8 \frac{\text{m}^3}{\text{s}}$$

Apply the equation for flow rate.

$$Q = Av$$

Divide both sides of the equation by the cross-sectional area.

$$v = \frac{Q}{A} = \frac{1.8}{0.0009\pi} = \frac{2000}{\pi} \text{ m/s}$$

The speed of flow through the pipe is $v = \frac{2000}{\pi}$ m/s. If you use a calculator, $v = 637$ m/s.

Example: A large storage tank is filled with water. If a hole is made in the tank a distance of 5.0 m below the water level, what is the speed of the water as it leaves the hole?

Apply Torricelli's law:[2]

$$v = \sqrt{2gh} = \sqrt{(2)(9.81)(5)} \approx \sqrt{(2)(10)(5)} = \sqrt{100} = 10 \text{ m/s}$$

The water escapes through the hole with a speed of $v = 10$ m/s.

[2] Torricelli's law equates to applying conservation of energy to a motion problem with $v_0 = 0$ (since the water at the top of the water level is stationary). We obtained this same equation, $v = \sqrt{2gh}$, in an example in Volume 1, Chapter 22 (conservation of energy) on page 211. Furthermore, once the water leaves the hole, it follows the shape of a **projectile** (Volume 1, Chapter 11).

Example: Consider the horizontal pipe illustrated below, where the diameter is reduced from 4.0 cm to 2.0 cm. (Such a horizontal **constricted** pipe is called a **Venturi tube**.) Water with a density of 1000 $\frac{\text{kg}}{\text{m}^3}$ flows through the pipe. The pressure is initially 40 kPa, but is reduced to 10 kPa in the constricted portion of the pipe. Determine the speed of flow in each section of the pipe.

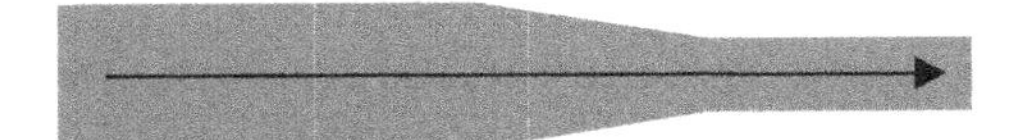

Identify the given information in SI units.

- The diameter is $D_1 = 0.040$ m at the first position.
- The diameter is $D_2 = 0.020$ m at the second position.
- The pressure is $P_1 = 40{,}000$ Pa at the first position.
- The pressure is $P_2 = 10{,}000$ Pa at the second position.
- The density of the water is $\rho = 1000 \ \frac{\text{kg}}{\text{m}^3}$.

Divide each diameter by 2 in order to find the corresponding radius.

$$R_1 = \frac{D_1}{2} = \frac{0.04}{2} = 0.020 \text{ m}$$
$$R_2 = \frac{D_2}{2} = \frac{0.02}{2} = 0.010 \text{ m}$$

Use the formula for the area of a circle to find the cross-sectional areas of the pipe.

$$A_1 = \pi R_1^2 = \pi(0.02)^2 = 0.0004\pi \text{ m}^2$$
$$A_2 = \pi R_2^2 = \pi(0.01)^2 = 0.0001\pi \text{ m}^2$$

Apply the continuity equation.

$$A_1 v_1 = A_2 v_2$$
$$v_2 = \frac{A_1}{A_2} v_1 = \frac{0.0004\pi}{0.0001\pi} v_1 = 4v_1$$

Apply Bernoulli's equation.

$$P_1 + \frac{1}{2}\rho v_1^2 + \rho g y_1 = P_2 + \frac{1}{2}\rho v_2^2 + \rho g y_2$$

Set $y_1 = y_2 = 0$ since the pipe is horizontal (the height doesn't change).

$$P_1 + \frac{1}{2}\rho v_1^2 = P_2 + \frac{1}{2}\rho v_2^2$$

Substitute the result from the continuity equation, $v_2 = 4v_1$, into Bernoulli's equation.

$$P_1 + \frac{1}{2}\rho v_1^2 = P_2 + \frac{1}{2}\rho (4v_1)^2$$

Note that $(4v_1)^2 = 4^2 v_1^2 = 16v_1^2$.

$$P_1 + \frac{1}{2}\rho v_1^2 = P_2 + \frac{1}{2}\rho 16 v_1^2$$
$$P_1 + \frac{1}{2}\rho v_1^2 = P_2 + 8\rho v_1^2$$

Combine like terms. Subtract P_2 and $\frac{1}{2}\rho v_1^2$ from both sides of the equation.

$$P_1 - P_2 = 8\rho v_1^2 - \frac{1}{2}\rho v_1^2$$

Combine fractions with a **common denominator**: $8 - \frac{1}{2} = \frac{16}{2} - \frac{1}{2} = \frac{15}{2}$.

$$P_1 - P_2 = \frac{15}{2}\rho v_1^2$$

Multiply both sides of the equation by 2 and divide by 15ρ.

$$v_1^2 = \frac{2(P_1 - P_2)}{15\rho}$$

Squareroot both sides of the equation.

$$v_1 = \sqrt{\frac{2(P_1 - P_2)}{15\rho}} = \sqrt{\frac{(2)(40{,}000 - 10{,}000)}{(15)(1000)}} = \sqrt{\frac{(2)(30{,}000)}{(15)(1000)}} = \sqrt{\frac{60}{15}} = \sqrt{4} = 2.0 \text{ m/s}$$

Plug this speed into the equation $v_2 = 4v_1$ that we previously obtained from the continuity equation.

$$v_2 = 4v_1 = (4)(2) = 8.0 \text{ m/s}$$

The speed of flow is $v_1 = 2.0$ m/s in the section of the pipe with a diameter of 4.0 cm, and is $v_2 = 8.0$ m/s in the section of the pipe with a diameter of 2.0 cm. Note that the pressure is smaller in the constricted part of the pipe, but that the speed of water flow is greater in the constricted part of the pipe.

51. Water flows through a pipe with circular cross section. The speed of flow is $\frac{16}{\pi}$ m/s in a section of the pipe that has a diameter of 5.0 cm.

(A) What is the flow rate in cubic meters per minute?

(B) What is the speed of flow in a section of the pipe that has a diameter of 2.0 cm?

52. A container is filled with water. The water level is 50 cm above the bottom of the container. A hole is made in the wall of the container a distance of 30 cm above the bottom of the container. The hole has a diameter of 1.0 mm.

(A) What is the speed of the water as it leaves the hole?

(B) What is the flow rate of water leaking through the hole in units of $\text{cm}^3/\text{min.}$?

Want help? Check the hints section at the back of the book.

Answers: #51. (A) 0.60 $\frac{\text{m}^3}{\text{min.}}$ (B) $\frac{100}{\pi}$ m/s = 32 m/s, #52. (A) 2.0 m/s (B) $30\pi \frac{\text{cm}^3}{\text{min.}} = 94 \frac{\text{cm}^3}{\text{min.}}$

53. Consider the Venturi tube illustrated below, where the diameter is reduced from $2\sqrt{2}$ cm to 2.00 cm. Water with a density of 1000 $\frac{\text{kg}}{\text{m}^3}$ flows through the tube. In the section of the tube with the larger diameter, the pressure is 82.5 kPa and the speed of flow is 5.0 m/s.

(A) What is the speed of flow in the constricted portion of the tube?

(B) What is the pressure in the constricted portion of the tube?

(C) What is the flow rate through the tube?

Want help? Check the hints section at the back of the book.

Answers: 10 m/s, 45 kPa, $\pi \times 10^{-3}\ \frac{\text{m}^3}{\text{s}} = 3.1 \times 10^{-3}\ \frac{\text{m}^3}{\text{s}}$

12 PASCAL'S LAW

Relevant Terminology

Fluid – a liquid or a gas.
Force – a push or a pull.
Mass – a measure of **inertia**. More massive objects are more difficult to accelerate as they have greater inertia to overcome.
Weight – the **gravitational force** that an object experiences. Weight equals mass times gravitational acceleration.
Volume – the amount of space that an object occupies. Also referred to as **capacity**.
Density – a measure of the compactness of a substance. Density is **mass per unit volume**.
Pressure – force per unit area.

Pascal's Law

Consider a liquid that fills some container. According to **Pascal's law**, if the pressure that is applied to the liquid changes, that change in applied pressure is transmitted throughout the entire liquid and even to the walls of the container. We see Pascal's law in the following equation from Chapter 9 for the pressure in a **fluid** (a liquid or gas). If the pressure at the surface, P_0, changes, the new value of P_0 affects the pressure at a depth h below the surface through the term P_0 in the equation.

$$P = P_0 + \rho g h$$

Recall from Chapter 9 that pressure is **force per unit area**.

$$P = \frac{F}{A}$$

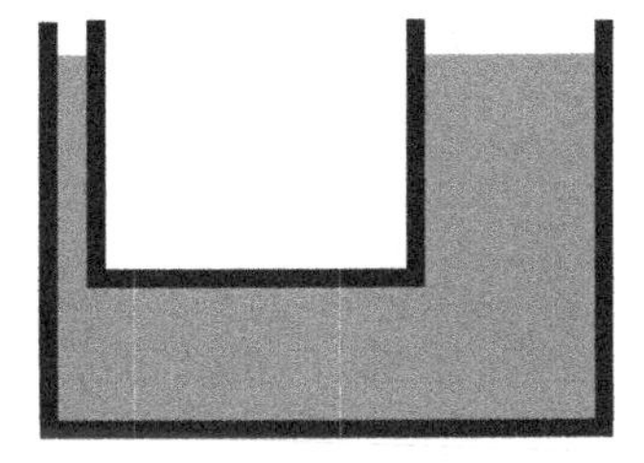

The **hydraulic press** shown above applies Pascal's law. The hydraulic press consists of a narrow tube and a wide tube joined together. A fluid resides inside. A force $\vec{\mathbf{F}}_1$ pushes downward on a piston in the narrow tube of cross-sectional area A_1. Pascal's law results in a much larger upward force $\vec{\mathbf{F}}_2$ in the wide tube of cross-sectional area A_2. This allows a small force $\vec{\mathbf{F}}_1$ to lift a heavy weight (such as a car or truck) placed on the wide tube.

$$\frac{F_1}{A_1} = \frac{F_2}{A_2}$$

Symbols and SI Units

Symbol	Name	SI Units
P	pressure	Pa
A_1	cross-sectional area of the narrow tube	m^2
A_2	cross-sectional area of the wide tube	m^2
F_1	downward force applied in the narrow tube	N
F_2	upward force produced in the wide tube	N
ρ	density	$\frac{kg}{m^3}$
g	gravitational acceleration	m/s^2
h	depth	m
R	radius	m
D	diameter	m
m_1	mass of object 1	kg
m_2	mass of object 2	kg

Hydraulic Press Strategy

If a problem involves a **hydraulic press** (like the one shown on page 141), follow these steps:

1. Apply **Pascal's law**.
$$\frac{F_1}{A_1} = \frac{F_2}{A_2}$$
2. You may need to substitute formulas for the cross-sectional area. For example, if the tubes have circular cross section, $A_1 = \pi R_1^2$ and $A_2 = \pi R_2^2$.
3. If either force equals the weight of an object, use the equation(s) $F_1 = m_1 g$ and/or $F_2 = m_2 g$.
4. Apply algebra to solve for the desired unknown.
5. If you need to solve for pressure, note that $P = \frac{F_1}{A_1}$ and $P = \frac{F_2}{A_2}$.

Example: An 8,000-kg car is to be lifted by a hydraulic press. The piston that raises the car has a diameter of 100 cm. What force must be applied to a smaller piston with a 20-cm diameter in order to lift the car?

First identify the given information in SI units.

- The diameter of the narrow piston is $D_1 = 0.20$ m. Use the subscript 1 for the narrow piston.
- The diameter of the wide piston is $D_2 = 1.00$ m. Use the subscript 2 for the wide piston.
- The mass of the car is $m_2 = 8{,}000$ kg. A hydraulic press provides a lift force in the wide piston, so we use subscript 2 for the mass of the car.
- You should also know that earth's surface gravity is $g = 9.81 \text{ m/s}^2 \approx 10 \text{ m/s}^2$.

Divide each diameter by 2 in order to find the corresponding radius.

$$R_1 = \frac{D_1}{2} = \frac{0.20}{2} = 0.10 \text{ m}$$

$$R_2 = \frac{D_2}{2} = \frac{1.00}{2} = 0.50 \text{ m}$$

Use the formula for the area of a circle to find the area of each piston.

$$A_1 = \pi R_1^2 = \pi(0.1)^2 = 0.01\pi \text{ m}^2$$

$$A_2 = \pi R_2^2 = \pi(0.5)^2 = 0.25\pi \text{ m}^2$$

Multiply the mass of the car by gravitational acceleration to find its weight.

$$F_2 = m_2 g = (8000)(9.81) \approx (8000)(10) = 80{,}000 \text{ N}$$

Apply **Pascal's law**.

$$\frac{F_1}{A_1} = \frac{F_2}{A_2}$$

Multiply both sides of the equation by A_1. Note that the π's cancel.

$$F_1 = \frac{A_1}{A_2} F_2 = \frac{0.01\pi}{0.25\pi}(80{,}000) = \frac{1}{25}(80{,}000) = 3200 \text{ N}$$

The needed force is $F_1 = 3200 \text{ N} = 3.2 \text{ kN}$.

The benefit of this hydraulic press is that a 3200 N force is able to lift an 80,000 N car. The ratio of the diameters is 5, but the ratio of the forces is 25 because radius is squared in the formula for area.

54. A hydraulic press consists of a 10-cm diameter piston and a 40-cm diameter piston. If a maximum force of 200 N can be applied to the 10-cm diameter piston, what is the maximum amount of mass that this hydraulic press can lift?

Want help? Check the hints section at the back of the book.

Answer: 320 kg

13 TEMPERATURE CONVERSIONS

Relevant Terminology

Temperature – a measure of the average kinetic energy of the molecules of a substance.

Temperature Conversions

The SI unit of temperature is the **Kelvin** (K). Many formulas in science, such as the ideal gas law (Chapter 16), only work when the temperature is expressed in Kelvin. When the temperature is expressed in Kelvin, it is called the **absolute temperature**. That's because a temperature of 0 K corresponds to zero kinetic energy, which establishes a minimum possible temperature according to the **third law of thermodynamics** (Chapter 18).

To convert from degrees **Celsius** (°C) to **Kelvin**[1] (K), add 273.15.

$$T_K = T_C + 273.15$$

To convert from **Kelvin** (K) to degrees **Celsius** (°C), subtract 273.15.

$$T_C = T_K - 273.15$$

It may help to remember that the temperature in Kelvin (K) is **always** a **larger** numerical value than the corresponding temperature in degrees Celsius (°C). The temperature in Celsius may be negative, whereas the temperature in Kelvin must be positive.

In this book, we will round 273.15 to 273 in order to work without a calculator.

To convert from degrees **Celsius** (°C) to degrees **Fahrenheit** (°F), first multiply by 9/5 and then add 32.

$$T_F = \frac{9}{5}T_C + 32$$

To convert from degrees **Fahrenheit** (°F) to degrees **Celsius** (°C), first subtract 32 and then multiply by 5/9.

$$T_C = \frac{5}{9}(T_F - 32)$$

To convert from **Kelvin** (K) to degrees **Fahrenheit** (°F), first convert from Kelvin to Celsius and then from Celsius to Fahrenheit. To convert from degrees **Fahrenheit** (°F) to **Kelvin** (K), first convert from Fahrenheit to Celsius and then from Celsius to Kelvin.

---

[1] Look closely. If you think the degree (°) is missing here, it's not a typo. We only use the degree symbol (°) for Celsius (°C) and Fahrenheit (°F). We don't use the degree symbol (°) for Kelvin (K). The reason for this is that the temperature in Kelvin is special: It is the **absolute** temperature.

Symbols and Units

Symbol	Name	Units
T_K	temperature in Kelvin (absolute temperature)	K
T_C	temperature in degrees Celsius	°C
T_F	temperature in degrees Fahrenheit	°F

Notes Regarding Units

The SI unit of temperature is the **Kelvin** (K). When solving physics problems, it's a good habit to convert temperature to Kelvin before plugging it into a formula. Some formulas, such as the ideal gas law (Chapter 16), only work when the temperature is expressed in Kelvin.

Note that we **don't** use a degree symbol (°) for Kelvin (K) because it is the **absolute** temperature. For example, compare 300 K to 27°C or 81°F.

Special Temperatures

It is handy to know the **freezing points** and **boiling points** for water in all three units of temperature. These are shown in the chart below.

Temperature Scale	Name	Value
Kelvin	freezing point of water	273.15 K
Celsius	freezing point of water	0°C
Fahrenheit	freezing point of water	32°F
Kelvin	boiling point of water	373.15 K
Celsius	boiling point of water	100°C
Fahrenheit	boiling point of water	212°F

Temperature Conversion Strategy

To perform a temperature conversion, follow these steps:

1. Identify the temperature that you would like to find (T_K, T_C, or T_F). You want to find this symbol on the **left**-hand side of the formula.
2. Identify the temperature that you already know (T_K, T_C, or T_F). You want to find this symbol on the **right**-hand side of the formula.
3. To convert from Celsius to Kelvin or vice-versa, use one of the formulas below.
$$T_K = T_C + 273.15$$
$$T_C = T_K - 273.15$$
4. To convert from Celsius to Fahrenheit or vice-versa, use one of the formulas below.
$$T_F = \frac{9}{5}T_C + 32$$
$$T_C = \frac{5}{9}(T_F - 32)$$
5. To convert from Kelvin to Fahrenheit or vice-versa, do it in **two** steps. Use one equation from Step 3 and one equation from Step 4, as illustrated in one of the examples that follow.
6. If possible, check that your answer makes sense.
   - If you solve for a temperature in Kelvin, make sure the answer is **positive**. A temperature in Kelvin **can't** be negative. (However, note that temperature **can** be negative in both the Celsius and Fahrenheit scales.)
   - When converting between Celsius and Kelvin, check that the numerical value is **larger** in **Kelvin** (T_K) than it is in Celsius (T_C). Don't ignore a minus sign. If the temperature in Celsius is negative, it is definitely smaller than the value in Kelvin (since the temperature in Kelvin can't be negative).
   - If you live in the United States, where Fahrenheit is common, you should have a feel for some values. For example, 72°F is a relatively comfortable room temperature, 50°F is chilly, below 32°F is freezing, 100°F is very hot outside, and water boils at 212°F.
   - If you live somewhere that Celsius is common, you should have a feel for some of these values. For example, 22°C is a relatively comfortable room temperature, 10°C is chilly, a negative temperature is freezing, 38°C is very hot outside, and water boils at 100°C.
7. Another way to help check your answer is to do the opposite conversion. For example, suppose that you converted 35°C to Fahrenheit and got 95°F. Now try converting from 95°F and see if you get 35°C.

Example: Convert 60°C to Kelvin.
Use the equation with T_K on the left (since we're solving for the temperature in Kelvin) and with T_C on the right (since we know the temperature in Celsius).

$$T_K = T_C + 273.15 = 60 + 273 = 333 \text{ K}$$

The answer, $T_K = 333$ K, corresponds to 60°C. As expected, the numerical value in Kelvin (333) is **larger** than the numerical value in degrees Celsius (60).

Example: Convert 59°F to Kelvin.
Convert from Fahrenheit to Kelvin in **two steps**. First convert from Fahrenheit to Celsius. Use the equation with T_C on the left (since we're initially solving for the temperature in Celsius) and with T_F on the right (since we know the temperature in Fahrenheit).

$$T_C = \frac{5}{9}(T_F - 32) = \frac{5}{9}(59 - 32) = \frac{5}{9}(27) = (5)(3) = 15°\text{C}$$

Note that it's often simpler to divide first: That is, $\frac{5}{9}(27) = \frac{27}{9}(5) = (3)(5) = 15$ is the same as $\frac{5}{9}(27) = \frac{135}{9} = 15$, but if you're not using a calculator, the arithmetic involves smaller numbers the first way.

Now that we know $T_C = 15°\text{C}$, convert from Celsius to Kelvin. Use the equation with T_K on the left (since we're solving for the temperature in Kelvin) and with T_C on the right (since we now know the temperature in Celsius).

$$T_K = T_C + 273.15 = 15 + 273 = 288 \text{ K}$$

The answer, $T_K = 288$ K, corresponds to 59°F. As expected, the numerical value in Kelvin (288) is **larger** than the numerical value in degrees Celsius (15).

Example: At what temperature does the numerical value in Celsius equal the numerical value in Fahrenheit?
Write down either formula for the conversion between Celsius to Fahrenheit and set the two temperatures equal to one another: Set $T_C = T_F$ in the equation $T_F = \frac{9}{5}T_C + 32$.

$$T_F = \frac{9}{5}T_F + 32$$

Combine like terms. Subtract T_F and 32 from both sides of the equation.

$$-32 = \frac{4}{5}T_F$$

Note that $\frac{9}{5} - 1 = \frac{9}{5} - \frac{5}{5} = \frac{9-5}{5} = \frac{4}{5}$. Multiply both sides of the equation by 5/4.

$$T_F = -\frac{(32)(5)}{4} = -40°\text{F}$$

The answer is -40 (note the **minus sign**) because $T_F = -40°\text{F} = T_C = -40°\text{C}$.

55. Perform the following temperature conversions.

(A) Convert 200 K to Celsius.

(B) Convert 200°C to Kelvin.

(C) Convert 14°F to Celsius.

(D) Convert −15°C to Fahrenheit.

(E) Convert 238 K to Fahrenheit.

(F) Convert 95°F to Kelvin.

Want help? Check the hints section at the back of the book.

Answers: (A) −73°C (B) 473 K (C) −10°C (D) 5.0°F (E) −31°F (F) 308 K

56. At what temperature does the numerical value in Kelvin equal the numerical value in Fahrenheit?

Want help? Check the hints section at the back of the book.

Answer: approximately 575 K or 575°F

14 THERMAL EXPANSION

Relevant Terminology

Temperature – a measure of the average kinetic energy of the molecules of a substance.

Thermal Expansion Equations

The formula for the thermal expansion of a liquid or solid depends on whether the expansion is characterized as a linear expansion (like a rod), a surface area expansion (like a plane or disc), or a volume expansion (like a cube or sphere).

$$\Delta L = \alpha L_0 \Delta T \quad \text{(linear)}$$
$$\Delta A = 2\alpha A_0 \Delta T \quad \text{(surface)}$$
$$\Delta V = 3\alpha V_0 \Delta T \quad \text{(volume)}$$

The symbol α (lowercase Greek letter alpha) represents the **coefficient of linear expansion.** The symbols L_0, A_0, and V_0 represent the length, area, or volume, respectively, at the reference temperature, T_0. The delta (Δ) means "change in." For example, $\Delta T = T - T_0$ is the change in temperature, and $\Delta L = L - L_0$ is the change in length. The third equation is sometimes written $\Delta V = \beta V_0 \Delta T$, where $\beta = 3\alpha$ is the coefficient of volume expansion. **Note**: Water is unusual in that it expands when it freezes into ice (whereas materials normally expand when heated, not cooled), which can cause pipes to burst in the winter.

Thermal Expansion Strategy

To solve a problem involving thermal expansion, follow these steps:

1. Determine whether you need the formula for a linear, surface, or volume expansion.
   - If the expansion is primarily along one dimension, $\Delta L = \alpha L_0 \Delta T$.
   - If the expansion is significant in two dimensions, $\Delta A = 2\alpha A_0 \Delta T$.
   - If the expansion is significant in three dimensions, $\Delta V = 3\alpha V_0 \Delta T$.
2. Determine which temperature corresponds to which length (or area or volume): T_0 must correspond to L_0 (or A_0 or V_0), and T must correspond to L (or A or V).
3. You may need to use $\Delta T = T - T_0$, $\Delta L = L - L_0$, $\Delta A = A - A_0$, or $\Delta V = V - V_0$.
4. If you're working with **surface area**, note that $A = L^2$ for a square, $A = LW$ for a rectangle, $A = \frac{1}{2}bh$ for a triangle, $A = \pi R^2$ for a circle, $A = 2\pi RL$ for the body of a cylinder, and $A = 4\pi R^2$ is the surface area of a sphere.
5. If you're working with **volume**, $V = L^3$ for a cube, $V = LWH$ for a cuboid (a rectangular box), $V = \pi R^2 L$ for a cylinder, and $V = \frac{4}{3}\pi R^3$ for a sphere.
6. For a textbook problem, you may need to look up the value of α in a table.
7. Apply algebra to solve for the desired unknown.

Symbols and SI Units

Symbol	Name	SI Units
T_0	reference temperature	K
T	temperature	K
ΔT	change in temperature	K
L_0	reference length	m
L	length	m
ΔL	change in length	m
A_0	reference surface area	m^2
A	surface area	m^2
ΔA	change in surface area	m^2
V_0	reference volume	m^3
V	volume	m^3
ΔV	change in volume	m^3
α	coefficient of linear expansion	$\frac{1}{K}$
β	coefficient of volume expansion	$\frac{1}{K}$
W	width	m
b	base	m
h or H	height	m
R	radius	m

Notes Regarding Units

It's okay to use degrees Celsius (°C) for ΔT or to use $\frac{1}{°C}$ for α because a **change in temperature** (ΔT) is the same in both Kelvin and Celsius. Since $T_K = T_C + 273.15$, it follows that $\Delta T_K = \Delta T_C$ (the 273.15 cancels out in the subtraction).

Example: A metal rod is 4.0000 m long at 20.000°C. The coefficient of linear expansion[1] for the metal is 6.0000×10^{-6} /°C. What is the length of the rod at 70.000°C?
First identify the given information. It's okay to use Celsius for thermal expansion.

- The coefficient of thermal expansion is $\alpha = 6.0000 \times 10^{-6}$ /°C.
- The length of the rod is $L_0 = 4.0000$ m when the temperature is $T_0 = 20.000$°C.
- We want to find the length L when the temperature is $T = 70.000$°C.

Apply the formula for **linear** expansion.

$$\Delta L = \alpha L_0 \Delta T$$

Note that $\Delta L = L - L_0$ and $\Delta T = T - T_0$.

$$L - L_0 = \alpha L_0 (T - T_0)$$

Add L_0 to both sides of the equation in order to solve for L.

$$L = L_0 + \alpha L_0 (T - T_0) = 4 + (6 \times 10^{-6})(4)(70 - 20) = 4 + (6 \times 10^{-6})(4)(50)$$

$$L = 4 + 1200 \times 10^{-6} = 4 + 1.2 \times 10^{-3} = 4 + 0.0012 = 4.0012 \text{ m}$$

The length of the rod is $L = 4.0012$ m at 70.000°C. The rod is now 1.2 mm longer.

Example: A rectangular plate has a length of 5.0000 m and a width of 2.0000 m at 0.000°C. The coefficient of linear expansion for the plate is 15.0000×10^{-6} /°C. How much would the area of the plate increase at 100.000°C?
First identify the given information. It's okay to use Celsius for thermal expansion.

- The coefficient of thermal expansion is $\alpha = 15.0000 \times 10^{-6}$ /°C.
- The length is $L_0 = 5.0000$ m and the width is $W_0 = 2.0000$ m when $T_0 = 0.000$°C.
- We want to find the change in area ΔA when $T = 100.000$°C.

Apply the formula for **surface** expansion.

$$\Delta A = 2\alpha A_0 \Delta T$$

Note that $A_0 = L_0 W_0$ and $\Delta T = T - T_0$.

$$\Delta A = 2\alpha L_0 W_0 (T - T_0) = 2(15 \times 10^{-6})(5)(2)(100 - 0) = 2(15 \times 10^{-6})(5)(2)(100)$$

$$\Delta A = 30{,}000 \times 10^{-6} = 3.0 \times 10^{-2} = 0.030 \text{ m}^2$$

The area of the plate increases by $\Delta A = 0.030 \text{ m}^2$ when the temperature is raised to 100.000°C.

[1] Note that $1/°\text{C} = \frac{1}{°\text{C}} = (°\text{C})^{-1} = 1/\text{K} = \frac{1}{\text{K}} = (\text{K})^{-1}$ are all equivalent units. Recall that α may be expressed in $\frac{1}{°\text{C}}$ or $\frac{1}{\text{K}}$ since $\Delta T = T - T_0$ is the same in both Celsius and Kelvin (the 273.15 cancels out in the subtraction).

57. A monkey measures the height of a metal pole using a tape measure. The coefficient of linear expansion for the metal pole is 25.000×10^{-6} /°C. Assume that the thermal expansion of the tape measure is negligible compared to the thermal expansion of the metal pole. On a day when the temperature is −20.000°C (note the **minus** sign), the monkey measures the height of the metal pole to be 30.000 m. What would the monkey measure the height of the metal pole to be on a day when the temperature is 40.000°C?

58. A monkey measures the height of a pole using a metal tape measure. The coefficient of linear expansion for the metal tape measure is 25.000×10^{-6} /°C. Assume that the thermal expansion of the pole is negligible compared to the thermal expansion of the metal tape measure. On a day when the temperature is −20.000°C (note the **minus** sign), the monkey measures the height of the pole to be 30.000 m. What would the monkey measure the height of the pole to be on a day when the temperature is 40.000°C?

Answers: 30.045 m, 29.955 m

59. A large, solid ball has a radius of 3.0000 m at 10.000°C. The coefficient of linear expansion for the ball is 40.000×10^{-6} /°C.

(A) Use the formula for volume expansion to find the volume of the ball at 35.000°C.

(B) Use the formula for linear expansion to find the radius of the ball at 35.000°C.

(C) Use the formula for the volume of a sphere to show that your answers to (A) and (B) are consistent. (A calculator is convenient for this part.)

Want help? Check the hints section at the back of the book.

Answers: (A) 36.108π m^3 = 113.44 m^3 (B) 3.0030 m

60. A solid cube has an edge length of 2.0000 m at 5.000°C. The coefficient of linear expansion for the cube is 30.000×10^{-6} /°C.

(A) Determine the surface area and volume of the cube at 5.000°C.

(B) Determine the surface area and volume of the cube at 55.000°C.

Want help? Check the hints section at the back of the book.

Answers: (A) 24.000 m^2, 8.0000 m^3 (B) 24.072 m^2, 8.0360 m^3

15 HEAT TRANSFER

Relevant Terminology

Temperature – a measure of the **average kinetic energy** of the molecules of a substance.
Heat – thermal energy that is **transferred** between substances.
Heat conduction – heat that is transferred through interatomic collisions. Heat conduction tends to be efficient in **metals**.
Convection – heat that is transferred through the movement of a **fluid**. Convection involves the rising and falling of regions of the fluid with varying density.
Radiation – heat that is transferred in the form of **electromagnetic** waves.
Thermal conductivity – a measure of how well a substance conducts heat.
Emissivity – a measure of how efficiently a substance emits (or absorbs) radiation.
Power – the rate at which work is done or the rate at which energy is transferred.
Intensity – power per unit area.
Heat capacity – a measure of how much heat a given substance needs to absorb in order to increase its temperature a specified amount.
Calorimeter – a device that measures the amount of heat exchanged. A simple calorimeter consists of a container of liquid with a thermometer.
Phase transition – when a substance changes from one phase (solid, liquid, or gas) to another.
Latent heat – the heat per unit mass that is absorbed or released when a substance undergoes a **phase transition**.
Freezing – a phase transition from the liquid state to the solid state.
Melting – a phase transition from the solid state to the liquid state.
Condensation – a phase transition from the gaseous state to the liquid state.
Boiling – a phase transition from the liquid state to the gaseous state.
Sublimation – a phase transition from the solid state directly to the gaseous state.
Deposition – a phase transition from the gaseous state directly to the solid state.
Evaporation – when the fastest molecules of a liquid escape in vapor form.
Exothermic – heat energy is **released** by the system.
Endothermic – heat energy is **absorbed** by the system.

Heat vs. Temperature

Temperature is a measure of the average kinetic energy of the molecules of a substance, whereas **heat** is a transfer of thermal energy between substances. Heating a substance often results in an increase in temperature, but that's not necessarily the case. For example, a substance can undergo a phase transition for which heat is absorbed or released, but for which the temperature of the substance doesn't change. Heat is only one form of energy that affects temperature.

Phase Transitions

A substance undergoes a **phase transition** when it changes from one phase (solid, liquid, or gas) to another. Six basic phase transitions are tabulated below.

Phase Transition	Process	Latent Heat
liquid to solid	freezing	heat of fusion
solid to liquid	melting	heat of fusion
gas to liquid	condensation	heat of vaporization
liquid to gas	boiling	heat of vaporization
gas to solid	deposition	heat of sublimation
solid to gas	sublimation	heat of sublimation

The **heat of transformation** is the amount of thermal energy needed for a substance of mass m to undergo a phase transition. The heat (Q) of transformation is related to the **latent heat** (L) by the following formula.

$$Q = mL$$

There are three forms of latent heat:

- The **heat of fusion** applies to melting and freezing.
- The **heat of vaporization** applies to boiling and condensation.
- The **heat of sublimation** applies to sublimation and deposition.

The latent heat has the same value for both melting and freezing, for example, but the **sign** of the heat transfer is different in the two cases because one absorbs energy while the other releases energy. Melting, boiling, and sublimation are **endothermic** processes: They **absorb** energy (since heat must be **added** to the substance, Q is **positive**). Freezing, condensation, and deposition are **exothermic** processes: They **release** energy (since the substance **releases** energy, Q is **negative**). The values of the latent heats of fusion and vaporization can be found in most standard textbooks for common substances.

Sign Convention for Heat Exchanges

Signs are important in thermal physics.

- Heat (Q) is **positive** when it is **added**.
- Heat (Q) is **negative** when it is **released**.

Heat Flow

The naturally tendency is for heat to flow from a **higher** temperature region to a **lower** temperature region. For example, if you touch an ice cube with your finger, heat flows from your finger to the ice cube, cooling your finger and warming the ice cube. As another example, if you touch a hot plate with your finger, heat flows from the plate to your finger, warming your finger and cooling the plate.

Heat Capacity

For a process by which the heat added to an object goes towards[1] increasing the object's temperature, the amount of **heat** (Q) needed to raise the object's temperature (T) by an amount $\Delta T = T - T_0$ is proportional to the heat capacity (C) by $Q = C\Delta T$. However, most problems involve **specific heat capacity** (C_V, C_P, c_V, or c_P) rather than heat capacity (C). There are actually four[2] kinds of specific heat capacity:

- Specific heat capacity (C_V or C_P) depends on the mass (m) of the substance.
  - C_V is the specific heat capacity at constant **volume**.
    $$Q = mC_V\Delta T \quad (V \text{ is constant})$$
  - C_P is the specific heat capacity at constant **pressure**.
    $$Q = mC_P\Delta T \quad (P \text{ is constant})$$
- **Molar** specific heat capacity (c_V or c_P) depends on the number of moles (n).
  - c_V is the molar specific heat capacity at constant **volume**.
    $$Q = nc_V\Delta T \quad (V \text{ is constant})$$
  - c_P is the molar specific heat capacity at constant **pressure**.
    $$Q = nc_P\Delta T \quad (P \text{ is constant})$$

Note: C_V and C_P differ noticeably for **gases** (Chapter 16), but not for liquids and solids.

Calorimeter

A calorimeter is a device used to measure how much heat is exchanged. A container of liquid with a thermometer serves as a simple calorimeter. For example, if a solid object is placed in the liquid, and if the solid and liquid are initially at different temperatures, heat is exchanged until the system attains **thermal equilibrium**, at which point the solid and liquid achieve the **same final temperature**.

---

[1] Not all of the thermal energy added in the form of heat always goes into increasing an object's temperature. The first law of thermodynamics (Chapter 18) accounts for all of the changes in thermal energy.

[2] Some first-year physics textbooks only use specific heat capacity, and don't mention molar specific heat capacity. Also, some textbooks don't distinguish between the specific heat capacity at constant pressure and volume until a later chapter on the kinetic theory of gases. When a textbook does use molar specific heat capacity, beware that it may not use exactly the same notation (we use lowercase "c" for molar specific heat).

Calorimetry

The first law of thermodynamics (Chapter 18) expresses conservation of energy, and accounts for internal energy changes, heat exchanges, and work done. For a perfectly insulated calorimeter with negligible volume changes, conservation of energy simplifies to state that the sum of the heat changes equals zero.

$$\sum Q_i = 0$$

This sum includes heat gained as well as heat lost by each substance. Note that heat **gained** is **positive** ($Q_{gained} > 0$) whereas heat **lost** is **negative** ($Q_{lost} < 0$). Apply the equation for **specific heat** at constant volume (since the volume change is negligible) to each substance: $Q = mC_V\Delta T$. If a phase transition occurs during the process, you must also include the heat associated with the phase transition: $Q = mL$, where L is the appropriate **latent heat** of transformation. See the examples.

Methods of Heat Transfer

Heat is transferred by three basic methods:

- Through **conduction**, heat that is transferred through interatomic collisions. Heat conduction tends to be efficient in **metals**.
- Through **convection**, heat that is transferred through the movement of a **fluid**.
- Through **radiation**, heat that is transferred in the form of **electromagnetic** waves.

Thermal Conduction

When two surfaces of a solid object are at different temperatures, heat may flow across the object from the higher temperature surface to the lower temperature surface via heat conduction. If the heat flows spontaneously (that is, instead of being forced to differ from its natural route) and if the **thermal conductivity** (k) is uniform, the instantaneous **rate of energy transfer** (P) is proportional to the cross-sectional area (A) and a derivative of the temperature (T) with respect to a coordinate (s) along the heat flow.

$$P = kA\left|\frac{dT}{ds}\right|$$

It is often convenient to express this equation as an integral.

$$\Delta T = \frac{P}{k}\int_i^f \frac{ds}{A}$$

Note that the **cross-sectional area** (A) is generally a function of the **heat flow coordinate** (s), in which case A may **not** come out of the integral. Instead, simply express A in terms of s, and then perform the integration.

The R Value

For a thermal conductor with a length L separating two surfaces at different temperatures, the R value is defined as:

$$R = \frac{L}{k}$$

Convection

When one portion of a **fluid** (a liquid or gas) is heated, that portion of the fluid expands, becoming less dense. The less dense portion of the fluid then rises upward according to **Archimedes' principle** (Chapter 10): **Buoyancy** is the reason that **heat rises**. After rising, that portion of the fluid cools, and then sinks. These recurring processes create convection currents, which eventually warm the entire fluid. When you boil water in a pot on the oven, the water in the pot is primarily heated through convection.

Stefan's Law

Objects constantly absorb and emit radiation in the form of **electromagnetic waves**. The **absorption** of radiation causes the object's temperature to **increase**, while the **emission** of radiation causes the object's temperature to **decrease**. When the absorption and emission rates are equal, the object is in **radiative equilibrium** and the object's temperature remains **constant**.

According to **Stefan's law**, the instantaneous **rate at which an object emits thermal radiation** (P_e) is proportional to the fourth power of the object's temperature (T) in Kelvin (K), while the instantaneous **rate at which an object absorbs thermal radiation** (P_a) is proportional to the fourth power of the temperature of the object's surroundings (T_{env}) in Kelvin (K).

$$P_e = \sigma \epsilon A T^4 \quad , \quad P_a = \sigma \epsilon A T_{env}^4$$

The **net power** radiated is the difference between these two rates.

$$P_{net} = P_e - P_a = \sigma \epsilon A (T^4 - T_{env}^4)$$

The **sign** of the net power is:

- **positive** if $P_e > P_a$, such that the object is **losing** thermal energy through radiation.
- **negative** if $P_e < P_a$, such that the object is **gaining** thermal energy through radiation.

The constant σ is the **Stefan-Boltzmann constant**: $\sigma = 5.67 \times 10^{-8} \frac{\text{W}}{\text{m}^2\text{K}^4}$. (The 5, 6, 7, and 8 make this number easy to remember.) We will round this to $\sigma \approx \frac{17}{3} \times 10^{-8} \frac{\text{W}}{\text{m}^2\text{K}^4}$ to solve problems without a calculator. The symbol ϵ is the **emissivity**: ϵ approaches 1 for a perfect absorber/emitter (called a **blackbody**) and approaches 0 for a perfect reflector. The symbol A is the **surface area** of the object.

Symbols and SI Units

Symbol	Name	SI Units
T	temperature	K
T_0	initial temperature	K
T_{env}	temperature of the surroundings (environment)	K
ΔT	change in temperature	K
Q	heat	J
L	latent heat of transformation in $\frac{\text{J}}{\text{kg}}$ (or length in m)	see text to left
C	heat capacity	$\frac{\text{J}}{\text{K}}$
C_V	specific heat capacity at constant volume	$\frac{\text{J}}{\text{kg}\cdot\text{K}}$
C_P	specific heat capacity at constant pressure	$\frac{\text{J}}{\text{kg}\cdot\text{K}}$
c_V	molar specific heat capacity at constant volume	$\frac{\text{J}}{\text{mol}\cdot\text{K}}$
c_P	molar specific heat capacity at constant pressure	$\frac{\text{J}}{\text{mol}\cdot\text{K}}$
C_ℓ	specific heat of a liquid	$\frac{\text{J}}{\text{kg}\cdot\text{K}}$
C_s	specific heat of a solid	$\frac{\text{J}}{\text{kg}\cdot\text{K}}$
P	the rate of energy transfer	W
P_e	rate of emission of thermal radiation	W
P_a	rate of absorption of thermal radiation	W
P_{net}	net power radiated	W
k	thermal conductivity	$\frac{\text{W}}{\text{m}\cdot\text{K}}$
A	cross-sectional area	m^2
s	heat flow coordinate	m

R	R value	$\frac{\text{m}^2\text{K}}{\text{W}}$
σ	Stefan-Boltzmann constant	$\frac{\text{W}}{\text{m}^2\text{K}^4}$
ϵ	emissivity	unitless
I	intensity	$\frac{\text{W}}{\text{m}^2}$

Notes Regarding Units

Recall from Volume 1 that the SI unit is the Joule for both work and energy (since energy is the ability to do work) and that the SI unit of power (which is the instantaneous rate at which work is done) is the Watt. Also recall that one Watt (W) equals a Joule (J) per second: $1 \text{ W} = 1 \text{ J/s}$. That's why we use the symbol P for the rate of energy transfer, and why its SI unit is the Watt (W).

The temperature T is the absolute temperature in Kelvin (K). Most of the formulas only work when the temperature is expressed in Kelvin. The exception is when a formula involves the change in temperature ($\Delta T = T - T_0$). The change in temperature may be expressed in either Kelvin or Celsius, since the 273.15 (from $T_K = T_C + 273.15$) will cancel out in the subtraction ($T - T_0$). Since the formula $\Delta T = \frac{P}{k}\int \frac{ds}{A}$ involves the change in temperature, the thermal conductivity (k) may be expressed in $\frac{\text{W}}{\text{m}\cdot\text{K}}$ or $\frac{\text{W}}{\text{m}\cdot{}^\circ\text{C}}$.

If you remember that the SI unit of temperature (T) is the Kelvin (K), the SI unit of heat (Q) is the Joule (J), the SI unit of distance (s, for example) is the meter (m), the SI unit of area (A) is the square meter (m^2), the SI unit of mass (m) is the kilogram (kg), the SI unit of mole number (n) is the mole (mol), the SI unit of power (P) is the Watt (W), and that the emissivity (ϵ) is unitless, you can determine the SI units of the other quantities in this chapter directly from the relevant formula.[3]

- $Q = mL$ shows that the SI units of latent heat (L) are $\frac{\text{J}}{\text{kg}}$.
- $Q = mC_V\Delta T$ and $Q = mC_P\Delta T$ show that the SI units of specific heat are $\frac{\text{J}}{\text{kg}\cdot\text{K}}$.
- $Q = nc_V\Delta T$ and $Q = nc_P\Delta T$ show that the SI units of molar specific heat are $\frac{\text{J}}{\text{mol}\cdot\text{K}}$.
- $P = kA\left|\frac{dT}{ds}\right|$ shows that the SI units of thermal conductivity (k) are $\frac{\text{W}}{\text{m}\cdot\text{K}}$.
- $P_e = \sigma\epsilon AT^4$ shows that the SI units of the Stefan-Boltzmann constant (σ) are $\frac{\text{W}}{\text{m}^2\text{K}^4}$.

[3] In general, if you know the base units like the kilogram, meter, second, and Ampère, you can determine the SI units of other quantities from the corresponding formulas.

Important Distinctions

In human experience, heat is often associated with temperature because heating an object often results in an increase in its temperature. However, **heat** and **temperature** have precise and different meanings, and there are cases where heat can be added to a substance without any change in temperature. For example, a substance can undergo a phase transition, for which the substance absorbs or releases heat without any change of temperature. The two terms heat and temperature are <u>not</u> interchangeable. Heat is just one form of energy that affects temperature.

- **Temperature** is a measure of the average kinetic energy of the molecules.
- **Heat** is a transfer of thermal energy between substances.

There are four kinds of specific **heat capacities**:

- A subscript "V" designates constant **volume**: C_V and c_V.
- A subscript "P" designates constant **pressure**: C_P and c_P.
- An uppercase "C" represents specific heat capacity: $Q = mC_V\Delta T$ and $Q = mC_P\Delta T$.
- A lowercase "c" represents **molar** specific heat capacity: $Q = nc_V\Delta T$ and $Q = nc_P\Delta T$.

However, for **liquids** and **solids**, you may write $Q = mC\Delta T$ and $Q = nc\Delta T$, as the values of C_V and C_P (or c_V and c_P) are about the same. The distinction between constant volume and constant pressure is significant for **gases**.

How you should interpret the symbols L and R depends on the context:

- L is latent heat in $\frac{\text{J}}{\text{kg}}$ in the formula $Q = mL$, but in the context of thermal conduction L is the length of the conductor in meters (m).
- The R value is in $\frac{\text{m}^2\text{K}}{\text{W}}$ for a thermal conductor. We're using a (<u>not</u> R) for radius in this chapter. (Also note that thermal resistance is different from the R value.)

Note the distinction between boiling and evaporation:

- Boiling changes a liquid into a gas, whereas in evaporation the fastest molecules escape from the liquid in vapor form.
- Boiling is a phase transition. Evaporation is <u>not</u> a phase transition.
- Boiling occurs when the temperature reaches the boiling point. Evaporation occurs at all temperatures, even when the liquid is much cooler than the boiling point.
- Boiling occurs throughout the volume of the liquid, whereas evaporation occurs at the surface of the liquid.

You must express the temperature (T) in **Kelvin** (K) when using **Stefan's law** for thermal radiation: $P_e = \sigma\epsilon A T^4$, $P_a = \sigma\epsilon A T_{env}^4$, and $P_{net} = \sigma\epsilon A(T^4 - T_{env}^4)$. However, you may use degrees Celsius (°C) or Kelvin (K) in the integral for **thermal conduction**, $\Delta T = \frac{P}{k}\int_i^f \frac{ds}{A}$, or any of the equations for specific heat capacity (such as $Q = mC_V\Delta T$) because these equations involve the change in temperature (ΔT). Since $T_K = T_C + 273.15$, the constant 273.15 cancels out in the subtraction $\Delta T = T - T_0$.

Strategy for Problems Involving Heat

How you solve a problem involving heat depends on which kind of problem it is:

- If a problem involves a **phase transition**, the **heat** (Q) is related to the **latent heat** (L) by the equation $Q = mL$. If the temperature of the object also changes in the problem, you must combine the equation $Q_1 = mL$ together with the equation $Q_2 = mC\Delta T$ (see the next bullet point): The equation $Q_1 = mL$ accounts for the heat required for the substance to undergo the phase transition, while the equation $Q_2 = mC\Delta T$ accounts for the heat required to change the substance's temperature. See the third bullet point regarding the **sign conventions**.
- If a problem involves **specific heat capacity**, use the appropriate equation.
    - For the specific heat capacity at constant **volume**: $Q = mC_V\Delta T$.
    - For the specific heat capacity at constant **pressure**: $Q = mC_P\Delta T$.
    - For the **molar** specific heat capacity at constant **volume**: $Q = nc_V\Delta T$.
    - For the **molar** specific heat capacity at constant **pressure**: $Q = nc_P\Delta T$.

  Note: For a **liquid** or **solid**, simply write $Q = mC\Delta T$ or $Q = nc\Delta T$ without a subscript on the specific heat capacity, since the distinction between C_V and C_P (or c_V and c_P) is significant for gases.
- It's important to get the **signs** right:
    - $Q = mL$ is **positive** for melting, boiling, and sublimation.
    - $Q = mL$ is **negative** for freezing, condensation, and deposition.
    - $Q = mC\Delta T$ and $Q = nc\Delta T$ are **positive** for an increase in temperature.
    - $Q = mC\Delta T$ and $Q = nc\Delta T$ are **negative** for a decrease in temperature.
- If a problem involves a **calorimeter** - which includes the simple case of a substance added to a liquid in a well-insulated container - relate the heat gained to the heat lost by each substance via the following equation. Note that heat **gained** is **positive** ($Q_{gained} > 0$) whereas heat **lost** is **negative** ($Q_{lost} < 0$). See the previous bullet point regarding the **sign conventions**.

  $$\sum Q_i = 0$$

  Apply the equation for **specific heat** at constant volume (since the volume change is negligible) to each substance: $Q = mC_V\Delta T$. If a phase transition occurs during the process, you must also include the heat associated with the phase transition: $Q = mL$ (see the first bullet point). Recall that $Q = mL$ **may be negative**. See the previous bullet point regarding the **sign conventions** for $Q = mL$. This technique is shown in the examples.

- If a problem involves thermal **conduction** with spontaneous heat flow through a substance with uniform[4] conductivity (k), perform the following integral, where P is the instantaneous rate of thermal energy transfer. Express the cross-sectional area (A) in terms of the heat flow coordinate (s). Note that s is along the heat flow.

$$\Delta T = \frac{P}{k} \int_i^f \frac{ds}{A}$$

For a thermal conductor with a length L separating two surfaces at different temperatures, the R value is defined as $R = \frac{L}{k}$.

- If a problem involves thermal **radiation**, apply **Stefan's law** to determine the instantaneous rate at which thermal energy is emitted or absorbed, where T is the temperature of the **object** in Kelvin (K) and T_{env} is the temperature of the **surroundings** in Kelvin (K).

$$P_e = \sigma \epsilon A T^4 \quad , \quad P_a = \sigma \epsilon A T_{env}^4 \quad , \quad P_{net} = P_e - P_a = \sigma \epsilon A (T^4 - T_{env}^4)$$

The **Stefan-Boltzmann constant** is $\sigma = 5.67 \times 10^{-8} \frac{\mathrm{W}}{\mathrm{m^2K^4}} \approx \frac{17}{3} \times 10^{-8} \frac{\mathrm{W}}{\mathrm{m^2K^4}}$. The **emissivity** ($\epsilon$) approaches 1 for a perfect absorber/emitter (called a **blackbody**) and approaches 0 for a perfect reflector. Note that **intensity** (I) is defined as pressure per unit area: $I = P/A$. The symbol A is the **surface area** of the object. It may help to recall the formula for the surface area of a few common geometries.
    - The area of a circle is $A = \pi R^2$.
    - The surface area of a sphere is $A = 4\pi R^2$.
    - The surface area of a right-circular cylinder is $A = 2\pi RL + 2\pi R^2$ (including the body and the two ends).

Note: Chapter 20 includes an example and problem involving the **intensity** of light (electromagnetic radiation).

- For a textbook problem, you may need to look up the latent heat of fusion, the latent heat of vaporization, the specific heat capacity, or the thermal conductivity for a specified material in a table.
- If a problem involves an **ideal gas**, see Chapter 16.
- If a problem involves a van der **Waals fluid**, see Chapter 17.
- If a problem involves the **laws of thermodynamics** or related quantities (such as **entropy**), see Chapter 18.
- If a problem involves a **heat engine**, see Chapter 19.

[4] If the thermal conductivity is non-uniform, you will need to put it inside of the integral.

Example: Water has a specific heat capacity of approximately $4200 \frac{\text{J}}{\text{kg}\cdot°\text{C}}$ and a latent heat of fusion of approximately $3.3 \times 10^5 \frac{\text{J}}{\text{kg}}$. If 5.0 g of water at 25°C is placed into a freezer and turns into an ice cube at 0°C, how much heat is released by the water?

This solution involves two parts:

- First the water cools from 25°C to 0°C. Use $Q_1 = mC\Delta T$ for this process. (Note that the distinction between C_V and C_P is significant for a gas. This is **liquid** water.)
- Next the water freezes into ice. Use $Q_2 = mL$ for this phase transition.
- The net heat equals $Q_{net} = Q_1 + Q_2$.

It's important to get the **signs** right:

- Cooling is exothermic: Water releases energy when cooling. The sign of Q_1 is **negative**. This sign comes naturally in $Q_1 = mC\Delta T$: The change in temperature, $\Delta T = T - T_0$, is negative since the final temperature ($T = 0°\text{C}$) is less than the initial temperature ($T_0 = 25°\text{C}$).
- Freezing is also exothermic: Water releases energy when it freezes into ice. The sign of Q_2 is **negative**. Take care to include this minus sign in $Q_2 = -mL$.

Convert the mass to SI units: $m = 5.0 \text{ g} = 0.0050 \text{ kg}$. It's okay to leave the temperature in Celsius (°C) since the equations for this solution involve ΔT. First determine the heat released by the water as it cools from $T_0 = 25°\text{C}$ to $T = 0°\text{C}$.

$$Q_1 = mC\Delta T = (0.005)(4200)(0 - 25) = (0.005)(4200)(-25) = -525 \text{ J}$$

Next determine the heat released by the water as it freezes into ice.

$$Q_2 = -mL = -(0.005)(3.3 \times 10^5) = -1650 \text{ J}$$

Combine these together to determine the net heat.

$$Q_{net} = Q_1 + Q_2 = -525 - 1650 = -2175 \text{ J} = -2.2 \text{ kJ}$$

The water releases 2.2 kJ (to two significant figures) of heat energy. (We interpret the **minus** sign to mean that heat is **released**.)

Example: A perfectly insulating container holds 300 g of water at 20°C. A 70-g aluminum sphere at 125°C is added to the water and the container is sealed. The specific heat capacity of aluminum is $900 \frac{\text{J}}{\text{kg}\cdot°\text{C}}$ and the specific heat capacity of water is approximately $4200 \frac{\text{J}}{\text{kg}\cdot°\text{C}}$. What equilibrium temperature will the system reach?

This is a **calorimetry** problem. First identify the given information in suitable units. As we will see, every term in the heat equation will include mass, so in this case it's okay to work with grams (g) instead of kilograms (kg) – it's the equivalent of multiplying **every** term in the equation by 1000. We will use the subscript "ℓ" for liquid (the water), "s" for solid (the aluminum), and "e" for equilibrium.

- The mass of the water is $m_\ell = 300$ g. (Grams are okay for calorimetry problems.)
- The mass of aluminum is $m_s = 70$ g. (Grams are okay for calorimetry problems.)
- The initial temperature of the water is $T_\ell = 20°\text{C}$.
- The initial temperature of the aluminum is $T_s = 125°\text{C}$.
- The specific heat of water is $C_\ell = 4200 \frac{\text{J}}{\text{kg}\cdot°\text{C}}$.
- The specific heat of aluminum is $C_s = 900 \frac{\text{J}}{\text{kg}\cdot°\text{C}}$.

In a calorimetry problem, it's important to express all of the ΔT's correctly.

- The equilibrium temperature is the **final** temperature (T_e).
- When we write ΔT, it's final minus initial.
- The temperature change for the water is $T_e - T_\ell$.
- The temperature change for aluminum is $T_e - T_s$.

In this problem, the **signs** come about **naturally**:

- Note that the equilibrium temperature (T_e) will be between 20°C and 125°C (the initial temperatures of the water and aluminum).
- For water, $Q_\ell = m_\ell C_\ell (T_e - T_\ell)$, and Q_ℓ is naturally **positive** since $T_e > T_\ell$. In this problem, Q_ℓ is positive because the water is **warmed** by the aluminum.
- For aluminum, $Q_s = m_s C_s (T_e - T_s)$, and Q_s is naturally **negative** since $T_e < T_s$. In this problem, Q_s is negative because the aluminum is **cooled** by the water.

The sum of the heat changes equals zero for a perfectly insulated calorimeter.

$$Q_\ell + Q_s = 0$$
$$m_\ell C_\ell (T_e - T_\ell) + m_s C_s (T_e - T_s) = 0$$
$$m_\ell C_\ell (T_e - T_\ell) = -m_s C_s (T_e - T_s)$$
$$(300)(4200)(T_e - 20) = -(70)(900)(T_e - 125)$$

Divide both sides of the equation by 1000. Cross out 3 zeroes on each side.

$$(3)(420)(T_e - 20) = -(7)(9)(T_e - 125)$$
$$1260(T_e - 20) = -63(T_e - 125)$$

Divide both sides of the equation by 63. Note that $\frac{1260}{63} = 20$.

$$20(T_e - 20) = -(T_e - 125)$$

Distribute the 20 and distribute the minus sign. Note that $(-1)(-125) = +125$.

$$20T_e - 400 = -T_e + 125$$

Combine like terms. Add T_e and 400 to both sides of the equation.

$$21T_e = 525$$

Divide both sides of the equation by 21. Note that $\frac{525}{21} = 25$.

$$T_e = \frac{525}{21} = 25°\text{C}$$

The equilibrium temperature is $T_e = 25°\text{C}$.

Example: A perfectly insulating container holds 265 g of water at 50°C. A 35-g ice cube at 0°C is added to the water and the container is sealed. Water has a specific heat capacity of approximately $4200 \frac{\text{J}}{\text{kg}\cdot°\text{C}}$ and a latent heat of fusion of approximately $3.3 \times 10^5 \frac{\text{J}}{\text{kg}}$. What equilibrium temperature will the system reach?

This is a **calorimetry** problem. It's similar to the previous example, except that the ice cube will **melt**, which means that we will need to account for the **phase transition**.

- The initial mass of the liquid water is $m_\ell = 265$ g (where "ℓ" is for liquid).
- The mass of the ice cube is $m_s = 35$ g (where "s" is for solid).
- The initial temperature of the water is $T_\ell = 50°\text{C}$.
- The initial temperature of the ice cube is $T_s = 0°\text{C}$.
- The specific heat of water is $C_\ell = 4200 \frac{\text{J}}{\text{kg}\cdot°\text{C}}$. (**Note**: We don't need the specific heat of ice since the ice transforms to water before it changes temperature.)
- The latent heat of fusion for water is $L = 3.3 \times 10^5 \frac{\text{J}}{\text{kg}} = 330{,}000 \frac{\text{J}}{\text{kg}}$.

It's important to get the **signs** right:

- Note that the equilibrium temperature (T_e) will be between 0°C and 50°C (the initial temperatures of the liquid water and ice cube).
- For water, $Q_\ell = m_\ell C_\ell (T_e - T_\ell)$, and Q_ℓ is naturally **negative** since $T_e < T_\ell$. In this problem, Q_ℓ is negative because the water is **cooled** by the ice cube.
- For the ice cube, $Q_s = m_s C_\ell (T_e - T_s)$, and Q_s is naturally **positive** since $T_e > T_s$. In this problem, Q_s is positive because the ice cube is **warmed** by the water. Note that we use C_ℓ for the ice cube because it **melts** into water <u>**before**</u> it changes temperature.
- As the ice cube **melts** into liquid water, $Q_{melt} = m_s L$ is **positive** because melting is an endothermic process (the ice cube absorbs heat from the liquid water as it melts).

The sum of the heat changes equals zero for a perfectly insulated calorimeter.

$$Q_\ell + Q_s + Q_{melt} = 0$$
$$m_\ell C_\ell (T_e - T_\ell) + m_s C_\ell (T_e - T_s) + m_s L = 0$$
$$(265)(4200)(T_e - 50) + (35)(4200)(T_e - 0) + (35)(330{,}000) = 0$$

Divide both sides of the equation by 100. Cross out 2 zeroes for each term.

$$(265)(42)(T_e - 50) + (35)(42)T_e + (35)(3300) = 0$$
$$11{,}130(T_e - 50) + 1470T_e + 115{,}500 = 0$$

The numbers get smaller if you divide every term by 30.

$$371(T_e - 50) + 49T_e + 3850 = 0$$
$$371T_e - 18{,}550 + 49T_e + 3850 = 0$$
$$420T_e = 14{,}700$$
$$T_e = \frac{14{,}700}{420} = 35°\text{C}$$

The equilibrium temperature is $T_e = 35°\text{C}$.

Example: The left face of the conducting cuboid (rectangular box) illustrated below has a higher temperature (T_h) than the right face (T_c), resulting in a steady rate of heat flow from left to right. The remaining four sides are thermally insulated. The conductor has uniform thermal conductivity.

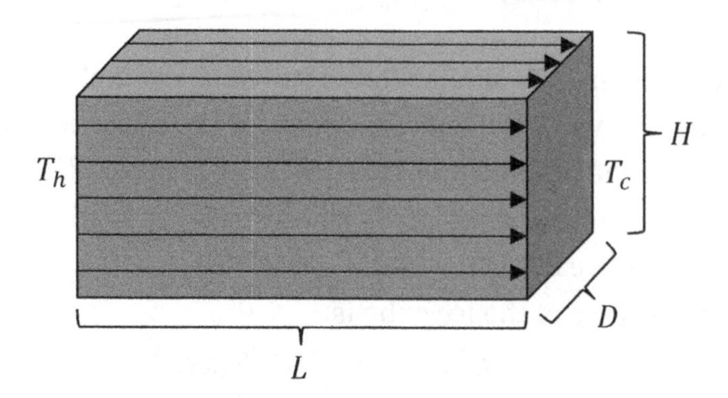

(A) Derive an equation for the rate of heat transfer in terms of k, L, D, H, and $\Delta T = T_h - T_c$. Begin with the integral for thermal conduction with uniform thermal conductivity.

$$\Delta T = \frac{P}{k}\int_{i}^{f}\frac{ds}{A}$$

The **heat flow coordinate**, which we call s in general, runs along the direction of the heat flow from the higher temperature surface (at T_h) to the lower temperature surface (at T_c). In this problem, the heat flows horizontally to the right. If we setup an (x, y, z) coordinate system with the origin at the left surface, in this example the Cartesian coordinate x is the same as the heat flow coordinate: $s = x$. The **cross-sectional area** (A) is perpendicular to the heat flow coordinate, as shown below. In this example, the cross-sectional area is the area of a rectangle with height H and depth D: $A = HD$.

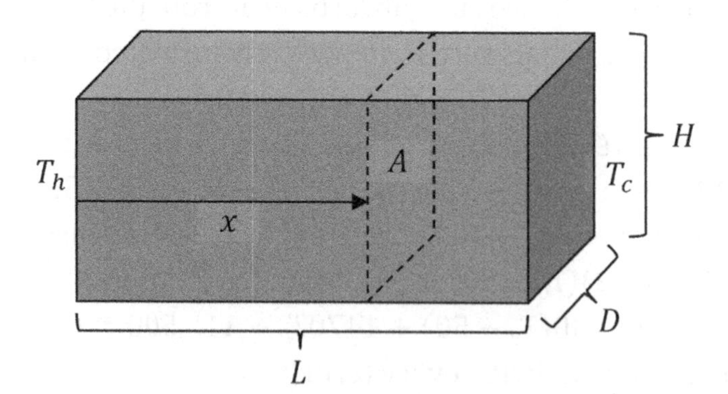

Plug $ds = dx$ and $A = HD$ into the thermal conduction integral. The limits are from $x = 0$ (at the left face with T_h) to $x = L$ (at the right face with T_c).

$$\Delta T = \frac{P}{k}\int_{x=0}^{L}\frac{dx}{HD}$$

Since H and D are both constants, we may pull them out of the integral.

$$\Delta T = \frac{P}{kHD}\int_{x=0}^{L} dx = \frac{P}{kHD}[x]_{x=0}^{L} = \frac{P}{kHD}(L-0) = \frac{PL}{kHD}$$

Solve for the rate of heat transfer (P). Multiply both sides by kHD and divide by L.

$$P = \frac{kHD\Delta T}{L}$$

The equation $P = \frac{kHD\Delta T}{L}$ is in the symbols that the question asked for. (Note, however, that it's common to write this equation in the form $P = \frac{kA\Delta T}{L}$, where we substituted $A = HD$.)

(B) Derive an equation for the rate of heat transfer in terms of L, D, H, $\Delta T = T_h - T_c$, and the R value.

Compare the instructions for parts (A) and (B). The only difference is that part (B) wants the answer to include the R value instead of the thermal conductivity (k). The R value is related to the thermal conductivity by $R = \frac{L}{k}$. Reciprocate this equation to see that $\frac{1}{R} = \frac{k}{L}$. The equation from part (A) has $\frac{k}{L}$ in it: $P = \frac{kHD\Delta T}{L}$. Since $\frac{1}{R} = \frac{k}{L}$, we may replace $\frac{k}{L}$ with $\frac{1}{R}$.

$$P = \frac{HD\Delta T}{R}$$

The equation $P = \frac{HD\Delta T}{R}$ is in the symbols that the question asked for. (Note, however, that it's common to write this equation in the form $P = \frac{A\Delta T}{R}$, where we substituted $A = HD$.)

Example: The three conducting slabs illustrated below are connected in series. The slabs make thermal contact at the shared walls. The far left wall has a higher temperature (T_h) than the far right wall (T_c), resulting in a steady rate of heat flow from left to right. The remaining sides of the slabs are thermally insulated. The slabs have uniform thermal conductivities. Derive an equation relating the R values for this series.

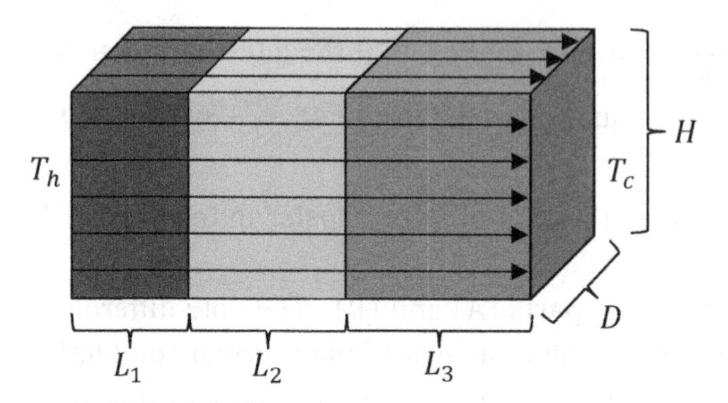

The overall temperature difference between the left and right walls is $\Delta T = T_h - T_c$. This must equal the sum of the temperature differences for the three slabs.

$$\Delta T = \Delta T_1 + \Delta T_2 + \Delta T_3$$

Each individual slab is just like the previous example,[5] where we showed that $P = \frac{kA\Delta T}{L}$. Multiply both sides of this equation by L and divide by kA to see that $\Delta T = \frac{PL}{kA}$. Apply this equation to each slab: $\Delta T_1 = \frac{P_1 L_1}{k_1 A_1}$, $\Delta T_2 = \frac{P_2 L_2}{k_2 A_2}$, and $\Delta T_3 = \frac{P_3 L_3}{k_3 A_3}$. Substitute these equations for temperature difference into the equation $\Delta T = \Delta T_1 + \Delta T_2 + \Delta T_3$.

$$\frac{P_s L_s}{k_s A_s} = \frac{P_1 L_1}{k_1 A_1} + \frac{P_2 L_2}{k_2 A_2} + \frac{P_3 L_3}{k_3 A_3}$$

Note that all of the slabs have the same cross-sectional area: $A_1 = A_2 = A_3 = A_s = HD$. Since all of the areas are the same, area cancels out.

$$\frac{P_s L_s}{k_s} = \frac{P_1 L_1}{k_1} + \frac{P_2 L_2}{k_2} + \frac{P_3 L_3}{k_3}$$

At the shared walls, where thermal energy is transferred from one slab to the next, $P_1 = P_2$ and $P_2 = P_3$. If these rates of thermal energy transfer weren't equal, then energy would either be gained or lost at the shared walls, which would violate the law of **conservation of energy**. Therefore, $P_1 = P_2 = P_3 = P_s$, and the rate of energy transfer cancels out.

$$\frac{L_s}{k_s} = \frac{L_1}{k_1} + \frac{L_2}{k_2} + \frac{L_3}{k_3}$$

Since the R value is defined as $R = \frac{L}{k}$, the previous equation can also be expressed as:

$$R_s = R_1 + R_2 + R_3$$

Note that this is just like the equation for series resistors in an electric circuit (compare with Volume 2, Chapter 13).

---

[5] See the note in parentheses at the end of part (A) of the previous example.

Example: The inner surface of the conducting spherical shell illustrated below has a higher temperature (T_h) than the outer surface (T_c), resulting in a steady rate of heat flowing outward. The conducting shell has uniform thermal conductivity. Derive an equation for the rate of heat transfer in terms of k, a, b, and $\Delta T = T_h - T_c$.

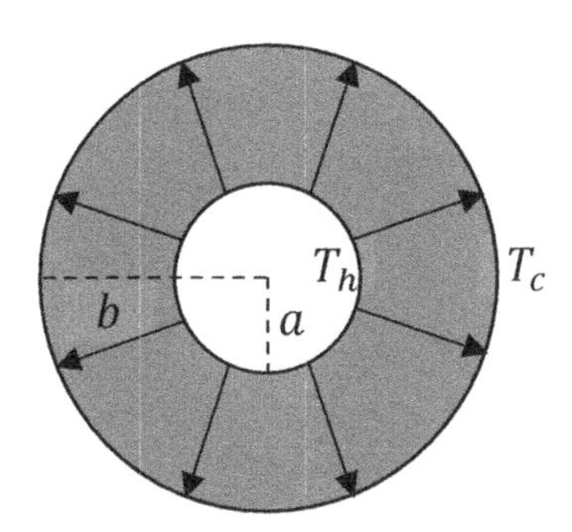

Begin with the integral for thermal conduction with uniform thermal conductivity.

$$\Delta T = \frac{P}{k} \int_{i}^{f} \frac{ds}{A}$$

The **heat flow coordinate**, which we call s in general, runs along the direction of the heat flow from the higher temperature surface (at T_h) to the lower temperature surface (at T_c). In this problem, the heat flows radially outward along r of spherical coordinates: $s = r$. The **cross-sectional area** (A) is perpendicular to the heat flow coordinate. In this example, the cross-sectional area is the surface area of a sphere with radius r. Recall that the surface area of a **sphere** (**not** the area of a circle) is $A = 4\pi r^2$. Plug $ds = dr$ and $A = 4\pi r^2$ into the thermal conduction integral. The limits are from $r = a$ (at the inner surface with T_h) to $r = b$ (at the outer surface with T_c).

$$\Delta T = \frac{P}{k} \int_{r=a}^{b} \frac{dr}{4\pi r^2} = \frac{P}{4\pi k} \int_{r=a}^{b} \frac{dr}{r^2} = \frac{P}{4\pi k} \int_{r=a}^{b} r^{-2}\, dr = \frac{P}{4\pi k}[-r^{-1}]_{r=a}^{b} = \frac{P}{4\pi k}\left[-\frac{1}{r}\right]_{r=a}^{b}$$

Note that $\frac{1}{r^2} = r^{-2}$ and $r^{-1} = \frac{1}{r}$.

$$\Delta T = -\frac{P}{4\pi k}\left[\frac{1}{r}\right]_{r=a}^{b} = -\frac{P}{4\pi k}\left(\frac{1}{b} - \frac{1}{a}\right)$$

Make a **common denominator** in order to subtract the fractions.

$$\Delta T = -\frac{P}{4\pi k}\left(\frac{a}{ab} - \frac{b}{ab}\right) = -\frac{P}{4\pi k}\left(\frac{a-b}{ab}\right) = \frac{P}{4\pi k}\left(\frac{b-a}{ab}\right)$$

Note that $-(a - b) = +(b - a)$. Multiply both sides of the equation by $4\pi k$.

$$4\pi k \Delta T = P\left(\frac{b-a}{ab}\right)$$

Multiply both sides of the equation by ab and divide by $(b - a)$.

$$P = 4\pi k \Delta T\left(\frac{ab}{b-a}\right)$$

61. Water has a specific heat capacity of approximately $4200 \frac{\text{J}}{\text{kg}\cdot{}^\circ\text{C}}$, a latent heat of fusion of approximately $3.3 \times 10^5 \frac{\text{J}}{\text{kg}}$, and a latent heat of vaporization of approximately $2.3 \times 10^6 \frac{\text{J}}{\text{kg}}$. A cup contains 500 g of water at 25°C.

(A) Determine the heat exchange involved in completely freezing all of the water into ice. Is this heat absorbed or released by the water?

(B) Determine the heat exchange involved in completely boiling all of the water into steam. Is this heat absorbed or released by the water?

Want help? Check the hints section at the back of the book.

Answers: -2.2×10^5 J (released), 1.3×10^6 J (absorbed)

62. A perfectly insulating container holds 400 g of a liquid at 30°C. A 100-g metal cube at 110°C is added to the liquid and the container is sealed. The specific heat capacity of the metal is $640 \frac{\text{J}}{\text{kg}\cdot°\text{C}}$ and the specific heat capacity of the liquid is $2400 \frac{\text{J}}{\text{kg}\cdot°\text{C}}$. What equilibrium temperature will the system reach?

Want help? Check the hints section at the back of the book.

Answer: 35°C

63. A perfectly insulating container holds 600 g of liquid at 50°C. A 50-g ice cube at 0°C is added to the liquid and the container is sealed. The liquid has a specific heat capacity of $2150 \frac{\text{J}}{\text{kg}\cdot°\text{C}}$, water has a specific heat capacity of approximately $4200 \frac{\text{J}}{\text{kg}\cdot°\text{C}}$, and water has a latent heat of fusion of approximately $3.3 \times 10^5 \frac{\text{J}}{\text{kg}}$. What equilibrium temperature will the system reach?

Want help? Check the hints section at the back of the book.

Answer: 32°C

64. A perfectly insulating container holds 200 g of water at 33°C. A 100-g ice cube at 0°C is added to the water and the container is sealed. Water has a specific heat capacity of approximately 4200 $\frac{\text{J}}{\text{kg}\cdot°\text{C}}$ and a latent heat of fusion of approximately $3.3 \times 10^5 \frac{\text{J}}{\text{kg}}$.

(A) What percentage of the ice melts?

(B) What equilibrium temperature will the system reach?

Want help? Check the hints section at the back of the book.

Answers: (A) 84% (B) 0°C

65. The left face of the conducting cylinder illustrated below has a higher temperature (T_h) than the right face (T_c), resulting in a steady rate of heat flow from left to right. The body of the cylinder is thermally insulated. The conductor has uniform thermal conductivity.

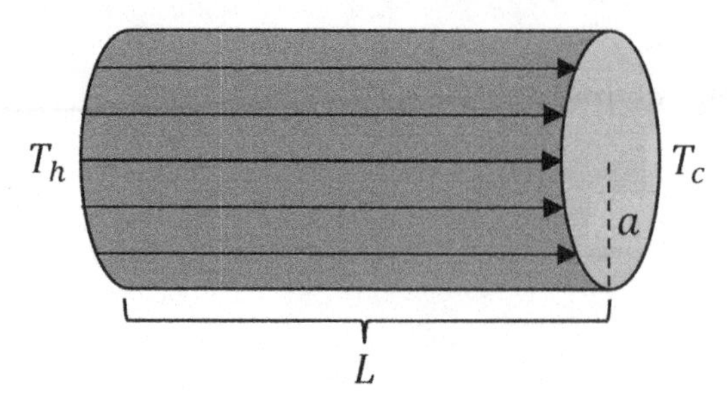

(A) Derive an equation for the rate of heat transfer in terms of k, L, a, and $\Delta T = T_h - T_c$.

(B) Derive an equation for the rate of heat transfer in terms of L, a, $\Delta T = T_h - T_c$, and the R value. **Note**: a is the radius; R is **not** the radius.

Want help? Check the hints section at the back of the book.

Answers: (A) $P = \frac{\pi k a^2 \Delta T}{L}$ (B) $P = \frac{\pi a^2 \Delta T}{R}$

66. The inner surface of the conducting cylindrical shell illustrated below has a higher temperature (T_h) than the outer surface (T_c), resulting in a steady rate of heat flowing outward. The conducting shell has uniform thermal conductivity. Derive an equation for the rate of heat transfer in terms of k, a, b, L, and $\Delta T = T_h - T_c$.

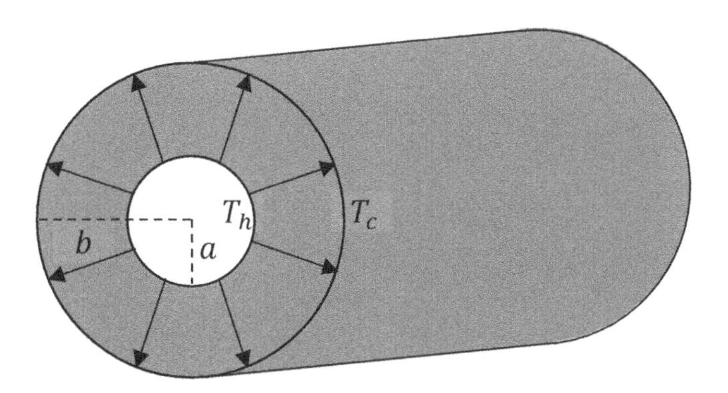

Want help? Check the hints section at the back of the book.

Answer: $P = \frac{2\pi k L \Delta T}{\ln\left(\frac{b}{a}\right)}$

67. A copper conducting slab and aluminum conducting slab of the same length and cross-sectional area are placed in series. The two slabs make thermal contact at the shared wall. The far end of the copper slab is at 65°C, while the far end of the aluminum slab is at 25°C. The remaining sides of the slabs are thermally insulated. The thermal conductivities of copper and aluminum are approximately $400 \frac{\text{W}}{\text{m}\cdot°\text{C}}$ and $240 \frac{\text{W}}{\text{m}\cdot°\text{C}}$, respectively. Find the temperature of the shared wall.

Want help? Check the hints section at the back of the book.

Answer: 50°C

16 IDEAL GASES

Relevant Terminology

Pressure – force per unit area.
Volume – the amount of space that an object occupies. Also referred to as **capacity**.
Temperature – a measure of the **average kinetic energy** of the molecules of a substance.
Heat – thermal energy that is **transferred** between substances.
Heat capacity – a measure of how much heat a given substance needs to absorb in order to increase its temperature a specified amount.
Internal energy – the total energy of the molecules of a substance, excluding their interactions with external fields (like earth's gravitational field). See Chapter 18.

The Ideal Gas Law

Gases that do not have a very high density obey the ideal gas law, where P is pressure, V is volume, T is the absolute temperature in **Kelvin** (K), n is the number of moles, and the **universal gas constant** equals $R = 8.314 \frac{\text{J}}{\text{mol}\cdot\text{K}}$ in physics units.[1] In this book, we will round R to $R \approx \frac{25}{3} \frac{\text{J}}{\text{mol}\cdot\text{K}}$ in order to solve problems without a calculator.

$$PV = nRT$$

An alternative way to express the ideal gas law uses the total number of **molecules** (N) rather than the number of **moles** (n) and Boltzmann's constant (k_B) instead of the universal gas constant (R). The number of molecules (N) equals the number of moles (N) times Avogadro's number (N_A): $N = nN_A$. **Avogadro's number** equals $N_A = 6.02 \times 10^{23}$ per mole. In this book, we will round Avogadro's number to $N_A \approx 6.0 \times 10^{23}$ per mole. The universal gas constant is related to Boltzmann's constant by: $R = N_A k_B$. Boltzmann's constant equals $k_B = 1.38 \times 10^{-23}$ J/K, which is approximately $k_B \approx \frac{7}{5} \times 10^{-23}$ J/K.

$$PV = Nk_BT$$

If the mole number (n) is constant (meaning that the system doesn't gain or lose particles[2]), the ideal gas law can also be expressed as the following ratio. This ratio comes in handy when you have information about the initial and final states of a gas (as we will see in the examples).

$$\frac{P_1V_1}{T_1} = \frac{P_2V_2}{T_2}$$

---

[1] In physics, we work with Joules, where it is convenient to work with $R = 8.314 \frac{\text{J}}{\text{mol}\cdot\text{K}}$. In chemistry, it is common to work with atmospheres and liters. In terms of those units, you get $R = 0.0821 \frac{\text{L}\cdot\text{atm}}{\text{mol}\cdot\text{K}}$.
[2] How can a gas gain or lose particles? One way is for the container to have a hole, such that some particles escape the volume of interest through the hole.

If the **temperature** (T) is constant, the ideal gas law reduces to **Boyle's law.**

$$P_1V_1 = P_2V_2$$

If the **pressure** (P) is constant, the ideal gas law reduces to **Charles's law.**

$$\frac{V_1}{T_1} = \frac{V_2}{T_2}$$

If the **volume** (V) is constant, the ideal gas law reduces to **Gay-Lussac's law.**

$$\frac{P_1}{T_1} = \frac{P_2}{T_2}$$

The root-mean-square[3] (rms) **speed** (v_{rms}) of the molecules in an ideal gas is related to the absolute temperature (T). This formula involves a squareroot because temperature is a measure of the average kinetic energy of the molecules of a substance, and the formula for kinetic energy is $KE = \frac{1}{2}mv^2$. There are two kinds of masses: In our notation, lowercase m represents the **total mass** of the gas in kilograms (kg), whereas uppercase M represents the **molar mass** of the substance in kilograms per mole (kg/mol).[4]

$$v_{rms} = \sqrt{\frac{3k_BT}{m}} = \sqrt{\frac{3RT}{M}}$$

The **molar specific heats**[5] at constant volume and constant pressure for an ideal gas depend on whether the gas is monatomic or diatomic.[6] See Chapter 5 for a list of elements that form monatomic or diatomic gases (and the footnote on page 75 for a tip on remembering them). The **adiabatic index** (γ) is the ratio of the molar specific heats (c_P over c_V).

$$c_V = \frac{3}{2}R \quad , \quad c_P = \frac{5}{2}R \quad , \quad \gamma = \frac{c_P}{c_V} = \frac{5}{3} \quad \text{(monatomic)}$$

$$c_V = \frac{5}{2}R \quad , \quad c_P = \frac{7}{2}R \quad , \quad \gamma = \frac{c_P}{c_V} = \frac{7}{5} \quad \text{(diatomic, moderate temp.)}$$

The molar specific heat at constant pressure is larger than at constant volume:

$$c_p = c_V + R$$

The **internal energy** (U) of an ideal gas is proportional to the absolute temperature.[6] We will learn about internal energy in Chapter 18.

$$U = \frac{3}{2}Nk_BT = \frac{3}{2}nRT \quad \text{(monatomic)} \quad , \quad U = \frac{5}{2}Nk_BT = \frac{5}{2}nRT \quad \text{(diatomic, mod. temp.)}$$

For an **adiabatic** process, an ideal gas obeys the following equation (in addition to the ideal gas law from the previous page). In this equation, volume is raised to the power of the **adiabatic index** (γ). We will explore adiabatic processes further in Chapter 18.

$$P_1V_1^\gamma = P_2V_2^\gamma$$

---

[3] We first encountered rms values in Volume 2, Chapter 31.

[4] Periodic tables generally provide the molar mass in g/mol, which requires converting to kg/mol.

[5] These are **molar** specific heats. See Chapter 15. The way to tell is that R has SI units of $\frac{\text{J}}{\text{mol}\cdot\text{K}}$, **not** $\frac{\text{J}}{\text{kg}\cdot\text{K}}$.

[6] The diatomic values assume that the temperature of the gas is moderate (not high, not very low).

Symbols and SI Units

Symbol	Name	SI Units
P	pressure	Pa
V	volume	m^3
n	number of moles	mol
N	number of molecules	unitless
N_A	Avogadro's number	$\frac{1}{\text{mol}}$
R	universal gas constant	$\frac{\text{J}}{\text{mol}\cdot\text{K}}$
k_B	Boltzmann's constant	J/K
T	absolute temperature	K
m	total mass	kg
M	mass per mole	kg/mol
Q	heat	J
c_V	molar specific heat capacity at constant volume	$\frac{\text{J}}{\text{mol}\cdot\text{K}}$
c_P	molar specific heat capacity at constant pressure	$\frac{\text{J}}{\text{mol}\cdot\text{K}}$
γ	adiabatic index	unitless
U	internal energy	J
v_{rms}	root-mean-square (rms) speed	m/s

Note Regarding Units

If you look up the molar mass (M) on a periodic table that gives values in grams per mole (g/mol), remember to convert to the SI units of kilograms per mole (kg/mol) by dividing by 1000.

Strategy for Problems Involving an Ideal Gas

How you solve a problem involving an ideal gas depends on which kind of problem it is:

- If an ideal gas problem gives you the temperature in degrees Celsius (°C), first convert the temperature to **Kelvin** (K): $T_K = T_C + 273.15 \approx T_C + 273$. If you are given the temperature in Fahrenheit, see Chapter 13 regarding the conversion.
- If a problem involves the number of **moles** (n), use the following formula:
$$PV = nRT$$
The universal gas constant is $R = 8.314 \frac{\text{J}}{\text{mol}\cdot\text{K}} \approx \frac{25}{3} \frac{\text{J}}{\text{mol}\cdot\text{K}}$.
- If a problem involves the number of **molecules** (N), use the following formula:
$$PV = Nk_BT$$
Boltzmann's constant is $k_B = 1.38 \times 10^{-23}$ J/K $\approx \frac{7}{5} \times 10^{-23}$ J/K.
- If a problem gives you information about initial and final states of an ideal gas, apply the following ratio. Usually, one of the three symbols (P, V, or T) is held constant: If one of the symbols is constant, it cancels out. See the examples.
$$\frac{P_1V_1}{T_1} = \frac{P_2V_2}{T_2}$$
- If a problem involves the root-mean-square (rms) speed of the molecules, use the following equation (where m is the **total** mass and M is the **molar** mass).
$$v_{rms} = \sqrt{\frac{3k_BT}{m}} = \sqrt{\frac{3RT}{M}}$$
- If an ideal gas problem involves **molar specific heat**, use the following equations.
$$c_V = \frac{3}{2}R \quad , \quad c_P = \frac{5}{2}R \quad , \quad \gamma = \frac{c_P}{c_V} = \frac{5}{3} \quad \text{(monatomic)}$$
$$c_V = \frac{5}{2}R \quad , \quad c_P = \frac{7}{2}R \quad , \quad \gamma = \frac{c_P}{c_V} = \frac{7}{5} \quad \text{(diatomic, moderate temp.)}$$
$$c_p = c_V + R$$
- If a problem involves the **internal energy** of an ideal gas, use the following equations. Also see Chapter 18.
$$U = \frac{3}{2}Nk_BT = \frac{3}{2}nRT \quad \text{(monatomic)}$$
$$U = \frac{5}{2}Nk_BT = \frac{5}{2}nRT \quad \text{(diatomic, mod. temp.)}$$
- If an ideal gas goes through an **adiabatic** process, apply the following equation. You may also need the equation $\frac{P_1V_1}{T_1} = \frac{P_2V_2}{T_2}$. See Chapter 18.
$$P_1V_1^\gamma = P_2V_2^\gamma$$
- If a problem involves **entropy**, see Chapter 18. For a **heat engine**, see Chapter 19.

Important Distinctions

There are two different kinds of masses:

- Lowercase m is the **total mass** of the gas in kilograms (kg).
- Uppercase M is the **molar mass** of the substance in kilograms per mole (kg/mol).

Boltzmann's constant is $k_B = 1.38 \times 10^{-23}$ J/K. You may recall the name Boltzmann from Chapter 15: The Stefan-Boltzmann constant is a different constant ($\sigma = 5.67 \times 10^{-8} \frac{\text{W}}{\text{m}^2\text{K}^4}$).

Example: An ideal gas has an initial temperature of 200 K. The volume of the gas triples while pressure remains constant. What is the final temperature of the gas?
Apply the ratio form of the ideal gas law.

$$\frac{P_1 V_1}{T_1} = \frac{P_2 V_2}{T_2}$$

Since pressure is held **constant,** $P_1 = P_2$ and pressure cancels out.

$$\frac{V_1}{T_1} = \frac{V_2}{T_2}$$

Solve for the final temperature: Multiply both sides by T_2 and T_1, and divide by V_1.

$$T_2 = \left(\frac{V_2}{V_1}\right) T_1$$

Since the volume of the gas triples, $V_2 = 3V_1$, which means that $\frac{V_2}{V_1} = 3$.

$$T_2 = 3T_1 = 3(200) = 600 \text{ K}$$

The final temperature of the gas is $T_2 = 600$ K.

Example: An ideal gas has an initial volume of 40 cc. The pressure of the gas halves under constant temperature. What is the final volume of the gas?
Apply the ratio form of the ideal gas law.

$$\frac{P_1 V_1}{T_1} = \frac{P_2 V_2}{T_2}$$

Since temperature is held **constant,** $T_1 = T_2$ and temperature cancels out.

$$P_1 V_1 = P_2 V_2$$

Solve for the final volume: Divide both sides by P_2.

$$V_2 = \left(\frac{P_1}{P_2}\right) V_1$$

Since the pressure halves, $P_2 = \frac{P_1}{2}$, which means that $\frac{P_1}{P_2} = 2$. (Put another way, $\frac{P_2}{P_1} = \frac{1}{2}$, but the above equation involves $\frac{P_1}{P_2}$, **not** $\frac{P_2}{P_1}$). You have to be careful with these ratios.

$$V_2 = 2V_1 = 2(40) = 80 \text{ cc}$$

The final volume of the gas is $V_2 = 80$ cc (that's in cubic centimeters, or cm^3).

68. An ideal gas has an initial pressure of 0.80 atm. The volume of the gas quadruples under constant temperature. What is the final pressure of the gas?

69. An ideal gas has an initial volume of 60 cc. The temperature of the gas doubles under constant pressure. What is the final volume of the gas?

70. An ideal gas has an initial temperature of 400 K. The pressure of the gas halves while the volume remains constant. What is the final temperature of the gas?

71. An ideal gas has an initial temperature of 27°C. The pressure of the gas halves while the volume of the gas triples. What is the final temperature of the gas?

Want help? Check the hints section at the back of the book.

Answers: 0.20 atm, 120 cc, 200 K, 177°C

17 VAN DER WAALS FLUIDS

Relevant Terminology

Pressure – force per unit area.
Volume – the amount of space that an object occupies. Also referred to as **capacity**.
Temperature – a measure of the **average kinetic energy** of the molecules of a substance.

Van der Waals Equation

The van der Waals equation[1] modifies the ideal gas law to better describe higher density gases. As the equation also shows the basic characteristics of a phase transition between the gaseous and liquid states, the van der Waals equation describes a **fluid**. The constants a and b depend on the specific fluid.

$$\left(P + \frac{an^2}{V^2}\right)\left(\frac{V}{n} - b\right) = RT$$

You can understand the van der Waals equation as follows:

- The pressure (P) is modified by $\frac{an^2}{V^2}$ to account for the fact that the molecules interact with one another, which impacts the pressure that the fluid exerts on the walls of the container. A more concentrated fluid has a larger value of $\frac{n}{V}$ (the number of moles divided by the volume provides a measure of concentration), which has a greater impact on the pressure.
- The volume of each mole $\left(\frac{V}{n}\right)$ is modified to account for the fact that the molecules have some size. The constant b is the volume that one mole of the molecules occupy. A more concentrated fluid has a larger value of b, which has a greater impact on the molar volume.

In the limit of a low-density gas, a and b have less of an effect on the pressure and volume: In the extreme case that a and b equal zero, you get $\frac{PV}{n} = RT$, which is the ideal gas law.

As the fluid has a higher concentration, a and b have a larger effect on the pressure and volume, and the distinction between a van der Waals fluid and an ideal gas becomes more important.

---

[1] The van der Waals equation may be written a little differently in some texts. For one, if another textbook defines the constants a and b differently, the equation will appear different. For another, some books work with molar volume rather than volume: Our V represents volume, but in some books it is instead volume divided by mole number. Yet another way that the equation may be different is in terms of structure: The equation is sometimes written with $\frac{P}{T}$ by itself on the left-hand side, for example.

Symbols and SI Units

Symbol	Name	SI Units
P	pressure	Pa
V	volume	m^3
n	number of moles	mol
R	universal gas constant	$\frac{\mathrm{J}}{\mathrm{mol{\cdot}K}}$
T	absolute temperature	K
a	a correction for intermolecular forces	$\frac{\mathrm{J{\cdot}m^3}}{\mathrm{mol^2}}$
b	the volume occupied by one mole of the molecules	$\frac{\mathrm{m^3}}{\mathrm{mol}}$

Note Regarding Units

The SI units of the constant a can be determined from the expression $P + \frac{an^2}{V^2}$ in the van der Waals equation. The term $\frac{an^2}{V^2}$ must have units of pressure, so one Pascal (Pa) must equal the SI units of $\frac{an^2}{V^2}$: $1 \text{ Pa} = \frac{[a][n^2]}{[V^2]}$. The brackets [ ] mean "the SI units of." Since the SI unit of n is the mole (mol) and the SI units of V are m^3, it follows that $1 \text{ Pa} = \frac{[a]\mathrm{mol}^2}{\mathrm{m}^6}$. Recall from Chapter 9 that one Pascal (Pa) equals a Newton (N) per square meter: $1 \text{ Pa} = 1\frac{\mathrm{N}}{\mathrm{m}^2}$. Plugging this into the previous equation, we get $1\frac{\mathrm{N}}{\mathrm{m}^2} = \frac{[a]\mathrm{mol}^2}{\mathrm{m}^6}$. Multiply both sides of the equation by m^6 and divide by mol^2 to see that $[a] = \frac{\mathrm{N{\cdot}m^4}}{\mathrm{mol^2}}$. Recall from Volume 1, Chapter 21 that one Joule (J) equals a Newton (N) times a meter: $1 \text{ J} = 1 \text{ Nm}$. Therefore, the SI units of a are $\frac{\mathrm{J{\cdot}m^3}}{\mathrm{mol^2}}$.

The SI units of the constant b can be determined from the expression $\frac{V}{n} - b$. The constant b must have the same units as $\frac{V}{n}$. Therefore, the SI units of b are $\frac{\mathrm{m}^3}{\mathrm{mol}}$.

If you look up the values of a and b in a table, you may need to convert them to SI units, as they might be given in terms of other units, such as atmospheres (atm) and liters (L).

Strategy for Problems Involving a van der Waals Fluid

If a problem involves a van der Waals fluid, follow these steps:

- If a problem gives you the temperature in degrees Celsius (°C), first convert the temperature to **Kelvin** (K): $T_K = T_C + 273.15 \approx T_C + 273$. If you are given the temperature in Fahrenheit, see Chapter 13 regarding the conversion.
- Apply the equation for a van der Waals fluid:

$$\left(P + \frac{an^2}{V^2}\right)\left(\frac{V}{n} - b\right) = RT$$

  The universal gas constant is $R = 8.314 \frac{\text{J}}{\text{mol·K}} \approx \frac{25}{3} \frac{\text{J}}{\text{mol·K}}$. For a textbook problem, you may need to look up the values of a and b in a table (if so, be sure to express them in SI units before plugging them into the equation). In the equation above, P is the pressure in Pascals (Pa), n is the mole number in moles (mol), V is the volume of the fluid in cubic meters (m^3), and T is the absolute temperature in Kelvin (K). If you are given the number of **molecules** (N) instead of the number of **moles** (n), note that $n = \frac{N}{N_A}$, where Avogadro's number equals $N_A = 6.02 \times 10^{23} \approx 6.0 \times 10^{23}$ per mole.
- If a problem asks you to compare the behavior of the van der Waals fluid to an **ideal gas**, compare with the equation for an ideal gas.

$$\frac{PV}{n} = RT$$

- If a problem involves **entropy**, **internal energy**, a P-V diagram, or a **specific process** (like an **isotherm** or **adiabat**), see Chapter 18. For a **heat engine**, see Chapter 19.

Example: For a particular van der Waals fluid, the molecules occupy 10% of the volume and the molecular interactions affect the measured pressure by 20%. If you were to use the ideal gas law to determine the temperature, by what percentage would you be off?

The van der Waals equation replaces $\frac{PV}{n}$ from the ideal gas law with $\left(P + \frac{an^2}{V^2}\right)\left(\frac{V}{n} - b\right)$, while both expressions equal RT in their respective equations. To answer the question, we just need to determine by what percentage $\left(P + \frac{an^2}{V^2}\right)\left(\frac{V}{n} - b\right)$ differs from $\frac{PV}{n}$. According to the question, $b = (10\%)\frac{V}{n} = 0.1\frac{V}{n}$ and $\frac{an^2}{V^2} = (20\%)P = 0.2P$. Note that $P + 0.2P = 1.2P$ and that $\frac{V}{n} - 0.1\frac{V}{n} = (1 - 0.1)\frac{V}{n} = 0.9\frac{V}{n}$.

$$\left(P + \frac{an^2}{V^2}\right)\left(\frac{V}{n} - b\right) = (P + 0.2P)\left(\frac{V}{n} - 0.1\frac{V}{n}\right) = (1.2P)\left(0.9\frac{V}{n}\right) = 1.08\frac{PV}{n}$$

In this problem, the van der Waals equation corrects the ideal gas law's prediction for the absolute temperature by 8% (since 1.08 is differs from 1 by 0.08 = 8%).

72. A van der Waals fluid exhibits the basic characteristics of a phase transition. The critical point for the phase transition can be found by applying the following equations (holding both temperature and mole number constant):

$$\frac{\partial P}{\partial V} = 0 \quad , \quad \frac{\partial^2 P}{\partial V^2} = 0$$

Apply the above equations to derive equations for the critical pressure, critical molar volume, and critical temperature of a van der Waals fluid in terms of the constants a, b, and R.

Note: It may help to review partial derivatives in Chapter 5 (see page 83).

Want help? Check the hints section at the back of the book.

Answers: $P_c = \frac{a}{27b^2}$, $\frac{V_c}{n} = 3b$, $T_c = \frac{8a}{27bR}$

18 THE LAWS OF THERMODYNAMICS

Relevant Terminology

Pressure – force per unit area.
Volume – the amount of space that an object occupies. Also referred to as **capacity**.
Temperature – a measure of the **average kinetic energy** of the molecules of a substance.
Entropy – a quantitative measure of the **statistical disorder** of a system.
Work – thermodynamic work is done when the volume changes under pressure.
Heat – thermal energy that is **transferred** between substances.
Heat capacity – a measure of how much heat a given substance needs to absorb in order to increase its temperature a specified amount.
Internal energy – the total energy of the molecules of a substance, excluding their interactions with external fields (like earth's gravitational field).
Isothermal – a process for which **temperature** remains constant.
Isobaric – a process for which **pressure** remains constant.
Isochoric – a process for which **volume** remains constant.
Adiabatic – a process for which <u>**no**</u> **heat** is absorbed or released.
Isentropic – a process for which the **entropy** of a system remains constant.
Steady-state – a process for which the **internal energy** of a system remains constant.
Quasistatic – a process that is carried out through a series of virtually infinitesimal disturbances from equilibrium, which occurs slowly enough that the system establishes equilibrium between the infinitesimal disturbances.
Spontaneous – a process that occurs without being driven by an energy source.
Reversible – a quasistatic process that can be carried out in reverse.
Irreversible – any process that is not quasistatic, or any process that features frictional or dissipative forces.

The Laws of Thermodynamics

According to the **zeroth** law of thermodynamics, if two objects, A and B, are in thermal equilibrium with a third object, C, then the two objects, A and B, are also in thermal equilibrium with each other. This fundamental law allows us to use a **thermometer** to measure temperature and to establish **thermal equilibrium** between multiple objects.

According to the **first** law of thermodynamics, the **heat** (dQ) absorbed by a system with constant mole numbers minus the **work** (dW) done by the system equals the change in the system's **internal energy** (dU). This law expresses **conservation of energy**.

$$dU = dQ - dW$$

Note the following **sign conventions:**

- The **heat** change (dQ) is **positive** if heat is **added** to (or absorbed by) the system and **negative** if it is released by the **system.**
- **Work** (dW) is **positive** if work is done **by** the system and **negative** if work is done **on** the system. Note how those tiny words "by" and "on" make a huge difference!

According to the **second** law of thermodynamics, the total entropy (statistical disorder) of the system plus the surroundings **cannot decrease** for a thermodynamic process (for a macroscopic system) that begins and ends in equilibrium states. It can remain constant or it may increase, but the overall entropy (if you include both the entropy of the system and the entropy of the surroundings) cannot decrease. (The entropy of the system can decrease provided that the entropy of the surroundings increases at least as much, and vice-versa.) One consequence of the second law of thermodynamics is that **it is impossible to construct a perfectly efficient heat engine.** We will consider heat engines in Chapter 19.

According to the **third** law of thermodynamics, it is impossible for any object to reach a temperature of exactly **absolute zero** Kelvin by performing a finite number of operations, regardless of what idealized process may be applied. In practice, we can get very close to absolute zero (a small fraction of one Kelvin), but we can never reach it exactly.[1]

Heat, Work, Entropy, and Internal Energy

Internal energy (U), **heat** (Q), and **work** (W) are related through the first law of thermodynamics, which expresses conservation of energy for thermal systems.

$$dU = dQ - dW$$

Work (W) is the integral of **pressure** (P) over **volume** (V).

$$dW = PdV \quad , \quad W = \int_{V=V_0}^{V} P\, dV$$

The sign of work is determined as follows:

- When work is done **by** the system, the **volume increases** and work is **positive.**
- When work is done **on** the system, the **volume decreases** and work is **negative.**

Heat (Q) exchanges (Chapter 15) may be expressed in terms of **specific heat** (C_V or C_P), **molar specific heat** (c_V or c_P), or **entropy** (S) – or for a phase transition, in terms of **latent heat** (L).

$$dQ = mC_V dT \quad , \quad dQ = mC_P dT \quad , \quad dQ = nc_V dT \quad , \quad dQ = nc_P dT \quad , \quad dQ = Ldm$$

$$dQ = TdS$$

[1] The first, second, and third laws of thermodynamics have been paraphrased as, "You can't get ahead," "You can't even break even," and "You can't even get back out of the game."

The sign of a heat exchange is determined as follows:

- When heat is **added** to (or absorbed by) the system, heat is **positive**.
- When heat is **released** by the system, heat is **negative**.

The **entropy** (S) is related to the heat and temperature by $dQ = TdS$. Entropy provides a measure of the **statistical disorder** of the system, and is valid only for equilibrium states. Entropy is path-independent (like internal energy in this regard, but unlike work and heat), meaning that the change in entropy can be computed for any reversible process that takes the system between the initial and final equilibrium states.

$$\Delta S = S - S_0 = \int_{Q=Q_0}^{Q} \frac{dQ}{T}$$

P-V Diagrams

There are two common ways to calculate the work done in thermal physics.

- Express pressure as a function of volume and integrate: $W = \int_{V=V_0}^{V} P\, dV$.
- Find the **area** under the curve for a P-V diagram.

(Recall from calculus that a definite integral equals the area under the curve.)

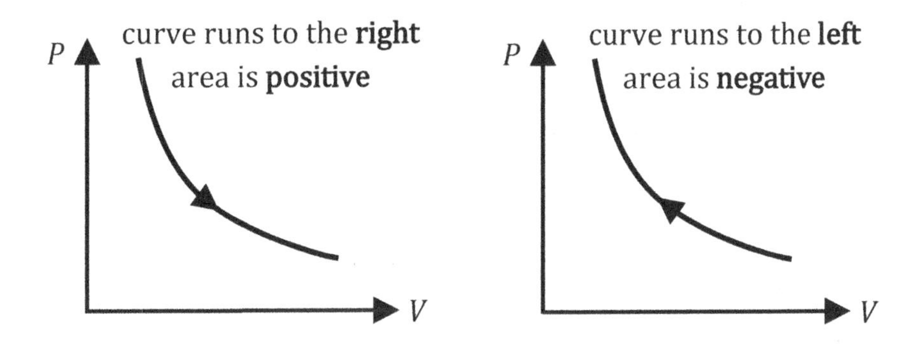

Note that area can be positive or negative:

- The area is **positive** when **volume increases** (the process goes to the **right**).
- The area is **negative** when **volume decreases** (the process goes to the **left**).

For a complete cycle (a closed path), the **net work** equals the area enclosed by the path.

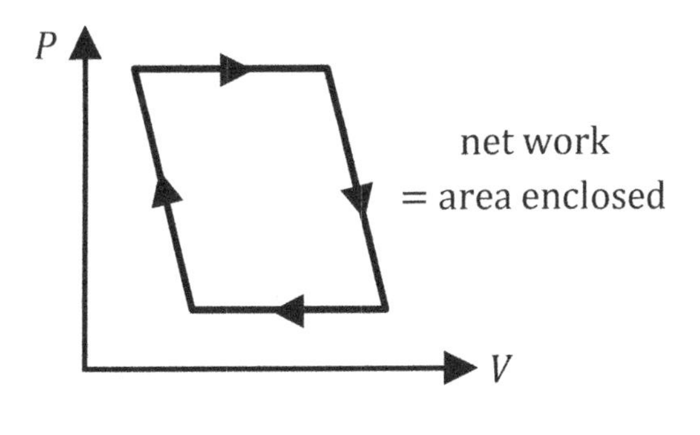

Thermodynamic Processes

Following are some common thermodynamic processes.

- **Volume** remains constant for an **isochoric** process.
  - Since the volume doesn't change ($\Delta V = 0$), <u>no</u> work is done ($W = 0$). An isochor is a **vertical** line on a P-V diagram.
  - Express the heat exchange in terms of the specific heat capacity at constant **volume**: $dQ = mC_V dT$ or $dQ = nc_V dT$. Since no work is done, the change in internal energy equals the heat exchanged: $\Delta U = Q$.
- **Pressure** remains constant for an **isobaric** process.
  - Since pressure is constant, it may come out of the work integral: $W = \int_{V=V_0}^{V} P\,dV = P\int_{V=V_0}^{V} dV = P(V - V_0) = P\Delta V$. An isobar is a **horizontal** line on a P-V diagram.
  - Express the heat exchange in terms of the specific heat capacity at constant **pressure**: $dQ = mC_P dT$ or $dQ = nc_P dT$. The internal energy change equals $\Delta U = Q - W$ according to the first law of thermodynamics.
- A **free expansion** occurs under **zero** pressure.
  - <u>No</u> work is done ($W = 0$) because the pressure equals zero. Since no work is done, the change in internal energy equals the heat exchanged: $\Delta U = Q$.
- **Temperature** remains constant for an **isothermal** process.
  - For an ideal gas or a van der Waals fluid, solve for pressure in terms of volume to perform the work integral: $W = \int_{V=V_0}^{V} P\,dV$. For example, for an ideal gas, $P = \frac{nRT}{V}$. **Temperature is a constant** for an isotherm.
  - Although temperature is constant for an isotherm, there is generally still heat exchanged ($Q \neq 0$). However, if the potential energy of the molecules is constant (it is for an **ideal gas**), the change in internal energy will be zero ($\Delta U = 0$) and the heat exchange will equal the work done: $Q = W$.
- <u>No</u> **heat** is absorbed or released by the system for an **adiabatic** process.
  - If the relationship between pressure and volume is known for the adiabat, express pressure in terms of volume to perform the work integral: $W = \int_{V=V_0}^{V} P\,dV$. For example, for an ideal gas, $P_0 V_0^\gamma = PV^\gamma$, or $P = \frac{P_0 V_0^\gamma}{V^\gamma}$.
  - Since no heat is exchanged, the change in internal energy equals the negative of the work done: $\Delta U = -W$.
- For a **complete cycle** (a **closed path** on a P-V diagram), the net internal energy change is zero ($\Delta U_{net} = 0$) for the complete cycle, although internal energy may change for parts of the cycle. The net heat exchange equals the net work done: $Q_{net} = W_{net}$.

Symbols and SI Units

Symbol	Name	SI Units
P	pressure	Pa
V	volume	m^3
n	number of moles	mol
R	universal gas constant	$\frac{\text{J}}{\text{mol}\cdot\text{K}}$
T	absolute temperature	K
m	total mass	kg
Q	heat	J
L	latent heat of transformation	$\frac{\text{J}}{\text{kg}}$
C_V	specific heat capacity at constant volume	$\frac{\text{J}}{\text{kg}\cdot\text{K}}$
C_P	specific heat capacity at constant pressure	$\frac{\text{J}}{\text{kg}\cdot\text{K}}$
c_V	molar specific heat capacity at constant volume	$\frac{\text{J}}{\text{mol}\cdot\text{K}}$
c_P	molar specific heat capacity at constant pressure	$\frac{\text{J}}{\text{mol}\cdot\text{K}}$
γ	adiabatic index	unitless
U	internal energy	J
W	work	J
S	entropy	J/K
a	a correction for intermolecular forces	$\frac{\text{J}\cdot\text{m}^3}{\text{mol}^2}$
b	the volume occupied by one mole of the molecules	$\frac{\text{m}^3}{\text{mol}}$

Note Regarding Units

From the equation $dQ = TdS$, it follows that the SI units of entropy (S) are J/K.

Where Do the Ideal Gas Equations $U = nc_V T$ and $c_P = c_V + R$ Come From?

We will begin with the ideal gas law, $PV = nRT$, which may be expressed in the form $\frac{P_0 V_0}{T_0} = \frac{PV}{T}$, and show where the equations on the bottom half of page 182 come from.

For an **isochor, volume** is constant, so __no__ work is done: $W = 0$. The corresponding heat change depends on the molar specific heat at constant **volume**: $Q = nc_V \Delta T$. According to the first law of thermodynamics, $\Delta U = Q - W$. For an isochor, this reduces to $\Delta U = Q = nc_V \Delta T$. Since ΔU is path-independent, and since the internal energy of an ideal gas depends only on temperature, $U = nc_V T$ and $\Delta U = nc_V \Delta T$ for __any__ process.

For an **isobar, pressure** is constant. We may pull pressure out of the work integral: $W = \int P\,dV = P \int dV = P\Delta V$. From the ideal gas law, $P\Delta V = nR\Delta T$ for an isobar, such that $W = nR\Delta T$. The corresponding heat change depends on the molar specific heat at constant **pressure**: $Q = nc_P \Delta T$. According to the first law of thermodynamics, $\Delta U = Q - W$. For an isobar, this becomes $\Delta U = nc_P \Delta T - nR\Delta T$. Recall that $\Delta U = nc_V \Delta T$ for an ideal gas for any process. Therefore, $nc_V \Delta T = nc_P \Delta T - nR\Delta T$. Divide both sides of the equation by $n\Delta T$. This shows that $c_V = c_P - R$, which is the same as $c_P = c_V + R$. The adiabatic index is then $\gamma = \frac{c_P}{c_V} = \frac{c_V + R}{c_V}$. For a **monatomic** ideal gas, $c_V = \frac{3}{2}R$, $c_P = \frac{5}{2}R$, and $\gamma = \frac{5}{3}$.

Why Does $P_0 V_0^\gamma = PV^\gamma$ Along an Adiabat for An Ideal Gas?

For an ideal gas, $PV = nRT$. Implicitly differentiate the ideal gas law: $d(PV) = d(nRT)$. Apply the product rule: $d(PV) = PdV + VdP$. Therefore, $PdV + VdP = nRdT$. Apply the first law of thermodynamics: $dU = dQ - dW$. For an adiabatic process, __no__ heat is exchanged: $dQ = 0$. The first law simplifies to $dU = -dW$. Recall that $dU = nc_V dT$ and $dW = PdV$. Plug these expressions into $dU = -dW$ to get $nc_V dT = -PdV$. Therefore, $dT = -\frac{PdV}{nc_V}$. Plug this into the equation $PdV + VdP = nRdT$ to get $PdV + VdP = -\frac{R}{c_V}PdV$. Multiply both sides of the equation by c_V to get $c_V PdV + c_V VdP = -RPdV$. Combine like differentials: $c_V VdP = -(c_V + R)PdV$. Recall that $c_P = c_V + R$. Thus, $c_V VdP = -c_P PdV$. Separate variables and integrate. Recall that $\ln(b) - \ln(a) = \ln\left(\frac{b}{a}\right)$ and $a\ln(x) = \ln(x^a)$.

$$c_V VdP = -c_P PdV$$

$$\int \frac{dP}{P} = -\frac{c_P}{c_V}\int \frac{dV}{V}$$

$$\ln(P) - \ln(P_0) = -\gamma[\ln(V) - \ln(V_0)]$$

$$\ln\left(\frac{P}{P_0}\right) = -\gamma \ln\left(\frac{V_0}{V}\right) = \ln\left(\frac{V_0^\gamma}{V^\gamma}\right)$$

Use the rule $e^{\ln(x)} = x$ to get $\frac{P}{P_0} = \frac{V_0^\gamma}{V^\gamma}$. Cross multiply: $PV^\gamma = P_0 V_0^\gamma$.

Strategy for Problems Involving Work, Heat, Entropy, or Internal Energy

How you solve a thermal physics problem involving work, heat, entropy, or internal energy depends on which kind of problem it is:

- There are two ways to calculate the **work** done for a thermodynamic process:
  - If you are given a P-V diagram, find the **area** under the curve. Area is **positive** if **volume increases** and **negative** if **volume decreases**. For a closed path, the net work equals the area enclosed by the path: In this case, area is **positive** for a **clockwise** path and **negative** for a **counterclockwise** path.
  - Otherwise, express pressure as a function of volume and integrate.

$$W = \int_{V=V_0}^{V} P\,dV$$

  Consult the notes on page 194 regarding common thermodynamic processes.
- There are a variety of ways to determine the **heat** exchanged.
  - For an **isobar, pressure** remains constant: $dQ = mC_P dT$ and $dQ = nc_P dT$.
  - For an **isochor, volume** remains constant: $dQ = mC_V dT$ and $dQ = nc_V dT$.
  - For other thermodynamic process (such as an isotherm or adiabat), the notes on page 194 will help you apply the first law of thermodynamics to relate the heat exchange to work and internal energy.
  - For a **phase transition,** $Q = mL$.
  - For heat conduction or thermal radiation, see Chapter 15.
- If a problem involves an **ideal gas**, you may apply the ideal gas law (Chapter 16). The **internal energy** and **molar specific heat capacities** of a monatomic or diatomic (at moderate temperature) ideal gas is given by the expressions below.

$$PV = nRT \quad , \quad \frac{P_0 V_0}{T_0} = \frac{PV}{T} \quad , \quad c_p = c_V + R$$

$$U = \frac{3}{2} N k_B T = \frac{3}{2} nRT \quad \text{(monatomic)}$$

$$U = \frac{5}{2} N k_B T = \frac{5}{2} nRT \quad \text{(diatomic, mod. temp.)}$$

$$c_V = \frac{3}{2} R \quad , \quad c_P = \frac{5}{2} R \quad , \quad \gamma = \frac{c_P}{c_V} = \frac{5}{3} \quad \text{(monatomic)}$$

$$c_V = \frac{5}{2} R \quad , \quad c_P = \frac{7}{2} R \quad , \quad \gamma = \frac{c_P}{c_V} = \frac{7}{5} \quad \text{(diatomic, moderate temp.)}$$

  To find the work done along an **isotherm**, solve for pressure in one of the equations below, plug it into the work integral, and treat T as a constant. For an **isothermal** process involving an ideal gas, the **internal energy** change is <u>**zero**</u>.

$$PV = nRT$$

$$P_0 V_0 = PV \quad \text{(isotherm)}$$

$$\Delta U = 0 \quad \text{(isotherm)}$$

To find the work done by an ideal gas along an **adiabat**, solve for pressure in the following equation and plug it into the work integral.

$$P_0 V_0^\gamma = P V^\gamma \quad \text{(adiabat)}$$

- If a problem involves a **van der Waals fluid**, you may apply the van der Waals equation (Chapter 17). To find the work done, substitute the equation for pressure (see below) into the work integral. Along an **isotherm**, T will be constant.

$$P = \frac{RT}{\frac{V}{n} - b} - \frac{an^2}{V^2}$$

- **Internal energy** changes, heat, and work can be related through the first law of thermodynamics.

$$dU = dQ - dW$$

- **Entropy** (S) is related to heat via:

$$dQ = TdS$$

The change in entropy ($\Delta S = S - S_0$) can be computed for any reversible process that takes the system between the initial and final equilibrium states, according to the following integral. See the second bullet point on the previous page regarding how to express heat as a function of temperature. Express the temperature in **Kelvin** (K) **after** integrating (that is, after evaluating the anti-derivative over the limits).

$$\Delta S = S - S_0 = \int_{Q=Q_0}^{Q} \frac{dQ}{T}$$

- If a problem involves a **heat engine**, see Chapter 19.

Important Distinctions

It is very important to remember which quantity is constant for an isothermal, isobaric, isochoric, or adiabatic process, since the solution to a problem is much different for each process.

- **Temperature** is constant along an **isotherm**.
- **Pressure** is constant along an **isobar**.
- **Volume** is constant along an **isochor**.
- No **heat** is exchanged along an **adiabat**.

Note that the math is much different for **isotherms** and **adiabats**, since **temperature** and **heat** mean two different things (Chapter 15). Heat is generally exchanged along an isotherm even though temperature remains constant, and temperature generally changes along an adiabat even though no heat is exchanged.

Example: A non-ideal gas expands isobarically from a volume of 3.0 m^3 to 5.0 m^3 under a constant pressure of 50 kPa, as the temperature of the gas changes from 150 K to 450 K. The molar specific heat at constant pressure equals $2R$. There are 30 moles of the gas.

(A) Determine the work done.
Perform the thermodynamic work integral.

$$W = \int_{V=V_0}^{V} P\,dV$$

Since this example involves an **isobar**, the **pressure is constant**. This means that we may pull the pressure out of the integral. (For a process that isn't isobaric, you can't do this. The next example involves such a process.)

$$W = P \int_{V=V_0}^{V} dV = P(V - V_0) = (50{,}000)(5 - 3) = (50{,}000)(2) = 100{,}000 \text{ J} = 100 \text{ kJ}$$

The gas does $W = 100{,}000 \text{ J} = 100 \text{ kJ}$ of work. The work done is **positive** (meaning that work is done **by** the gas) because the **volume increased**.

(B) Determine the amount of heat exchanged.
Since this example involves an **isobar**, the **pressure is constant**. The heat can be found from the molar specific heat at constant pressure.

$$dQ = nc_P dT$$

$$Q = \int_{T=T_0}^{T} nc_P\,dT$$

The problem states that $c_P = 2R$ for this gas, where R is the universal gas constant. Recall that $R = 8.314 \frac{\text{J}}{\text{mol}\cdot\text{K}} \approx \frac{25}{3} \frac{\text{J}}{\text{mol}\cdot\text{K}}$.

$$Q = \int_{T=T_0}^{T} n2R\,dT = 2nR \int_{T=T_0}^{T} dT = 2nR(T - T_0)$$

$$Q = 2(30)(8.314)(450 - 150) \approx 2(30)\left(\frac{25}{3}\right)(300) = 150{,}000 \text{ J} = 150 \text{ kJ}$$

The gas **absorbs** (since Q is **positive**) $Q = 150{,}000 \text{ J} = 150 \text{ kJ}$ of heat energy.

(C) Determine the change in the internal energy of the gas.
Apply the **first law** of thermodynamics, using the answers to (A) and (B).

$$\Delta U = Q - W = 150{,}000 - 100{,}000 = 50{,}000 \text{ J} = 50 \text{ kJ}$$

The internal energy of the gas increases by $\Delta U = 50{,}000 \text{ J} = 50 \text{ kJ}$. **Note**: You **can't** use the equation $U = nc_V T$ for this problem because this is **not** an ideal gas.

Example: A monatomic ideal gas is compressed isothermally from a volume of 4.0 m^3 to 1.0 m^3. The initial pressure is 20 kPa.

(A) Determine the work done.
Perform the thermodynamic work integral.

$$W = \int_{V=V_0}^{V} P\,dV$$

Unlike the previous example, this process is <u>**not**</u> isobaric, so pressure is <u>**not**</u> constant and may <u>**not**</u> come out of the integral. However, since this examples involves an **ideal gas**, we may use the equation $\frac{P_0V_0}{T_0} = \frac{PV}{T}$. Furthermore, since this process is **isothermal, temperature** is constant: $T = T_0$, so T cancels out in $\frac{P_0V_0}{T_0} = \frac{PV}{T}$, reducing the equation to $P_0V_0 = PV$. Solve for P to get $P = \frac{P_0V_0}{V}$. Substitute this expression for pressure into the work integral.

$$W = \int_{V=V_0}^{V} \frac{P_0V_0}{V}\,dV$$

The initial pressure and volume are constants, so we may pull P_0V_0 out of the integral.

$$W = P_0V_0 \int_{V=V_0}^{V} \frac{dV}{V}$$

Recall from calculus that $\int \frac{dx}{x} = \ln(x)$. Also recall the logarithm identity $\ln(b) - \ln(a) = \ln\left(\frac{b}{a}\right)$. (For a quick review of **logarithms**, see Volume 2, Chapter 17.)

$$W = P_0V_0[\ln(V)]_{V=V_0}^{V} = P_0V_0[\ln(V) - \ln(V_0)] = P_0V_0 \ln\left(\frac{V}{V_0}\right)$$

$$W = (20{,}000)(4)\ln\left(\frac{1}{4}\right) = (80{,}000)\ln\left(\frac{1}{4}\right)\ \text{J} = -(80{,}000)\ln(4)\ \text{J} = -80\ln(4)\ \text{kJ}$$

Recall the logarithm identity $\ln\left(\frac{1}{x}\right) = -\ln(x)$. The work done is $W = -(80{,}000)\ln(4)\ \text{J} = -80\ln(4)\ \text{kJ}$. If you use a calculator, this works out to $W = -111$ kJ. The work done is **negative** (meaning that work is done <u>**on**</u> the gas) because the **volume decreased**. (Work is done "by" the gas when it expands, and "on" the gas when it is compressed. These two-letter words "by" and "on" make a sign difference.)

(B) Determine the change in the internal energy of the gas.
Since this example involves a **monatomic ideal gas**, the internal energy equals $U = \frac{3}{2}nRT$. Since this process is isothermal, T is constant, such that $\Delta T = 0$ and $\Delta U = 0$. The internal energy of the gas <u>**doesn't**</u> change: $\Delta U = 0$.

(C) Determine the amount of heat exchanged.
Apply the **first law** of thermodynamics, using the answers to (A) and (B).

$$\Delta U = Q - W$$

Add work to both sides of the equation.

$$Q = \Delta U + W = 0 + [-(80{,}000)\ln(4)] = -(80{,}000)\ln(4)\ \text{J} = -80\ln(4)\ \text{kJ}$$

The heat exchange equals $Q = -(80{,}000)\ln(4)$ J $= -80\ln(4)$ kJ. If you use a calculator, this works out to $Q = -110$ kJ. The gas **releases** heat (since Q is **negative**).

Example: A fluid expands from an initial volume of 6.0 m^3 to 12.0 m^3. The initial pressure is 25 kPa. The fluid obeys the equation $V\sqrt{P} = \text{const.}$ Determine the work done.

Perform the thermodynamic work integral.

$$W = \int_{V=V_0}^{V} P\,dV$$

Note that the pressure is **not** constant in this problem, so we may **not** pull it out of the integral. Instead, we must express the pressure in terms of the volume. We can write the given equation, $V\sqrt{P} = \text{const.}$ (where "const." is short for "constant"), as $V_0\sqrt{P_0} = V\sqrt{P}$ (since both $V_0\sqrt{P_0}$ and $V\sqrt{P}$ must equal the same constant). Square both sides to get $V_0^2 P_0 = V^2 P$ and then divide both sides by V^2 to find that $P = \frac{P_0 V_0^2}{V^2}$. Plug this expression into the work integral.

$$W = \int_{V=V_0}^{V} \frac{P_0 V_0^2}{V^2}\,dV$$

The initial pressure and volume are constants, so we may pull $P_0 V_0^2$ out of the integral.

$$W = P_0 V_0^2 \int_{V=V_0}^{V} \frac{dV}{V^2} = P_0 V_0^2 \int_{V=V_0}^{V} V^{-2}\,dV = P_0 V_0^2 [-V^{-1}]_{V=V_0}^{V} = P_0 V_0^2 \left[-\frac{1}{V}\right]_{V=V_0}^{V}$$

Note that $\frac{1}{V^2} = V^{-2}$ and $V^{-1} = \frac{1}{V}$.

$$W = -P_0 V_0^2 \left[\frac{1}{V}\right]_{V=V_0}^{V} = -P_0 V_0^2 \left(\frac{1}{V} - \frac{1}{V_0}\right) = -(25{,}000)(6)^2 \left(\frac{1}{12} - \frac{1}{6}\right)$$

It's convenient to distribute the $(6)^2$.

$$W = -(25{,}000)\left(\frac{6^2}{12} - \frac{6^2}{6}\right) = -(25{,}000)(3 - 6) = -(25{,}000)(-3) = 75{,}000\ \text{J} = 75\ \text{kJ}$$

The gas does $W = 75{,}000$ J $= 75$ kJ of work. The work done is **positive** (meaning that work is done **by** the gas) because the **volume increased**. (The two minus signs made a plus sign.)

Example: Determine the work done along the straight line path in the graph below.

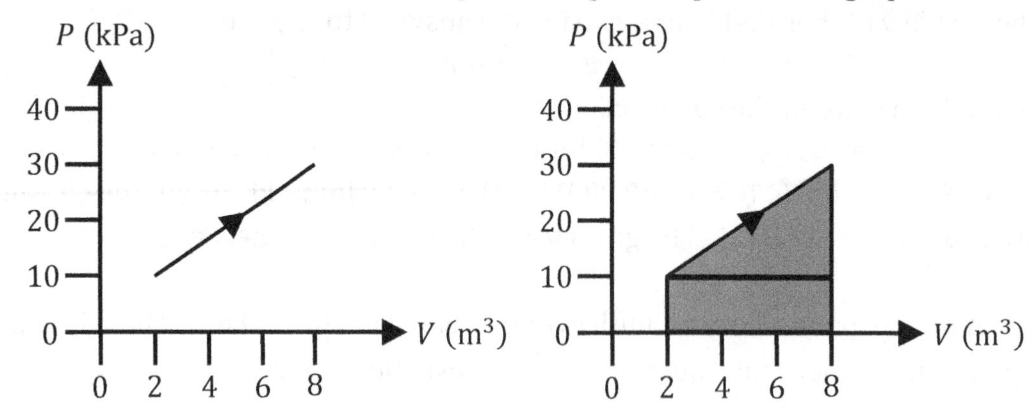

The work done equals the **area** under the P-V curve (which in this case is a straight line). On the diagram on the right, you can see that the area under the straight line path can be divided up into a triangle and rectangle. (It may help to review Volume 1, Chapter 6.) The work done equals the area of the triangle plus the area of the rectangle. In this example, the work is **positive** because the **volume increases** (the path goes to the **right**).

$$W = A_{tri} + A_{rect} = \frac{1}{2}bh + LW$$

The triangle has a base of $b = 8.0 - 2.0 = 6.0$ m^3 and a height of $h = 30 - 10 = 20$ kPa, while the rectangle has dimensions of $L = 8.0 - 2.0 = 6.0$ m^3 and $W = 10 - 0 = 10$ kPa.

$$W = \frac{1}{2}(6)(20{,}000) + (6)(10{,}000) = 60{,}000 + 60{,}000 = 120{,}000 \text{ J} = 120 \text{ kJ}$$

The work done is $W = 120{,}000$ J $= 120$ kJ.

Example: Determine the net work done for the complete cycle shown below.

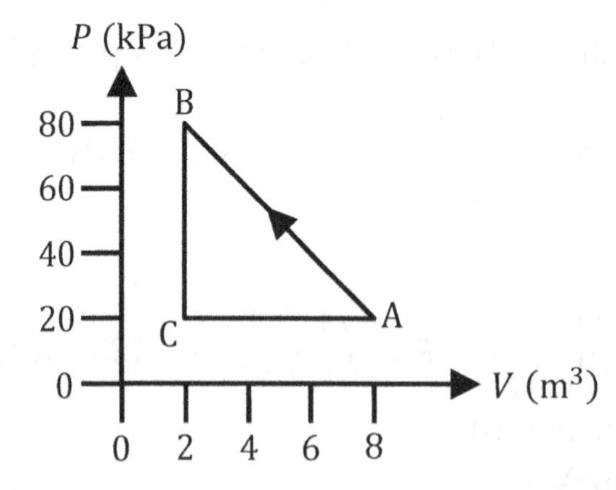

The __**net**__ work equals the **area** of the triangle. The net work is **negative** because the path is **counterclockwise**. The triangle has a base of $b = 8.0 - 2.0 = 6.0$ m^3 and a height of $h = 80 - 20 = 60$ kPa.

$$W_{net} = -A_{tri} = -\frac{1}{2}bh = -\frac{1}{2}(6)(60{,}000) = -180{,}000 \text{ J} = -180 \text{ kJ}$$

The __**net**__ work done is $W_{net} = -180{,}000$ J $= -180$ kJ.

Example: A monatomic ideal gas completes the thermodynamic cycle shown below, where path AB is an isotherm.

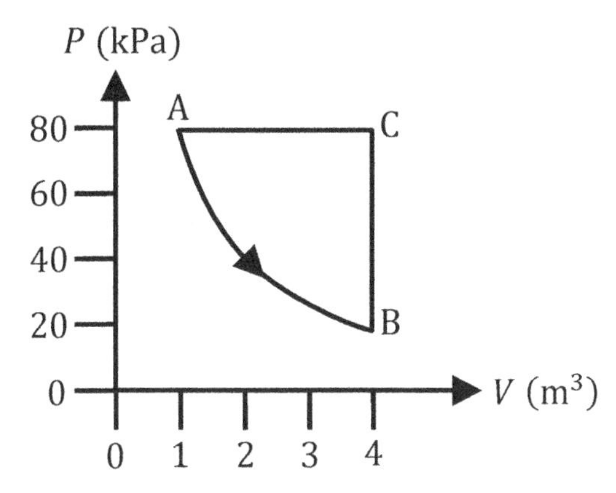

(A) Determine the work, heat, and internal energy change for path AB.
Unlike the two previous examples, path AB is a curve.[2] We can't get an exact answer by finding areas of triangles and rectangles for the curve running from A to B. However, we can get an exact answer by performing the thermodynamic work integral for this path.

$$W_{AB} = \int_{V=V_A}^{V_B} P\,dV$$

Since this example involves an **ideal gas**, we may use the equation $\frac{P_A V_A}{T_A} = \frac{PV}{T}$. Furthermore, since this process is **isothermal**, **temperature** is constant: $T = T_A$, so T cancels out in $\frac{P_A V_A}{T_A} = \frac{PV}{T}$, reducing the equation to $P_A V_A = PV$. Solve for P to get $P = \frac{P_A V_A}{V}$. Substitute this for pressure in the work integral.

$$W_{AB} = \int_{V=V_A}^{V_B} \frac{P_A V_A}{V}\,dV$$

The initial pressure and volume are constants, so we may pull $P_A V_A$ out of the integral.

$$W_{AB} = P_A V_A \int_{V=V_A}^{V_B} \frac{dV}{V}$$

Recall from calculus that $\int \frac{dx}{x} = \ln(x)$. Also recall the logarithm identity $\ln(b) - \ln(a) = \ln\left(\frac{b}{a}\right)$. (For a quick review of **logarithms**, see Volume 2, Chapter 17.)

$$W_{AB} = P_A V_A [\ln(V)]_{V=V_A}^{V_B} = P_A V_A [\ln(V_B) - \ln(V_A)] = P_A V_A \ln\left(\frac{V_B}{V_A}\right)$$

Read the initial and final values directly from the graph: $V_A = 1.0 \text{ m}^3$, $P_A = 80 \text{ kPa}$, $V_B = 4.0 \text{ m}^3$, and $P_B = 20 \text{ kPa}$.

[2] Although it may look like it in this picture, the equation for the isotherm is <u>not</u> a circular arc. It's actually a **hyperbola**: Since $P_A V_A = PV$ for this isotherm, $P = \frac{P_A V_A}{V}$ is a hyperbola (of the form $y = \frac{c}{x}$).

$$W_{AB} = (80{,}000)(1)\ln\left(\frac{4}{1}\right) = (80{,}000)\ln(4)\ \text{J} = 80\ln(4)\ \text{kJ}$$

The work done is $W_{AB} = (80{,}000)\ln(4)\ \text{J} = 80\ln(4)\ \text{kJ}$. If you use a calculator, this works out to $W_{AB} = 111$ kJ. The work done is **positive** because the **volume increased** (path AB goes to the **right**).

Since this example involves a **monatomic ideal gas**, the internal energy equals $U = \frac{3}{2}nRT$. Since this process is isothermal, T is constant, such that $\Delta T_{AB} = 0$ and $\Delta U_{AB} = 0$. The internal energy of the gas **doesn't** change: $\Delta U_{AB} = 0$.

Apply the **first law** of thermodynamics.

$$\Delta U_{AB} = Q_{AB} - W_{AB}$$

Add work to both sides of the equation.

$$Q_{AB} = \Delta U_{AB} + W_{AB} = 0 + (80{,}000)\ln(4) = (80{,}000)\ln(4)\ \text{J} = 80\ln(4)\ \text{kJ}$$

The heat exchange equals $Q_{AB} = (80{,}000)\ln(4)\ \text{J} = 80\ln(4)\ \text{kJ}$. If you use a calculator, this works out to $Q_{AB} = 111$ kJ. The gas **absorbs** (since Q_{AB} is **positive**) heat.

(B) Determine the work, heat, and internal energy change for path BC.
Path BC is an **isochor**: The **volume** doesn't change ($V_C = V_B$). Therefore, **no** work is done along path BC: $W_{BC} = 0$.

Since the **volume is constant**, use the formula for molar specific heat at constant volume.

$$dQ = nc_V dT$$

$$Q_{BC} = \int_{T=T_B}^{T_C} nc_V\, dT$$

Since this problem involves a **monatomic ideal gas**, $c_V = \frac{3}{2}R$ (see page 197), where R is the universal gas constant.

$$Q_{BC} = \int_{T=T_B}^{T_C} n\frac{3}{2}R\, dT = \frac{3}{2}nR\int_{T=T_B}^{T_C} dT = \frac{3}{2}nR(T_C - T_B)$$

We don't know the temperature, but that's okay. Since this is an **ideal gas**, $P_B V_B = nRT_B$ and $P_C V_C = nRT_C$. This means that $nR(T_C - T_B) = P_C V_C - P_B V_B$. Read the initial and final values directly from the graph: $V_B = 4.0\ \text{m}^3$, $P_B = 20$ kPa, $V_C = 4.0\ \text{m}^3$, and $P_C = 80$ kPa.

$$Q_{BC} = \frac{3}{2}nR(T_C - T_B) = \frac{3}{2}P_C V_C - \frac{3}{2}P_B V_B = \frac{3}{2}(80{,}000)(4) - \frac{3}{2}(20{,}000)(4)$$

$$Q_{BC} = 480{,}000 - 120{,}000 = 360{,}000\ \text{J} = 360\ \text{kJ}$$

The gas **absorbs** (since Q_{BC} is **positive**) $Q_{BC} = 360{,}000\ \text{J} = 360$ kJ of heat energy.

Apply the **first law** of thermodynamics.

$$\Delta U_{BC} = Q_{BC} - W_{BC} = 360{,}000 - 0 = 360{,}000 \text{ J} = 360 \text{ kJ}$$

The internal energy of the gas increases by $\Delta U_{BC} = 360{,}000 \text{ J} = 360 \text{ kJ}$.

(C) Determine the work, heat, and internal energy change for path CA.

Path CA is an **isobar**: The **pressure** doesn't change ($P_A = P_C$). Perform the thermodynamic work integral.

$$W_{CA} = \int_{V=V_C}^{V_A} P\, dV$$

Since the **pressure is constant**, we may pull the pressure out of the integral.

$$W_{CA} = P_C \int_{V=V_C}^{V_A} dV = P_C(V_A - V_C)$$

Read the initial and final values directly from the graph: $V_C = 4.0 \text{ m}^3$, $P_C = 80 \text{ kPa}$, $V_A = 1.0 \text{ m}^3$, and $P_A = 80 \text{ kPa}$.

$$W_{CA} = (80{,}000)(1 - 4) = (80{,}000)(-3) = -240{,}000 \text{ J} = -240 \text{ kJ}$$

The work done is $W_{CA} = -240{,}000 \text{ J} = -240 \text{ kJ}$. The work done is **negative** because the **volume decreased** (path CA goes to the **left**).

Since the **pressure is constant**, use the formula for molar specific heat at constant pressure.

$$dQ = nc_P dT$$

$$Q_{CA} = \int_{T=T_C}^{T_A} nc_P\, dT$$

Since this problem involves a **monatomic ideal gas**, $c_P = \frac{5}{2}R$ (see page 197), where R is the universal gas constant.

$$Q_{CA} = \int_{T=T_C}^{T_A} n\frac{5}{2}R\, dT = \frac{5}{2}nR \int_{T=T_C}^{T_A} dT = \frac{5}{2}nR(T_A - T_C)$$

We don't know the temperature, but that's okay. Since this is an **ideal gas**, $P_A V_A = nRT_A$ and $P_C V_C = nrT_C$. This means that $nR(T_A - T_C) = P_A V_A - P_C V_C$. Read the initial and final values directly from the graph: $V_C = 4.0 \text{ m}^3$, $P_C = 80 \text{ kPa}$, $V_A = 1.0 \text{ m}^3$, and $P_A = 80 \text{ kPa}$.

$$Q_{CA} = \frac{5}{2}nR(T_A - T_C) = \frac{5}{2}P_A V_A - \frac{5}{2}P_C V_C = \frac{5}{2}(80{,}000)(1) - \frac{5}{2}(80{,}000)(4)$$

$$Q_{CA} = 200{,}000 - 800{,}000 = -600{,}000 \text{ J} = -600 \text{ kJ}$$

The heat exchange is $Q_{CA} = -600{,}000 \text{ J} = -600 \text{ kJ}$. Since Q_{CA} is **negative**, the gas **releases** heat energy.

Apply the **first law** of thermodynamics, using the answers to (A) and (B).

$$\Delta U_{CA} = Q_{CA} - W_{CA} = -600{,}000 - (-240{,}000) = -600{,}000 + 240{,}000$$

$$\Delta U_{CA} = -360{,}000 \text{ J} = -360 \text{ kJ}$$

The internal energy change is $\Delta U_{CA} = -360{,}000 \text{ J} = -360 \text{ kJ}$.

(D) Determine the net work, heat, and internal energy change for the complete cycle.
Add up the work done by each process:

$$W_{net} = W_{AB} + W_{BC} + W_{CA} = (80{,}000)\ln(4) + 0 - 240{,}000$$

$$W_{net} = (80{,}000)\ln(4) \text{ J} - 240{,}000 \text{ J} = 80\ln(4) \text{ kJ} - 240 \text{ kJ}$$

The net work is $W_{net} = (80{,}000)\ln(4) \text{ J} - 240{,}000 \text{ J} = 80\ln(4) \text{ kJ} - 240 \text{ kJ}$. If you use a calculator, this comes out to $W_{net} = -129$ kJ. As expected, the **net** work is **negative** because the path is **counterclockwise**.

Add up the heat changes for each process.

$$Q_{net} = Q_{AB} + Q_{BC} + Q_{CA} = (80{,}000)\ln(4) + 360{,}000 - 600{,}000$$

$$Q_{net} = (80{,}000)\ln(4) \text{ J} - 240{,}000 \text{ J} = 80\ln(4) \text{ kJ} - 240 \text{ kJ}$$

The net heat change is $Q_{net} = (80{,}000)\ln(4) \text{ J} - 240{,}000 \text{ J} = 80\ln(4) \text{ kJ} - 240 \text{ kJ}$. If you use a calculator, this comes out to $Q_{net} = -129$ kJ.

Add up the internal energy changes for each process.

$$\Delta U_{net} = \Delta U_{AB} + \Delta U_{BC} + \Delta U_{CA} = 0 + 360{,}000 - 360{,}000 = 0$$

The net internal energy change is **zero**: $\Delta U_{net} = 0$. In fact, this will be true for the net internal energy of **any complete cycle** (closed path). Internal energy (unlike work and heat) is path-independent, so it only depends on the initial and final points: For a complete cycle, the initial and final positions are the same, so $\Delta U_{net} = 0$.

You can verify that the **first law** of thermodynamics, $\Delta U_{net} = Q_{net} - W_{net}$, is satisfied for the complete cycle: The left-hand side is $\Delta U_{net} = 0$, and the right-hand side is also zero because $Q_{net} = W_{net}$.

Example: One mole of an ideal gas at 300 K expands to twice its initial volume during a reversible isothermal process. Determine the change in entropy for the system.

Perform the integral for entropy change for this reversible process.

$$\Delta S = \int_{Q=Q_0}^{Q} \frac{dQ}{T}$$

From the **first law** of thermodynamics, $dU = dQ - dW$. For an ideal gas, $dU = nc_V dT$. For an **isothermal** process, **temperature** remains constant. Therefore, $dU = 0$ (internal energy doesn't change) and $0 = dQ - dW$, such that $dQ = dW$. Recall that $dW = PdV$. The upper limit of integration (which is over volume after substitution) is $2V_0$ because the problem states that the gas expands to twice its initial volume.

$$\Delta S = \int_{Q=Q_0}^{Q} \frac{dQ}{T} = \int_{W=W_0}^{W} \frac{dW}{T} = \int_{V=V_0}^{2V_0} \frac{P}{T} dV$$

For an ideal gas, $PV = nRT$ such that $\frac{P}{T} = \frac{nR}{V}$.

$$\Delta S = \int_{V=V_0}^{2V_0} \frac{nR}{V} dV = nR \int_{V=V_0}^{2V_0} \frac{dV}{V} = nR\,[\ln(V)]_{V=V_0}^{2V_0} = nR[\ln(2V_0) - \ln(V_0)]$$

Recall that $\ln(b) - \ln(a) = \ln\left(\frac{b}{a}\right)$.

$$\Delta S = nR \ln\left(\frac{2V_0}{V_0}\right) = nR \ln(2)$$

Recall that the universal gas constant is $R = 8.314 \frac{\text{J}}{\text{mol}\cdot\text{K}} \approx \frac{25}{3} \frac{\text{J}}{\text{mol}\cdot\text{K}}$. The problem states that there is just one mole ($n = 1.0$ mol).

$$\Delta S = (1)(8.314)\ln(2) \approx (1)\left(\frac{25}{3}\right)\ln(2) = \frac{25}{3}\ln(2)\ \text{J/K}$$

The entropy of the system increases by $\Delta S \approx \frac{25}{3}\ln(2)$ J/K. If you use a calculator, this works out to $\Delta S = 5.8$ J/K.

73. A monatomic ideal gas expands isobarically from a volume of 3.0 m^3 to 8.0 m^3. The initial pressure is 40 kPa. The gas then goes through an isochoric process until the pressure reaches 80 kPa.

(A) Determine total the work done.

(B) Determine the total amount of heat exchanged.

(C) Determine the total change in the internal energy of the gas.

Want help? Check the hints section at the back of the book.

Answers: (A) 200 kJ (B) 980 kJ (C) 780 kJ

74. A monatomic ideal gas expands adiabatically from a volume of 1.0 m^3 to 8.0 m^3. The initial pressure is 16 kPa.

(A) Determine the work done.

(B) Determine the amount of heat exchanged.

(C) Determine the change in the internal energy of the gas.

Want help? Check the hints section at the back of the book.

Answers: (A) 18 kJ (B) 0 (C) −18 kJ

75. A fluid expands from an initial volume of 3.0 m^3 to 6.0 m^3. The initial pressure is 90 kPa. The fluid obeys the equation $PV^2 = \text{const}$. Determine the work done.

Want help? Check the hints section at the back of the book.

Answer: 135 kJ

76. Determine the work done along the straight line path in the graph below.

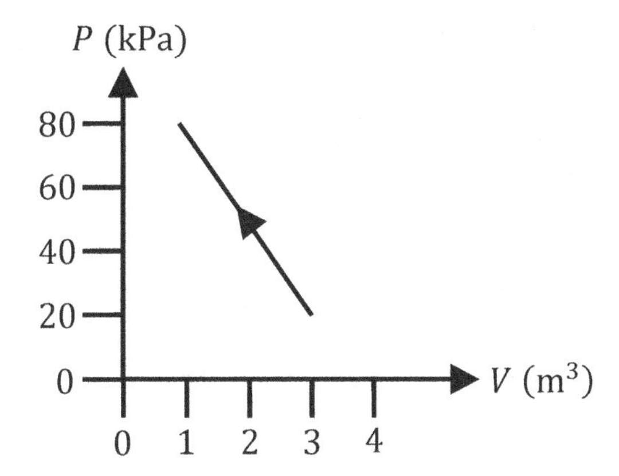

Want help? Check the hints section at the back of the book.

Answer: -100 kJ

77. Determine the net work done for the complete cycle shown below.

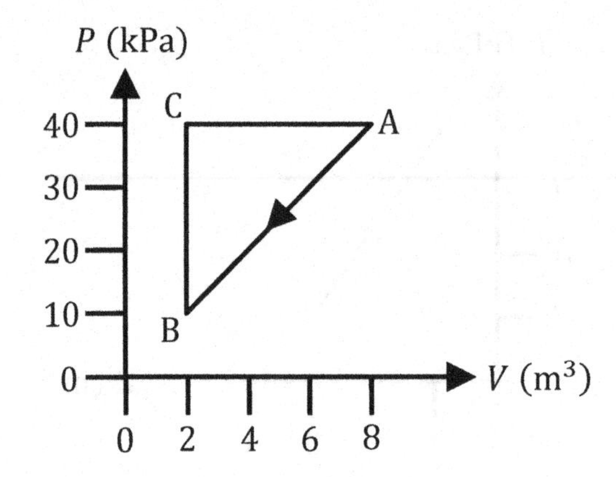

Want help? Check the hints section at the back of the book.

Answer: 90 kJ

78. A monatomic ideal gas completes the thermodynamic cycle shown below.

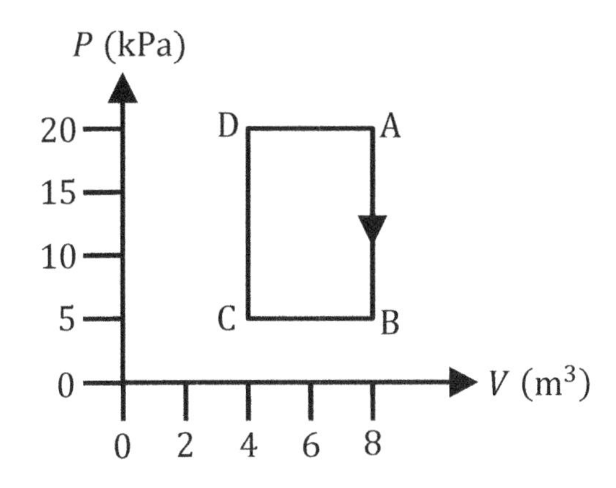

(A) Determine the work, heat, and internal energy change for path AB.

(B) Determine the work, heat, and internal energy change for path BC.

Note: This problem is **continued** on the next page.

(C) Determine the work, heat, and internal energy change for path CD.

(D) Determine the work, heat, and internal energy change for path DA.

(E) Determine the net work, heat, and internal energy change for the complete cycle.

Want help? Check the hints section at the back of the book.

Answers: (A) 0, −180 kJ, −180 kJ (B) −20 kJ, −50 kJ, −30 kJ (C) 0, 90 kJ, 90 kJ (D) 80 kJ, 200 kJ, 120 kJ (E) 60 kJ, 60 kJ, 0

79. A monatomic ideal gas completes the thermodynamic cycle shown below, where path AB is an adiabat. **Note**: Points B and C lie **above** the horizontal axis. (It is **not** necessary to interpolate the graph to determine the pressure at these points. You can use **math**.)

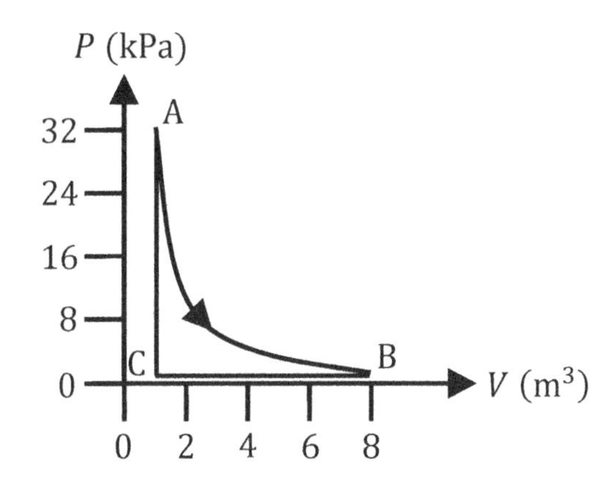

(A) Determine the work, heat, and internal energy change for path AB.

(B) Determine the work, heat, and internal energy change for path BC.

Note: This problem is **continued** on the next page.

(C) Determine the work, heat, and internal energy change for path CA.

(D) Determine the net work, heat, and internal energy change for the complete cycle.

Want help? Check the hints section at the back of the book.

Answers: (A) 36 kJ, 0, −36 kJ (B) −7.0 kJ, −17.5 kJ, −10.5 kJ
(C) 0, 46.5 kJ, 46.5 kJ (D)29 kJ, 29 kJ, 0

80. A 910-g ice cube slowly melts at 0°C. Water has a latent heat of fusion of approximately $3.3 \times 10^5 \frac{\text{J}}{\text{kg}}$. Determine the entropy change for the ice cube.

81. Water with a mass of 500 g is quasistatically cooled from 87°C to 27°C. Water has a specific heat capacity of approximately $4200 \frac{\text{J}}{\text{kg}\cdot{}^\circ\text{C}}$. Determine the entropy change for the water.

Want help? Check the hints section at the back of the book.

Answers: 1100 J/K, $-2100 \ln(1.2)$ J/K ≈ -383 J/K

82. An ideal gas expands to twice its initial volume during a reversible isobaric process. There are 6.0 moles of the gas. Determine the change in entropy for the system.

Want help? Check the hints section at the back of the book.

Answer: $125 \ln(2)$ J/K ≈ 87 J/K

19 HEAT ENGINES

Relevant Terminology

Heat engine – a device that adapts the natural flow of **heat** from a high-temperature thermal reservoir to a low-temperature thermal reservoir in order to perform mechanical **work**.
Heat pump – a device that pumps heat from a low-temperature thermal reservoir to a high-temperature thermal reservoir.
Thermal reservoir – a heat source that offers a virtually **limitless supply of thermal energy** (or which may receive a virtually limitless amount of thermal energy), with virtually no effect on the temperature of the heat source. Also called a **heat reservoir**.
Efficiency – a measure of the effectiveness of a **heat engine** at performing useful mechanical work. The efficiency of a heat engine equals the ratio of the net work output to the heat input.
Coefficient of performance – a measure of the thermal effectiveness of a **heat pump**.
Pressure – force per unit area.
Volume – the amount of space that an object occupies. Also referred to as **capacity**.
Temperature – a measure of the **average kinetic energy** of the molecules of a substance.
Work – thermodynamic work is done when the volume changes under pressure.
Heat – thermal energy that is **transferred** between substances.
Heat capacity – a measure of how much heat a given substance needs to absorb in order to increase its temperature a specified amount.
Internal energy – the total energy of the molecules of a substance, excluding their interactions with external fields (like earth's gravitational field).
Entropy – a quantitative measure of the **statistical disorder** of a system.
Isothermal – a process for which **temperature** remains constant.
Isobaric – a process for which **pressure** remains constant.
Isochoric – a process for which **volume** remains constant.
Adiabatic – a process for which **no heat** is absorbed or released.
Isentropic – a process for which the **entropy** of a system remains constant.
Steady-state – a process for which the **internal energy** of a system remains constant.
Quasistatic – a process that is carried out through a series of virtually infinitesimal disturbances from equilibrium, which occurs slowly enough that the system establishes equilibrium between the infinitesimal disturbances.
Spontaneous – a process that occurs without being driven by an energy source.
Reversible – a quasistatic process that can be carried out in reverse.
Irreversible – any process that is not quasistatic, or any process that features frictional or dissipative forces.
Compression ratio – the factor by which a working substance's volume is compressed.
Cutoff ratio – the factor by which the volume changes during combustion.

Heat Engines

Heat spontaneously flows from high temperature to low temperature. A **heat engine** is a device that utilizes this natural **heat** flow to do some useful mechanical **work**. The general features of a heat engine are:

- The system absorbs **heat**, $Q_{in} > 0$, from a **high-temperature** thermal reservoir. The subscript "in" is short for "input."
- The system uses this absorbed heat to perform useful mechanical **work**, $W_{out} > 0$. The subscript "out" is short for "output." This is the desired output.
- A fraction of the heat isn't converted to mechanical work: $Q_{out} < 0$. This heat is rejected to a **low-temperature** thermal reservoir. It is called the **exhaust**.

The work output equals the net work done by a heat engine in one complete cycle: $W_{out} = W_{net}$. The net heat for one cycle includes both heat input and heat output (exhaust): $Q_{net} = Q_{in} + Q_{out}$. The heat input is positive while the heat output is negative, such that we may write $Q_{net} = Q_{in} - |Q_{out}|$. The net internal energy change equals zero for a complete cycle: $\Delta U_{net} = 0$. From the first law of thermodynamics, $\Delta U_{net} = Q_{net} - W_{net}$. Since $\Delta U_{net} = 0$, this means that $Q_{net} = W_{net}$. Recall that $W_{net} = W_{out}$ and that $Q_{net} = Q_{in} + Q_{out}$. Therefore, the equation $Q_{net} = W_{net}$ may be written as $W_{out} = Q_{in} + Q_{out}$. Since Q_{out} is negative, we can write $W_{out} = Q_{in} - |Q_{out}|$.

The **efficiency** of a heat engine measures the fraction of the heat that is converted into mechanical work (rather than exhaust). **The efficiency of a heat engine is always less than 100% due to the second law of thermodynamics.**

$$e = \frac{W_{out}}{Q_{in}} = \frac{Q_{in} + Q_{out}}{Q_{in}} = 1 + \frac{Q_{out}}{Q_{in}} = 1 - \frac{|Q_{out}|}{Q_{in}}$$

No heat engine can be more efficient than a Carnot engine. For the Carnot cycle, the efficiency is given by the following formula. **Only** use the formula below to find the efficiency of a **Carnot** engine or to compare the efficiency of another heat engine to the **maximum** possible efficiency of any heat engine. **Don't** use the following formula to find the efficiency of another heat engine. Here, "ℓ" is for "low" while "h" is for "hot."

$$e_C = 1 - \frac{T_\ell}{T_h} \quad \text{(Carnot cycle)}$$

A **heat pump** involves the flow of heat in the opposite direction - that is, from low temperature to high temperature. Examples of heat pumps include heaters, refrigerators, and air conditioners (but not heat engines). The coefficient of performance (COP) provides a measure of the thermal effectiveness of a heat pump.

$$COP_{cooling} = -\frac{Q_{in}}{W_{in}}$$

$$COP_{heating} = \frac{Q_{out}}{W_{in}}$$

Thermodynamic Processes

Following are some common thermodynamic processes.

- **Volume** remains constant for an **isochoric** process.
  - Since the volume doesn't change ($\Delta V = 0$), **no** work is done ($W = 0$). An isochor is a **vertical** line on a P-V diagram.
  - Express the heat exchange in terms of the specific heat capacity at constant **volume**: $dQ = mC_V dT$ or $dQ = nc_V dT$. Since no work is done, the change in internal energy equals the heat exchanged: $\Delta U = Q$.
- **Pressure** remains constant for an **isobaric** process.
  - Since pressure is constant, it may come out of the work integral: $W = \int_{V=V_0}^{V} P\, dV = P \int_{V=V_0}^{V} dV = P(V - V_0) = P\Delta V$. An isobar is a **horizontal** line on a P-V diagram.
  - Express the heat exchange in terms of the specific heat capacity at constant **pressure**: $dQ = mC_P dT$ or $dQ = nc_P dT$. The internal energy change equals $\Delta U = Q - W$ according to the first law of thermodynamics.
- **Temperature** remains constant for an **isothermal** process.
  - For an ideal gas or a van der Waals fluid, solve for pressure in terms of volume to perform the work integral: $W = \int_{V=V_0}^{V} P\, dV$. For example, for an ideal gas, $P = \frac{nRT}{V}$. **Temperature is a constant** for an isotherm.
  - Although temperature is constant for an isotherm, there is generally still heat exchanged ($Q \neq 0$). However, if the potential energy of the molecules is constant (it is for an **ideal gas**), the change in internal energy will be zero ($\Delta U = 0$) and the heat exchange will equal the work done: $Q = W$.
- **No heat** is absorbed or released by the system for an **adiabatic** process.
  - If the relationship between pressure and volume is known for the adiabat, express pressure in terms of volume to perform the work integral: $W = \int_{V=V_0}^{V} P\, dV$. For example, for an ideal gas, $P_0 V_0^\gamma = PV^\gamma$, or $P = \frac{P_0 V_0^\gamma}{V^\gamma}$.
  - Since no heat is exchanged, the change in internal energy equals the negative of the work done: $\Delta U = -W$.
- For a **complete cycle** (a **closed path** on a P-V diagram), the net internal energy change is zero ($\Delta U_{net} = 0$) for the complete cycle, although internal energy may change for parts of the cycle. The net heat exchange equals the net work done: $Q_{net} = W_{net}$.

Symbols and SI Units

Symbol	Name	SI Units
P	pressure	Pa
V	volume	m^3
n	number of moles	mol
R	universal gas constant	$\frac{J}{mol \cdot K}$
T	absolute temperature	K
T_ℓ	low temperature (of the low-temperature reservoir)	K
T_h	high temperature (of the high-temperature reservoir)	K
Q	heat	J
Q_{net}	the net heat exchange for one cycle	J
Q_{in}	the heat input	J
Q_{out}	the heat output (exhaust)	J
C_V	specific heat capacity at constant volume	$\frac{J}{kg \cdot K}$
C_P	specific heat capacity at constant pressure	$\frac{J}{kg \cdot K}$
c_V	molar specific heat capacity at constant volume	$\frac{J}{mol \cdot K}$
c_P	molar specific heat capacity at constant pressure	$\frac{J}{mol \cdot K}$
γ	adiabatic index	unitless
U	internal energy	J
ΔU_{net}	the net internal energy change for one cycle (it equals zero)	J
W	work	J
W_{net}	the net work done in one cycle	J
W_{out}	the work output	J

e	efficiency	unitless
e_C	the efficiency of a Carnot cycle (maximum efficiency of any heat engine)	unitless
$COP_{cooling}$	the coefficient of performance for cooling	unitless
$COP_{heating}$	the coefficient of performance of heating	unitless
V_{min}	minimum volume	m^3
V_{max}	maximum volume	m^3
V_c	final volume of the combustion reaction	m^3
V_{c0}	initial volume of the combustion reaction	m^3
r	compression ratio	unitless
r_c	cutoff ratio	unitless
a	a correction for intermolecular forces	$\frac{\text{J}\cdot\text{m}^3}{\text{mol}^2}$
b	the volume occupied by one mole of the molecules	$\frac{\text{m}^3}{\text{mol}}$

Important Distinctions

The equation $e = \frac{W_{out}}{Q_{in}} = \frac{Q_{in}+Q_{out}}{Q_{in}} = 1 + \frac{Q_{out}}{Q_{in}} = 1 - \frac{|Q_{out}|}{Q_{in}}$ is true for any heat engine, whereas the equation $e_C = 1 - \frac{T_\ell}{T_h}$ applies **only** to a **Carnot** cycle. The efficiency of the Carnot cycle is important as it establishes a maximum possible limit for any heat engine. Thus, a problem that doesn't involve a Carnot cycle may ask you to compare the actual efficiency for a particular heat engine to the efficiency of the Carnot engine.

A **heat pump** is different from a **heat engine**. The heat flows in the opposite direction. For a heat engine, heat flows from the high-temperature thermal reservoir to the low-temperature thermal reservoir, whereas in a heat pump, the heat is forced to flow in the opposite direction. The **efficiency** (e) applies to a **heat engine**, whereas the **coefficient of performance** (COP) applies to a **heat pump**.

Strategy for Problems Involving Heat Engines or Heat Pumps

To solve a problem involving a heat engine or heat pump, follow these steps:

1. Draw a *P-V* diagram for the heat engine cycle. Label each process. Label the vertices where one process ends and another process begins. There are often four vertices: a, b, c, and d. Use two subscripts to represent a process. For example, W_{bc} means the work done along the process from point b to point c.
2. To find the work done for each process:
   - **Isochor** (constant **volume**) – **no** work is done: $W = 0$.
   - **Isobar** (constant **pressure**): $W = \int_{V=V_0}^{V} P\, dV = P \int_{V=V_0}^{V} dV = P(V - V_0)$.
   - **Isotherm** (constant **temperature**): Express P in terms of T, then integrate. For example, for an **ideal gas**, $P = \frac{nRT}{V}$ and $W = nRT \int_{V=V_0}^{V} \frac{dV}{V} = nRT \ln\left(\frac{V}{V_0}\right)$.
   - **Adiabat** (no **heat** exchange): Express P in terms of V (but **not** with T), then integrate. Alternatively, if the equation for the internal energy is known, apply the **first law** of thermodynamics: $\Delta U = Q - W = -W$ (since $Q = 0$ for an adiabat). For example, for a **monatomic** ideal gas, $U = \frac{3}{2}nRT$, such that $W = -\Delta U = -\frac{3}{2}nR\Delta T$.

   If you **only** need the net work, W_{net}, based on what the question asked, and if the complete cycle happens to be a polygon (like a triangle or rectangle) on the *P-V* diagram, then you can simply find the area of the polygon to get W_{net}. (However, most heat engine problems involve an isotherm or adiabat, which is **curved**.)
3. Determine the **work output** (net work) done by adding up the work done along each process. If the *P-V* diagram has four processes, $W_{out} = W_{ab} + W_{bc} + W_{cd} + W_{da}$.
4. To find the heat change for each process:
   - **Isochor** (constant **volume**): $Q = nc_V\Delta T$. For a **monatomic** ideal gas, $c_V = \frac{3}{2}R$.
   - **Isobar** (constant **pressure**): $Q = nc_P\Delta T$. For a **monatomic** ideal gas, $c_P = \frac{5}{2}R$.
   - **Isotherm** (constant **temperature**): Apply the **first law** of thermodynamics: $\Delta U = Q - W$. For example, for an **ideal gas**, $\Delta U = 0$ along an isotherm, such that $Q = W$. Step 2 above tells you how to find W.
   - **Adiabat** (no **heat** exchange): $Q = 0$.
5. The **heat input** (Q_{in}) is the **positive** heat change from Step 4. The **heat output** (Q_{out}) – or exhaust – is the **negative** heat change from Step 4.
6. To find the **efficiency**, use one of the following equations:

$$e = \frac{W_{out}}{Q_{in}} = \frac{Q_{in} + Q_{out}}{Q_{in}} = 1 + \frac{Q_{out}}{Q_{in}} = 1 - \frac{|Q_{out}|}{Q_{in}}$$

7. To find the **coefficient of performance**, see page 220.
8. To find **entropy**, see Chapter 18.

Example: The Carnot cycle involves an isothermal expansion, an adiabatic expansion, an isothermal compression, and an adiabatic compression. Derive an equation for the efficiency of the Carnot engine using a monatomic ideal gas as the working substance.

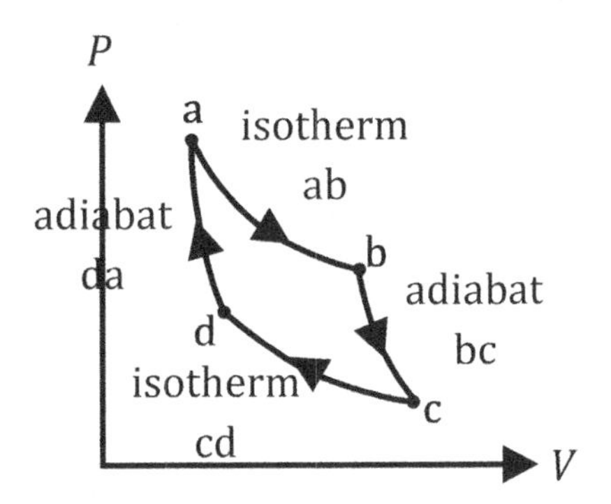

Find the work done for each process:

- **Isotherm** ab: For an ideal gas, $P = \frac{nRT}{V}$. Note that T is constant along the isotherm: $T_b = T_a$. Perform the thermodynamic work integral.
$$W_{ab} = nRT_a \int_{V=V_a}^{V_b} \frac{dV}{V} = nRT_a \ln\left(\frac{V_b}{V_a}\right)$$
- **Adiabat** bc: For a **monatomic** ideal gas, $U = \frac{3}{2}nRT$. Since $Q_{bc} = 0$ along the adiabat, the **first law** of thermodynamics gives $\Delta U_{bc} = Q_{bc} - W_{bc} = -W_{bc}$. The work is:
$$W_{bc} = -\Delta U_{bc} = -\frac{3}{2}nR(T_c - T_b) = -\frac{3}{2}nR(T_c - T_a) = \frac{3}{2}nR(T_a - T_c)$$
Recall that $T_b = T_a$. Note that $-(T_c - T_a) = (T_a - T_c)$.
- **Isotherm** cd: For an ideal gas, $P = \frac{nRT}{V}$. Note that T is constant along the isotherm: $T_d = T_c$. Perform the thermodynamic work integral.
$$W_{cd} = nRT_c \int_{V=V_c}^{V_d} \frac{dV}{V} = nRT_c \ln\left(\frac{V_d}{V_c}\right)$$
- **Adiabat** da: For a **monatomic** ideal gas, $U = \frac{3}{2}nRT$. Since $Q_{da} = 0$ along the adiabat, the **first law** of thermodynamics gives $\Delta U_{da} = Q_{da} - W_{da} = -W_{da}$. The work is:
$$W_{da} = -\Delta U_{da} = -\frac{3}{2}nR(T_a - T_d) = -\frac{3}{2}nR(T_a - T_c)$$
Recall that $T_d = T_c$.

The work output equals the sum of the work done by each process:
$$W_{out} = W_{ab} + W_{bc} + W_{cd} + W_{da}$$
$$W_{out} = nRT_a \ln\left(\frac{V_b}{V_a}\right) + \frac{3}{2}nR(T_a - T_c) + nRT_c \ln\left(\frac{V_d}{V_c}\right) - \frac{3}{2}nR(T_a - T_c)$$
The second and fourth terms **cancel** out.
$$W_{out} = nRT_a \ln\left(\frac{V_b}{V_a}\right) + nRT_c \ln\left(\frac{V_d}{V_c}\right)$$

Find the heat change for each process:

- **Isotherm** ab: For a **monatomic** ideal gas, $U = \frac{3}{2}nRT$. Since T is constant along the isotherm ($T_b = T_a$), $\Delta U_{ab} = 0$. The first law of thermodynamics, $\Delta U_{ab} = Q_{ab} - W_{ab}$, simplifies to $Q_{ab} = W_{ab}$. Recall that $W_{ab} = nRT_a \ln\left(\frac{V_b}{V_a}\right)$. Thus, $Q_{ab} = nRT_a \ln\left(\frac{V_b}{V_a}\right)$.
- **Adiabat** bc: $Q_{bc} = 0$ along the adiabat.
- **Isotherm** cd: For a **monatomic** ideal gas, $U = \frac{3}{2}nRT$. Since T is constant along the isotherm ($T_d = T_c$), $\Delta U_{cd} = 0$. The first law of thermodynamics, $\Delta U_{cd} = Q_{cd} - W_{cd}$, simplifies to $Q_{cd} = W_{cd}$. Recall that $W_{cd} = nRT_c \ln\left(\frac{V_d}{V_c}\right)$. Thus, $Q_{cd} = nRT_c \ln\left(\frac{V_d}{V_c}\right)$.
- **Adiabat** da: $Q_{da} = 0$ along the adiabat.

Recall the property[1] of logarithms that $\ln(x)$ is positive if $x > 1$ and negative if $x < 1$. Look at the P-V diagram for the Carnot cycle:

- Since $V_b > V_a$, it follows that $\frac{V_b}{V_a} > 1$, such that $\ln\left(\frac{V_b}{V_a}\right) > 0$ and $Q_{ab} > 0$. Since Q_{ab} is **positive**, Q_{ab} is the **heat input**: $Q_{in} = Q_{ab} = nRT_a \ln\left(\frac{V_b}{V_a}\right)$.
- Since $V_d < V_c$, it follows that $\frac{V_d}{V_c} < 1$, such that $\ln\left(\frac{V_d}{V_c}\right) < 0$ and $Q_{cd} < 0$. Since Q_{cd} is **negative**, Q_{cd} is the **heat output** (or exhaust): $Q_{out} = Q_{cd} = nRT_c \ln\left(\frac{V_d}{V_c}\right)$.

Use the equation for the **efficiency** of a heat engine:

$$e = \frac{W_{out}}{Q_{in}} = \frac{nRT_a \ln\left(\frac{V_b}{V_a}\right) + nRT_c \ln\left(\frac{V_d}{V_c}\right)}{nRT_a \ln\left(\frac{V_b}{V_a}\right)}$$

Apply the rule from algebra that $\frac{x+y}{z} = \frac{x}{z} + \frac{y}{z}$.

$$e = \frac{nRT_a \ln\left(\frac{V_b}{V_a}\right)}{nRT_a \ln\left(\frac{V_b}{V_a}\right)} + \frac{nRT_c \ln\left(\frac{V_d}{V_c}\right)}{nRT_a \ln\left(\frac{V_b}{V_a}\right)}$$

The first term equals one (everything cancels out), while nR cancels in the second term.

$$e = 1 + \frac{T_c \ln\left(\frac{V_d}{V_c}\right)}{T_a \ln\left(\frac{V_b}{V_a}\right)} = 1 - \frac{T_c \ln\left(\frac{V_c}{V_d}\right)}{T_a \ln\left(\frac{V_b}{V_a}\right)}$$

Since $V_d < V_c$, we applied the identity $\ln(x) = -\ln\left(\frac{1}{x}\right)$ to write $\ln\left(\frac{V_d}{V_c}\right) = -\ln\left(\frac{V_c}{V_d}\right)$. Consider the equations for the adiabats.

- **Adiabat** bc: For an ideal gas, $P_b V_b^\gamma = P_c V_c^\gamma$, which can be written $V_c^\gamma = \frac{P_b V_b^\gamma}{P_c}$.
- **Adiabat** da: For an ideal gas, $P_d V_d^\gamma = P_a V_a^\gamma$, which can be written $V_d^\gamma = \frac{P_a V_a^\gamma}{P_d}$.

---

[1] Using a calculator, you can see that $\ln(0.99) \approx -0.01005$ is negative, whereas $\ln(1.01) \approx 0.00995$ is positive. For a quick review of logarithms, see Volume 2, Chapter 17.

Divide these two equations. To divide by a fraction, multiply by its **reciprocal**.

$$\frac{V_c^\gamma}{V_d^\gamma} = \frac{P_b V_b^\gamma}{P_c} \div \frac{P_a V_a^\gamma}{P_d} = \frac{P_b V_b^\gamma}{P_c} \times \frac{P_d}{P_a V_a^\gamma} = \frac{P_b P_d}{P_c P_a} \frac{V_b^\gamma}{V_a^\gamma}$$

Now consider the equations for the isotherms.

- **Isotherm** ab: For an ideal gas, $\frac{P_a V_a}{T_a} = \frac{P_b V_b}{T_b}$. Since $T_b = T_a$ for this isotherm, T cancels out such that $P_a V_a = P_b V_b$, which can be written as $\frac{P_b}{P_a} = \frac{V_a}{V_b}$.
- **Isotherm** cd: For an ideal gas, $\frac{P_c V_c}{T_c} = \frac{P_d V_d}{T_d}$. Since $T_d = T_c$ for this isotherm, T cancels out such that $P_c V_c = P_d V_d$, which can be written as $\frac{P_d}{P_c} = \frac{V_c}{V_d}$.

Plug $\frac{P_b}{P_a} = \frac{V_a}{V_b}$ and $\frac{P_d}{P_c} = \frac{V_c}{V_d}$ into the equation $\frac{V_c^\gamma}{V_d^\gamma} = \frac{P_b P_d}{P_c P_a} \frac{V_b^\gamma}{V_a^\gamma}$ from the top of this page.

$$\frac{V_c^\gamma}{V_d^\gamma} = \frac{V_a}{V_b} \frac{V_c}{V_d} \frac{V_b^\gamma}{V_a^\gamma}$$

Divide both sides of the equation by V_c and multiply by V_d. Group symbols together.

$$\frac{V_c^\gamma}{V_c} \frac{V_d}{V_d^\gamma} = \frac{V_a}{V_a^\gamma} \frac{V_b^\gamma}{V_b}$$

Note that $\frac{V_c^\gamma}{V_c} = V_c^{\gamma-1}$, $\frac{V_d}{V_d^\gamma} = \frac{1}{V_d^{\gamma-1}}$, $\frac{V_a}{V_a^\gamma} = \frac{1}{V_a^{\gamma-1}}$, and $\frac{V_b^\gamma}{V_b} = V_b^{\gamma-1}$ according to the rule $\frac{x^m}{x^n} = x^{m-n}$.

$$\frac{V_c^{\gamma-1}}{V_d^{\gamma-1}} = \frac{V_b^{\gamma-1}}{V_a^{\gamma-1}}$$

Raise both sides of the equation to the power of $\frac{1}{\gamma-1}$. Note, for example, that $\left(V_c^{\gamma-1}\right)^{\frac{1}{\gamma-1}} = V_c$ according to the rule $(x^n)^{\frac{1}{n}} = x^1 = x$, which is a special case of $(x^m)^n = x^{mn}$.

$$\frac{V_c}{V_d} = \frac{V_b}{V_a}$$

Since we have just proven that $\frac{V_c}{V_d} = \frac{V_b}{V_a}$, it follows that $\frac{\ln\left(\frac{V_c}{V_d}\right)}{\ln\left(\frac{V_b}{V_a}\right)} = 1$, which means that the logarithms in our previous expression for efficiency cancel out.

$$e = 1 - \frac{T_c}{T_a}$$

For the Carnot cycle, temperature T_a (which equals T_b) is the **high** temperature, while temperature T_c (which equals T_d) is the **low** temperature: $T_a > T_c$. If we use the symbol T_h for the high temperature and T_ℓ for the low temperature, $T_h = T_a$ and $T_\ell = T_c$, we get:

$$e = 1 - \frac{T_\ell}{T_h}$$

Since $T_\ell < T_h$ and $T_\ell > 0$ K, the efficiency will always be less than 1.

83. The Otto cycle involves an adiabatic compression, an isochoric pressurization, an adiabatic expansion, and an isochoric depressurization. Use a monatomic ideal gas as the working substance.

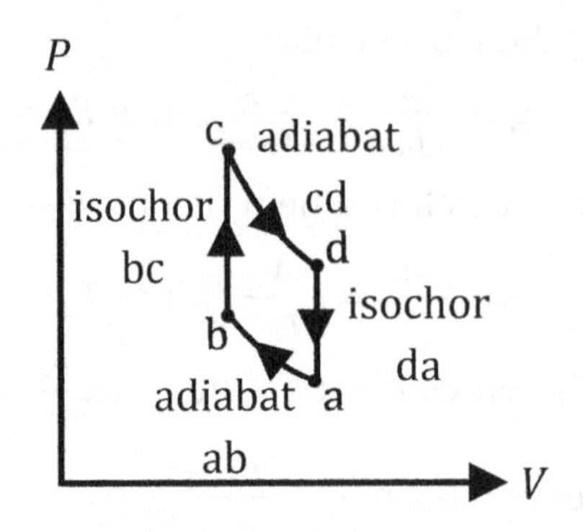

(A) Express the work done for each process in terms of n, R, T_a, T_b, T_c, and T_d.

(B) Express the heat change for each process in terms of n, R, T_a, T_b, T_c, and T_d.

Note: This problem is **continued** on the next page.

(C) Express the work output for the cycle in terms of n, R, T_a, T_b, T_c, and T_d.

(D) Express the heat input for the cycle in terms of n, R, T_a, T_b, T_c, and T_d.

(E) Express the heat output for the cycle in terms of n, R, T_a, T_b, T_c, and T_d.

Note: This problem is **continued** on the next page.

(F) Use the ideal gas law to show that $P_b = r\frac{T_b}{T_a}P_a$ and $P_c = r\frac{T_c}{T_d}P_d$, where $r = \frac{V_a}{V_b}$ is the compression ratio.

(G) Use the equations for the adiabats to show that $P_b = r^{5/3}\,P_a$ and $P_c = r^{5/3}\,P_d$.

(H) Use your answers to parts (F) and (G) to show that $T_b = r^{2/3}T_a$ and $T_c = r^{2/3}T_d$.

Note: This problem is **continued** on the next page.

(I) Use your answers to parts (D), (E), and (H) to express the efficiency of the heat engine in terms of r **only**. No other symbols may appear in your final answer.

(J) Use your answers to parts (H) and (I) to express the efficiency of the heat engine in terms of T_a and T_b. No other symbols may appear in your final answer.

(K) Use your answers to parts (H) and (I) to express the efficiency of the heat engine in terms of T_c and T_d. No other symbols may appear in your final answer.

Note: This problem is **continued** on the next page.

(L) Show that the highest temperature is $T_h = T_c$ and the lowest temperature is $T_\ell = T_a$.

(M) Use your answers to (J), (K), and (L) to show that the Otto heat engine is less efficient than a Carnot heat engine operating between the same extreme temperatures.

Want help? Check the hints section at the back of the book.

Answers: (A) $\frac{3nR}{2}(T_a - T_b), 0, \frac{3nR}{2}(T_c - T_d), 0$ (B) $0, \frac{3nR}{2}(T_c - T_b), 0, \frac{3nR}{2}(T_a - T_d)$
(C) $\frac{3nR}{2}(T_a - T_b + T_c - T_d)$ (D) $\frac{3nR}{2}(T_c - T_b)$ (E) $\frac{3nR}{2}(T_a - T_d)$
(I) $1 - \left(\frac{1}{r}\right)^{2/3}$ (J) $1 - \frac{T_a}{T_b}$ (K) $1 - \frac{T_d}{T_c}$

84. The heat engine graphed below involves an isochoric pressurization, an isobaric expansion, an isochoric depressurization, and an isobaric compression. Use a monatomic ideal gas as the working substance.

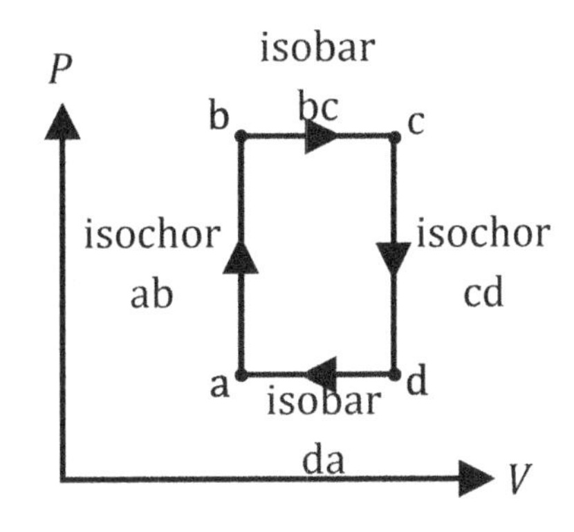

(A) Express the work done for each process in terms of n, R, T_a, T_b, T_c, and T_d.

(B) Express the heat change for each process in terms of n, R, T_a, T_b, T_c, and T_d.

Note: This problem is **continued** on the next page.

(C) Express the work output for the cycle in terms of n, R, T_a, T_b, T_c, and T_d.

(D) Express the heat input for the cycle in terms of n, R, T_a, T_b, T_c, and T_d.

(E) Express the heat output for the cycle in terms of n, R, T_a, T_b, T_c, and T_d.

(F) Use the equations for the isobars to show that $T_b = \frac{T_c}{r}$ and $T_d = rT_a$, where $r = \frac{V_c}{V_a}$ is the compression ratio.

(G) Use your answers to part (F) to express the efficiency of the heat engine in terms of r, T_a, and T_c. No other symbols may appear in your final answer.

Note: This problem is **continued** on the next page.

(H) Use your answers to part (F) to show that $\frac{T_c}{T_a} = r^2 \frac{T_b}{T_d}$.

(I) Use the equations for the isochors to show that $T_b = r_p T_a$ and $T_d = \frac{T_c}{r_p}$, where $r_p = \frac{P_b}{P_a}$ is the pressure ratio.

(J) Use your answers to part (I) to show that $\frac{T_c}{T_a} = r_p^2 \frac{T_d}{T_b}$.

(K) Use your answers to parts (H) and (J) to show that $T_c = r r_p T_a$.

(L) Use your answers to parts (G) and (K) to express the efficiency of the heat engine in terms of r and r_p. No other symbols may appear in your final answer.

Note: This problem is **continued** on the next page.

(M) Show that the highest temperature is $T_h = T_c$ and the lowest temperature is $T_\ell = T_a$.

(N) If $r = 2$ and $r_p = 4$, compare the efficiency of this heat engine with the efficiency of a Carnot heat engine operating between the same extreme temperatures.

Want help? Check the hints section at the back of the book.

Answers: (A) $0, nR(T_c - T_b), 0, nR(T_a - T_d)$
(B) $\frac{3nR}{2}(T_b - T_a), \frac{5nR}{2}(T_c - T_b), \frac{3nR}{2}(T_d - T_c), \frac{5nR}{2}(T_a - T_d)$
(C) $nR(T_a - T_b + T_c - T_d)$ (D) $\frac{nR}{2}(-3T_a - 2T_b + 5T_c)$ (E) $\frac{nR}{2}(5T_a - 3T_c - 2T_d)$
(G) $e = 1 - \frac{3T_c+(2r-5)T_a}{\left(\frac{5r-2}{r}\right)T_c-3T_a}$ (L) $1 - \frac{3rr_p+2r-5}{5rr_p-2r_p-3}$ (N) 21%, 88%

20 LIGHT WAVES

Relevant Terminology

Wavelength – the **horizontal** distance between two consecutive crests.
Period – the time it takes for the wave to complete exactly one oscillation.
Frequency – the number of oscillations completed per second.
Wave speed – how fast the wave travels.
Vacuum – a region of space completely devoid of matter. There is not even air.
Transverse – a wave for which the amplitude of oscillation is perpendicular to the direction that the wave propagates.
Propagation – the transmission of a wave through a medium (often, with the sense of spreading out, like the ripples of water wave – but not necessarily).
Power – the instantaneous rate at which work is done.
Intensity – power per unit area.

Essential Concepts

Light is an **electromagnetic wave**. As the wave propagates to the right in the illustration below, the electric field ($\vec{\mathbf{E}}$) oscillates up and down and the magnetic field ($\vec{\mathbf{B}}$) oscillates into and out of the page. Since the oscillation is perpendicular to the direction of propagation, light is a **transverse** wave (Chapter 5).

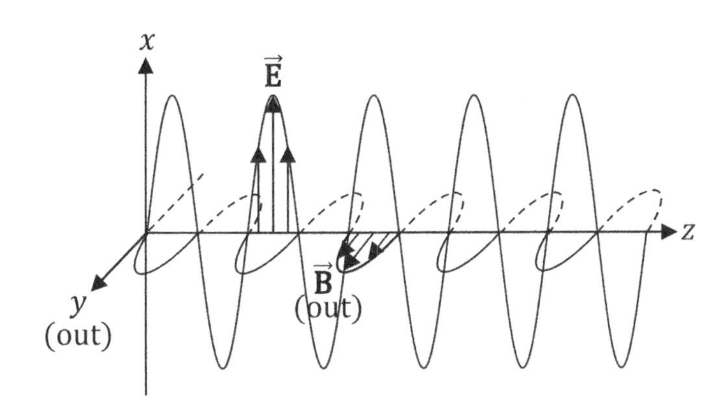

The electromagnetic spectrum consists of radio waves, microwaves, infrared, visible light, ultraviolet, x-rays, and gamma rays, as shown on the next page. The visible spectrum is a just a very narrow slice (much narrower than it would appear by looking at the chart on the following page – the blocks do <u>**not**</u> really have equal width when drawn to scale) of the full electromagnetic spectrum. The acronym Roy G. Biv stands for Red Orange Yellow Green Blue Indigo Violet, and helps to remember the order of the visible colors in increasing frequency (which corresponds to decreasing wavelength).

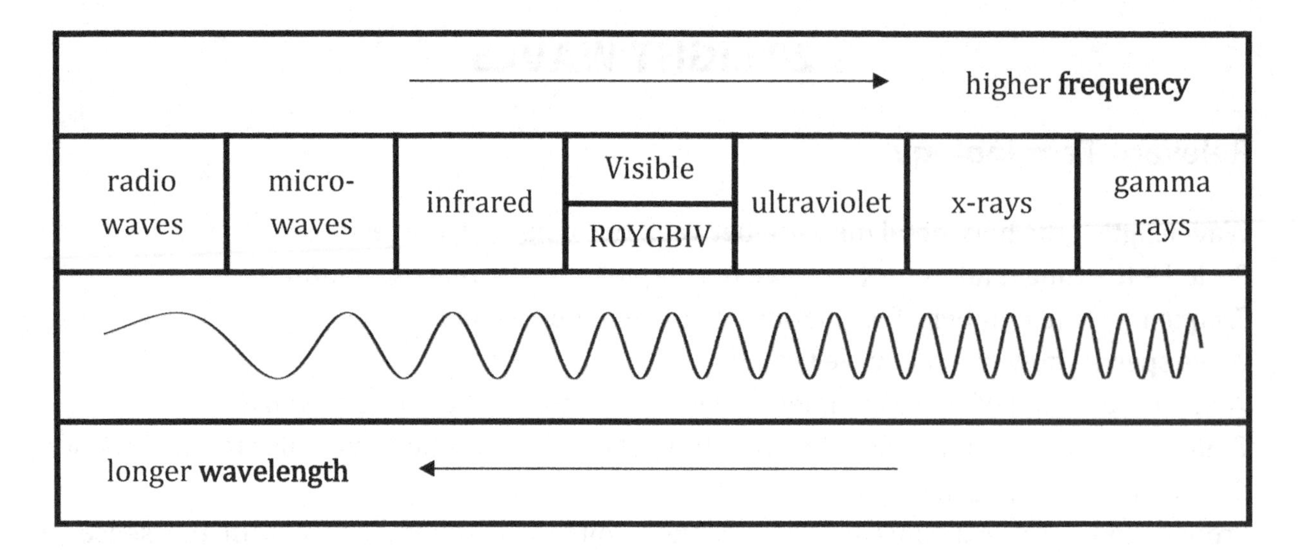

Relevant Equations

Recall the following wave quantities from Chapter 5:

- **Wavelength** (λ) is a **horizontal distance** from crest to crest.
- **Period** (T) is the **time** it takes to complete one oscillation.
- **Frequency** (f) is the **number of cycles per second**.

Wavelength and **frequency** are inversely related: Higher frequency corresponds to shorter wavelength, while lower frequency corresponds to longer wavelength. The wavelength (λ) and frequency (f) are related to the **wave speed** (v).

$$v = \lambda f \quad , \quad \lambda = \frac{v}{f} \quad , \quad f = \frac{v}{\lambda}$$

Recall from Chapter 5 that **frequency** (f) and **period** (T) share a reciprocal relationship.

$$f = \frac{1}{T} \quad , \quad T = \frac{1}{f}$$

Wave speed can be expressed in terms of frequency or period.

$$v = \lambda f \quad , \quad v = \frac{\lambda}{T}$$

We use the symbol v for the speed of light in a medium, but change it to c for the speed of light in vacuum. For example, we write $v = \lambda f$ when light travels through a medium such as water or glass, but write $c = \lambda f$ when light travels through vacuum. The speed of light in **vacuum** equals $c = 3.00 \times 10^8$ m/s to three significant figures.

The **intensity** (I) of a wave equals power (P) per unit area (A): $I = \frac{P}{A}$. For a wave that propagates radially outward (like the light radiated by a star), the area is the surface area of a sphere: $A = 4\pi r^2$. This shows that the intensity of the light decreases as an inverse square (since $I = \frac{P}{4\pi r^2}$ is proportional to $\frac{1}{r^2}$).

Symbols and SI Units

Symbol	Name	SI Units
v	speed of light in a medium	m/s
c	speed of light in vacuum	m/s
λ	wavelength	m
T	period	s
f	frequency	Hz
I	intensity	$\frac{\text{W}}{\text{m}^2}$
P	power	W
A	area	m^2
r	radius	m

Important Distinctions

We use the symbol c for the speed of light in **vacuum**, and the symbol v for the speed of light in a **medium** (such as glass or water). The equations $v = \lambda f$ and $c = \lambda f$ are basically the same formula, except that the second formula is specifically for vacuum.

In physics, period and wavelength are two different quantities:

- **Period** is the horizontal **time** between two crests. You determine the period from a graph that has time (t) on the horizontal axis.
- **Wavelength** is the horizontal **distance** between two crests. You determine the wavelength from a graph that has position (x) on the horizontal axis.

Metric Prefixes and the Angstrom

One **Angstrom** (Å) equals 10^{-10} m, which equates to 0.1 nm.

Prefix	Name	Power of 10
μ	micro	10^{-6}
n	nano	10^{-9}

Strategy for Problems Involving the Intensity or Wavelength of Light

To solve a problem involving the intensity or wavelength of light, pick the right equation:

- Relate wavelength, frequency, period, and wave speed with the following equations. The units can help you tell the quantities apart: The SI unit of **wavelength** (λ) is the meter (m), the SI unit of **frequency** (f) is the Hertz (Hz), the SI unit of **period** (T) is the second (s), and the SI units of **wave speed** (v) are meters per second (m/s).
$$v = \frac{\lambda}{T} = \lambda f$$
If light travels through a medium (such as glass or water), use the symbol v for wave speed. If instead light travels through **vacuum**, use the symbol c for wave speed. The speed of light in **vacuum** equals $c = 3.00 \times 10^8$ m/s to three significant figures. It's approximately the same (only slightly less) in **air**.
- If you are given wavelength in **Angstroms** (Å) or **nanometers** (nm), first convert the wavelength to meters using 1 Å $= 1 \times 10^{-10}$ m or 1 nm $= 1 \times 10^{-9}$ m.
- When light travels from one medium to another, its **frequency** (f) remains the same while its **wavelength** (λ) and **wave speed** (v) change. We will explore the change in wave speed in detail in Chapter 22 (see the index of refraction). In this chapter, you can relate the wavelengths and wave speeds via the following equations.
$$v_1 = \lambda_1 f \quad , \quad v_2 = \lambda_2 f$$
- Relate the intensity of a wave to power and area via the following equation. The units can help you tell the quantities apart: The SI units of **intensity** (I) are Watts per square meter ($\frac{\text{W}}{\text{m}^2}$), the SI unit of **power** (P) is the Watt (W), and the SI units of **area** (A) are square meters (m^2).
$$I = \frac{P}{A}$$
- For a wave that propagates radially outward (like the light radiated by a star), the area is the surface area of a sphere: $A = 4\pi r^2$. For light radiated by a star, power is constant, while intensity and area change.
$$I_1 = \frac{P}{4\pi r_1^2} \quad , \quad I_2 = \frac{P}{4\pi r_2^2}$$
- For interference, diffraction, or polarization, see Chapters 27-30.

Example: Orange light has a wavelength of 600 nm in vacuum. What is its frequency?
First convert the wavelength from nanometers (nm) to meters using 1 nm $= 1 \times 10^{-9}$ m. The wavelength is $\lambda = 600$ nm $= 600 \times 10^{-9}$ m $= 6.00 \times 10^{-7}$ m. Use the formula for wave speed with the speed of light in vacuum: $c = 3.00 \times 10^8$ m/s.

$$f = \frac{c}{\lambda} = \frac{3 \times 10^8}{6 \times 10^{-7}} = \frac{3}{6} \times 10^{8-(-7)} = \frac{1}{2} \times 10^{8+7} = 0.500 \times 10^{15} = 5.00 \times 10^{14} \text{ Hz}$$

The frequency is $f = 5.00 \times 10^{14}$ Hz.

Example: Red light with a wavelength of 720 nm in air slows down to 2.00×10^8 m/s when it enters a block of glass. What is the wavelength of the light in the glass?

The main idea is that the frequency (f) remains the same in both air and glass, whereas the speed (v) and wavelength (λ) change when the ray of light enters the glass. List the given quantities in SI units, and identify the desired unknown.

- The speed of light in air is about the same as in vacuum: $v_a = 3.00 \times 10^8$ m/s.
- The speed of light in the block of glass is $v_g = 2.00 \times 10^8$ m/s.
- The wavelength in air is $\lambda_a = 7.20 \times 10^{-7}$ m since 1 nm $= 1 \times 10^{-9}$ m.
- We wish to find the wavelength in the glass (λ_g).

Use the equation for wave speed, using subscripts for wavelength and wave speed.

$$v_a = \lambda_a f \quad , \quad v_g = \lambda_g f$$

Solve for frequency in each equation. Divide both sides by wavelength.

$$f = \frac{v_a}{\lambda_a} = \frac{v_g}{\lambda_g}$$

Cross multiply.

$$v_a \lambda_g = v_g \lambda_a$$

Divide both sides of the equation by v_a.

$$\lambda_g = \frac{v_g \lambda_a}{v_a} = \frac{(2 \times 10^8)(7.2 \times 10^{-7})}{3 \times 10^8} = \frac{(2)(7.2)}{3} \frac{10^8 \times 10^{-7}}{10^8} = 4.8 \times 10^{-7} \text{ m} = 480 \text{ nm}$$

The wavelength is $\lambda_g = 480$ nm in the glass.

Example: The intensity of sunlight reaching earth is about $1350 \frac{\text{W}}{\text{m}^2}$. What is the intensity of sunlight reaching Neptune? Neptune's orbital radius is 30 times earth's orbital radius.

The power (P) output of the sun is constant, while intensity varies with distance from the sun. Intensity (I) is power per unit area (A). Since the sun's light radiates outward from the sun, the area is the surface area of a sphere ($A = 4\pi r^2$) a distance r from the sun. Use subscripts for intensity, area, and radius.

$$I_e = \frac{P}{A_e} = \frac{P}{4\pi r_e^2} \quad , \quad I_n = \frac{P}{A_n} = \frac{P}{4\pi r_n^2}$$

Solve for power in each equation.

$$P = I_e 4\pi r_e^2 = I_n 4\pi r_n^2$$

Divide both sides by $4\pi r_n^2$. According to the problem, $r_n = 30 r_e$ and $I_e = 1350 \frac{\text{W}}{\text{m}^2}$.

$$I_n = \frac{4\pi r_e^2}{4\pi r_n^2} I_e = \frac{r_e^2}{r_n^2} I_e = \left(\frac{r_e}{r_n}\right)^2 I_e = \left(\frac{1}{30}\right)^2 (1350) = \frac{1350}{30^2} = \frac{1350}{900} = 1.5 \frac{\text{W}}{\text{m}^2}$$

The intensity of sunlight reaching Neptune is about $I_n = 1.5 \frac{\text{W}}{\text{m}^2}$.

85. A ray of violet light has a wavelength of 400 nm in air.

(A) What is the frequency of the ray of violet light?

(B) If the ray of violet light travels 1.80×10^8 m/s in a medium, what is its wavelength in that medium?

86. The light produced by a star has an intensity of $1800 \frac{\text{W}}{\text{m}^2}$ at one of its planets. What is the intensity of the light at another planet that is three times as far away from the star?

Want help? Check the hints section at the back of the book.

Answers: #85. (A) 7.5×10^{14} Hz (B) 240 nm. #86. $200 \frac{\text{W}}{\text{m}^2}$

21 REFLECTION AND REFRACTION

Relevant Terminology

Reflection – the ray of light that travels back into the same medium upon striking a surface.
Refraction – the bending of a ray of light associated with its change in speed as it passes from one medium into another medium.
Vacuum – a region of space completely devoid of matter. There is not even air.
Index of refraction – the ratio of the speed of light in vacuum to the speed of light in a medium. A **larger** index of refraction corresponds to a **slower** speed in the medium.
Normal – an imaginary line that is **perpendicular** to the surface.[1]
Incident – a ray of light that strikes the boundary between two different media.
Angle of incidence – the angle between the **incident** ray and the **normal**.
Angle of reflection – the angle between the **reflected** ray and the **normal**.
Angle of refraction – the angle between the **refracted** ray and the **normal**.
Transparent – light passes through the substance such that objects can be seen distinctly by looking through the substance.
Translucent – light passes through the substance, but the light is diffused such that objects can't be distinguished clearly by looking through the substance.
Opaque – light does not pass through the substance.

Essential Concepts

When a ray of light travels through a medium and strikes the boundary between that medium and an adjacent medium (or even a vacuum), the ray of light generally does two things:

- Part or all of the ray of light **reflects** back into the same medium.
- Part of the ray may also **refract** into the adjacent medium.

This is illustrated in the diagram below, where an incident ray both reflects and refracts.

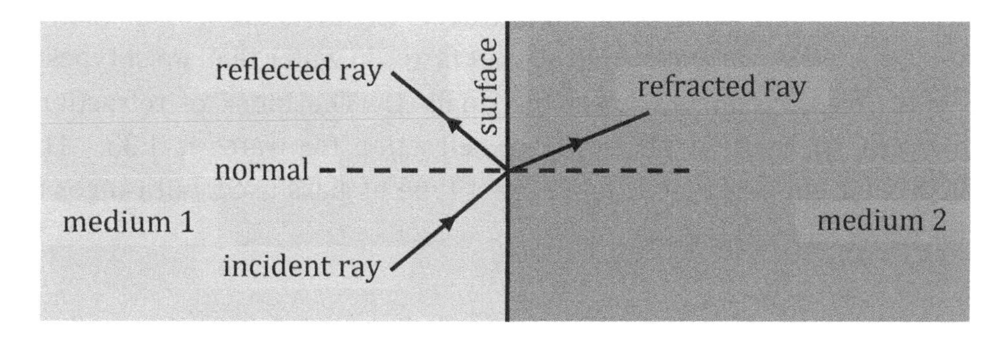

[1] You should remember from Volume 1 that "normal force" is perpendicular to the surface. If so, this may help reinforce that normal means perpendicular.

We measure angles relative to the normal. The **normal** is a line that is **perpendicular** to the surface. In a reflection or refraction problem, remember to always draw the normal and find the angle that the ray makes with the normal. There are three important angles:

- The angle of **incidence** (θ_i) is the angle between the **incident** ray and the **normal**.
- The angle of **reflection** (θ_r) is the angle between the **reflected** ray and the **normal**.
- The angle of **refraction**[2] (θ_t) is the angle between the **refracted** ray and the **normal**.

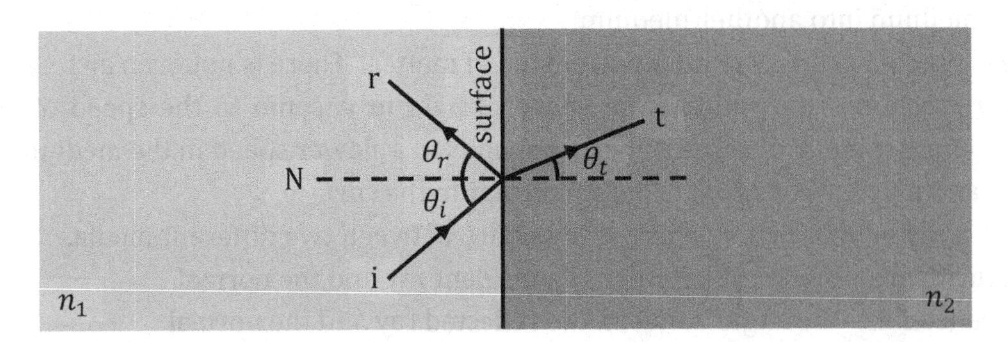

The **law of reflection** states that the **reflected** angle (θ_r) equals the incident angle (θ_i).

$$\theta_r = \theta_i$$

The **law of refraction (Snell's law)** relates the **refracted** angle (θ_t) to the incident angle (θ_i) via trig functions.

$$n_i \sin\theta_i = n_t \sin\theta_t$$

The symbol n represents the **index of refraction** of the medium. It is defined as the ratio of the speed of light in **vacuum** (c) to the speed of light in the medium (v).

$$n = \frac{c}{v}$$

We will explore the math in Chapter 22. In this chapter, we will focus on understanding reflection and refraction conceptually. Specifically, we will apply the following concepts.

- A medium with a **higher** index of refraction (n) has a **slower** speed (v).
- A medium with a **lower** index of refraction (n) has a **faster** speed (v).
- A ray of light refracts **towards** the normal if it **slows down** (n is getting **larger**).
- A ray of light refracts **away from** the normal if it **speeds up** (n is getting **smaller**).
- Light travels fastest in vacuum. Light travels slightly slower in air than in vacuum. Light travels slower in water than air, but faster in water than most types of glass.[3]
- The index of refraction for vacuum equals 1. The index of refraction for air is slightly greater than 1. The index of refraction for water is 1.33. The index of refraction for glass depends on the exact type of glass used, but ranges from 1.4 to 1.7 for many types of glass.

---

[2] Unfortunately, "reflection" and "refraction" both begin with the letter "r." It would be confusing to use the same symbol, θ_r, for both reflection and refraction. We will use θ_r for reflection and θ_t for refraction. It might help to remember the "t" if you think of the work "transmitted."

[3] Remember VAWG: vacuum, air, water, glass.

Symbols and Units

Symbol	Name	Units
v	speed of light in a medium	m/s
c	speed of light in vacuum	m/s
i	the incident ray	
r	the reflected ray	
t	the refracted ray	
N	the normal line	
n	index of refraction	unitless
n_i	index of refraction of the incident medium	unitless
n_t	index of refraction of the refracting medium	unitless
θ_i	angle of incidence	° or rad
θ_r	angle of reflection	° or rad
θ_t	angle of refraction	° or rad

Important Distinctions

The words reflection and refraction differ by only two letters, yet it's important to tell them apart. When you view a mirror, you experience **reflection**. When you look through a window and see objects on the other side, you experience **refraction**. Sometimes, when you look through a window, you experience **both** reflection and refraction – seeing objects on the same side and on the other side at once.

Note the distinction between transparent, translucent, and opaque substances:

- A substance is **transparent** if objects can be seen distinctly by looking through the substance.
- A substance is **translucent** if objects can be seen, but **can't** be distinguished clearly by looking through the substance (due to diffusion)
- A substance is **opaque** if light does not pass through the substance.

Reflection and Refraction Ray Drawing Strategy

To <u>draw</u> the reflected and refracted rays, follow these steps:

1. If the incident ray is not already drawn such that its tip meets the boundary between the two media, extend the incident ray until it does. Label the incident ray with the letter "i."
2. Draw the **normal** where the incident ray meets the boundary between two media. The normal is **perpendicular** to the boundary (or surface, or interface). Label the normal with the letter "N."
3. Label the angle between the **incident** ray (i) and the normal (N) as the angle of **incidence** (θ_i).
4. Draw the **reflected** ray back into the incident medium such that the angle of reflection (θ_r) equals the angle of incidence (θ_i). Label the reflected ray with the letter "r."
5. Label the angle between the **reflected** ray (r) and the normal (N) as the angle of **reflection** (θ_r).
6. Determine whether the refracted ray is slowing down or speeding up.
    - If the index of refraction of the second medium is **larger** than the index of refraction of the incident medium ($n_t > n_i$), the ray of light is **slowing down**.
    - If the index of refraction of the second medium is **smaller** than the index of refraction of the incident medium ($n_t < n_i$), the ray of light is **speeding up**.
    - Of vacuum, air, water, and glass, light travels fastest in **vacuum**, then **air**, then **water**, and slowest in (most types of) **glass**.
7. Draw the **refracted** ray into the new medium as follows.
    - First visualize the incident ray's initial path (imagine extending it), and then bend the refracted ray **closer to the normal** (without crossing[4] the normal) if the ray of light is **slowing down** (as determined in Step 6).
    - First visualize the incident ray's initial path (imagine extending it), and then bend the refracted ray **further from the normal** (but on the same side[4] of the normal) if the ray of light is **speeding up** (as determined in Step 6).

   Label the refracted ray with the letter "t."
8. Label the angle between the **refracted** ray (t) and the normal (N) as the angle of **refraction** (θ_t).

---

[4] Identify 4 Quadrants (like the 4 Quadrants from trig) as follows: Find the boundary (or surface, or interface) between the two media and find the normal. These two perpendicular lines define 4 Quadrants. The incident ray lies in one of these 4 Quadrants. The refracted ray always lies in the opposite Quadrant (diagonally from the incident Quadrant). Look at your diagram after you draw it: If the refracted and incident rays are in adjacent Quadrants, you made a big mistake. (The reflected ray is different: It will always be in the adjacent Quadrant compared to both the incident and refracted rays, and the reflected ray and incident ray will always lie in the same medium.) In addition to this, you must also be able to determine whether the ray of light is slowing down or speeding up so that you can bend the refracted ray correctly.

Example: Draw and label the reflected and refracted rays for the diagram below.

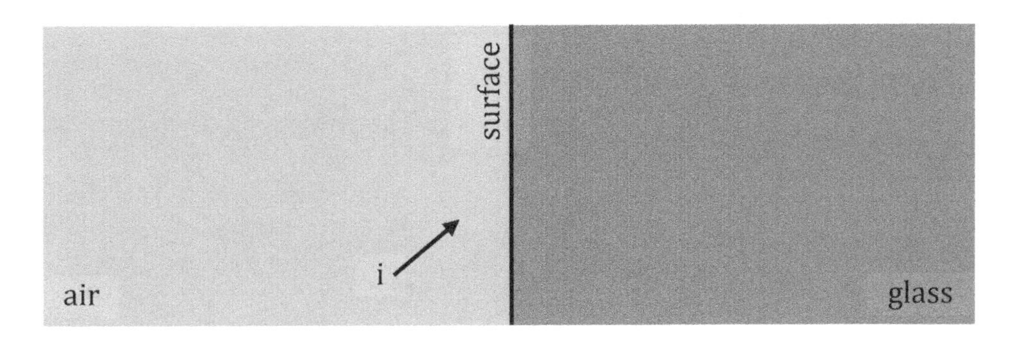

Draw the ray diagram as follows.

- Extend the incident ray (i) until it reaches the boundary between the air and glass.
- Since the boundary between the air and glass is vertical, the normal (N) is a **horizontal** line (since the normal must be **perpendicular** to the boundary). Label the normal "N."
- Draw the reflected ray coming back into the air, such that $\theta_r = \theta_i$.
- The refraction occurs from air to glass in this example.
- The refracted ray is **slowing down**: Light travels slower in glass than air.
- Bend the refracted ray **towards the normal** since it is slowing down.
- Label the reflected ray "r" and the refracted ray "t."
- Label θ_i, θ_r, and θ_t from the normal.

The completed ray diagram is shown below.

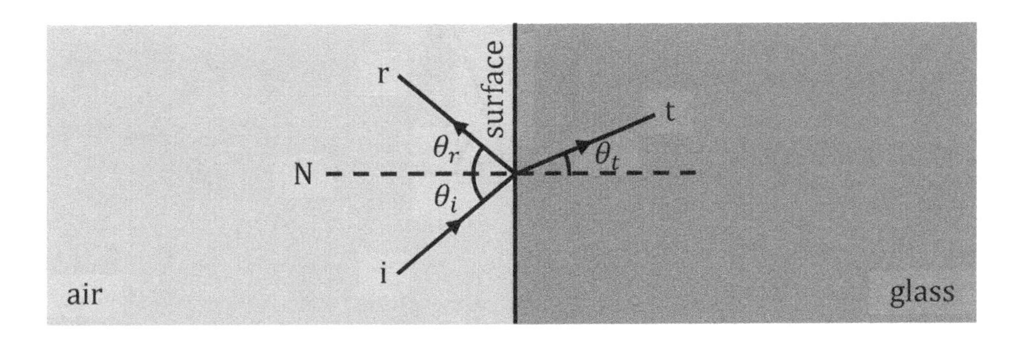

Example: Draw and label the reflected and refracted rays for the diagram below.

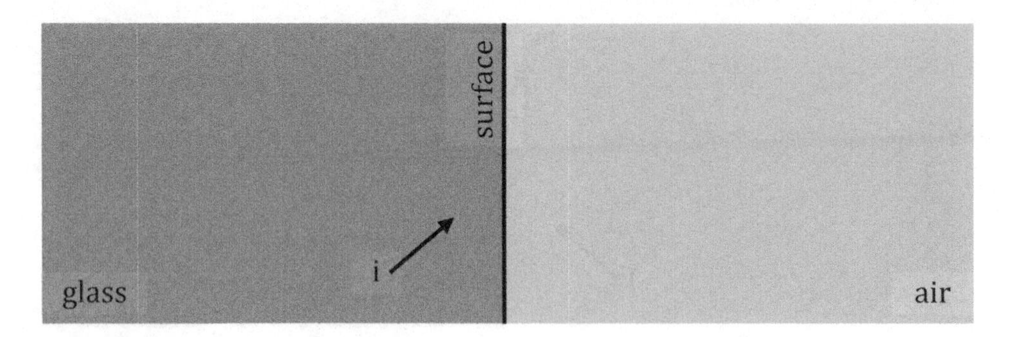

Draw the ray diagram as follows.

- Extend the incident ray (i) until it reaches the boundary between the glass and air.
- Since the boundary between the glass and air is vertical, the normal (N) is a **horizontal** line (since the normal must be **perpendicular** to the boundary). Label the normal "N."
- Draw the reflected ray coming back into the glass, such that $\theta_r = \theta_i$.
- The refraction occurs from glass to air in this example.
- The refracted ray is **speeding up**: Light travels faster in air than glass.
- Bend the refracted ray **away from the normal** since it is speeding up.
- Label the reflected ray "r" and the refracted ray "t."
- Label θ_i, θ_r, and θ_t from the normal.

The completed ray diagram is shown below.

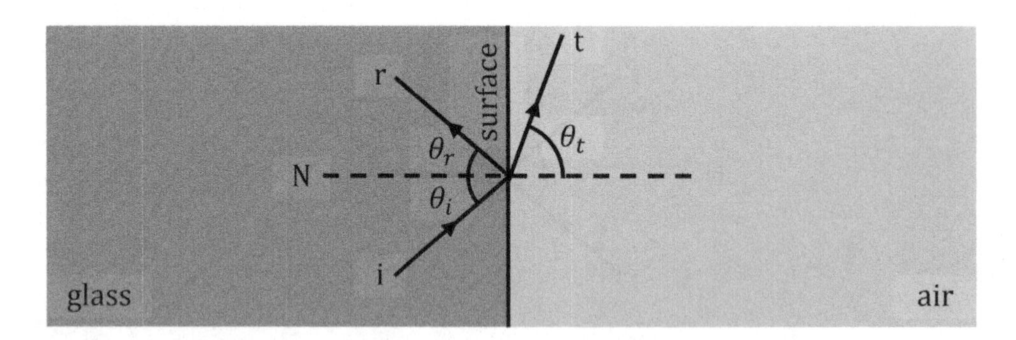

Note: It is instructive to compare this example with the previous example.

Example: Draw and label the reflected and refracted rays for the diagram below.

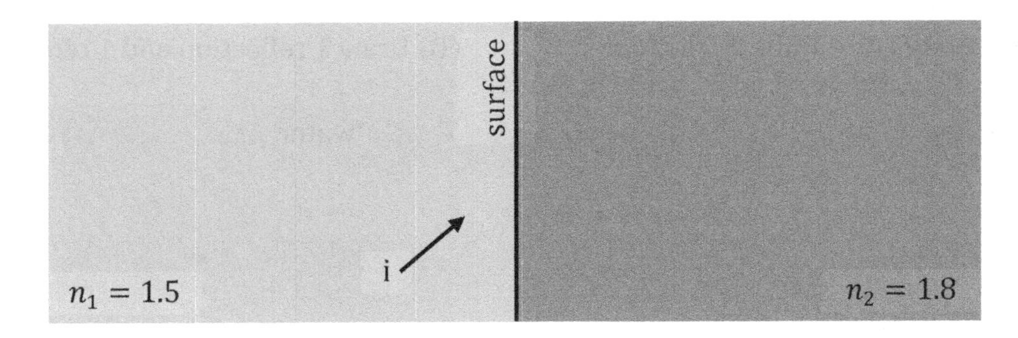

Draw the ray diagram as follows.

- Extend the incident ray (i) until it reaches the boundary between the two media.
- Since the boundary between the two media is vertical, the normal (N) is a **horizontal** line (since the normal must be **perpendicular** to the boundary). Label the normal "N."
- Draw the reflected ray coming back into the medium with $n_1 = 1.5$, such that $\theta_r = \theta_i$.
- The refraction occurs from the medium with $n_1 = 1.5$ to the medium with $n_2 = 1.8$ in this example.
- The refracted ray is **slowing down**: Since the index of refraction is **increasing** (from 1.5 to 1.8), the refracted ray travels slower than the incident ray.
- Bend the refracted ray **towards the normal** since it is slowing down.
- Label the reflected ray "r" and the refracted ray "t."
- Label θ_i, θ_r, and θ_t from the normal.

The completed ray diagram is shown below.

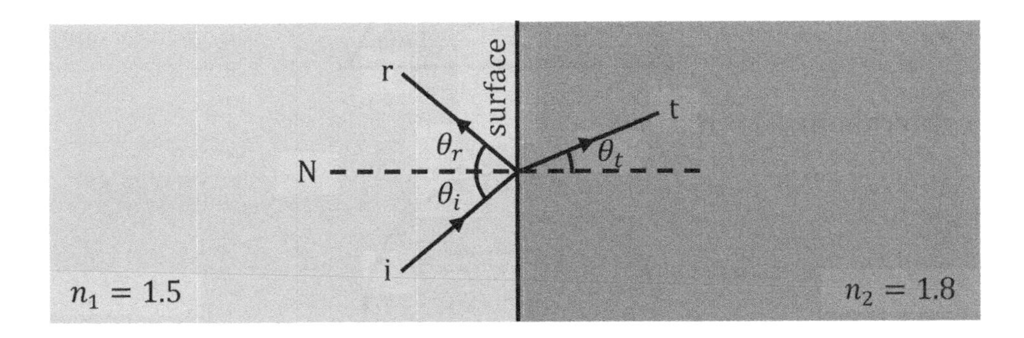

87. Draw and label the reflected and refracted rays for each diagram below.

(A) Draw 1 reflection and 1 refraction.

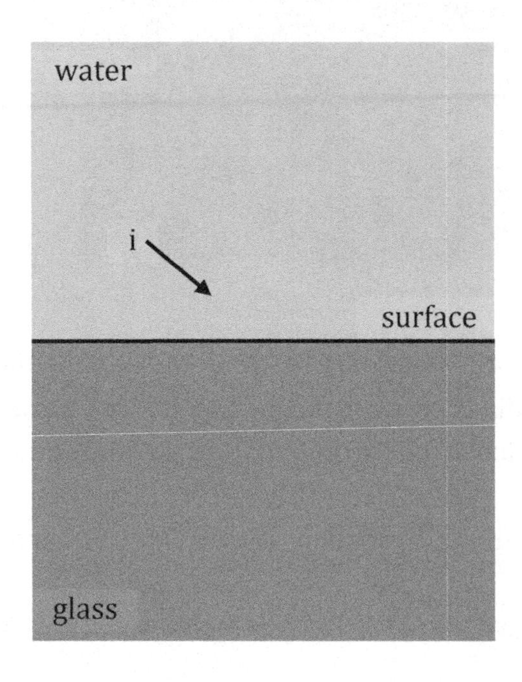

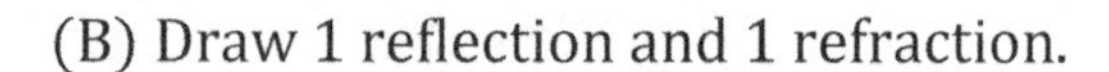

(B) Draw 1 reflection and 1 refraction.

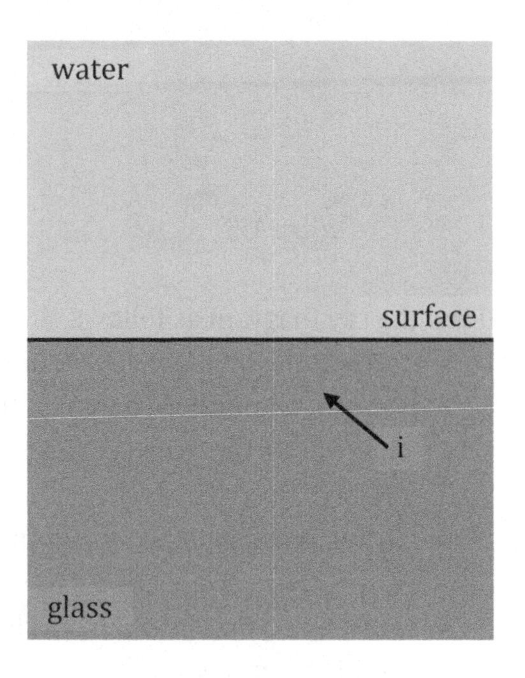

(C) Draw 2 reflections and 2 refractions. (Why 2 of each? Think about it.)

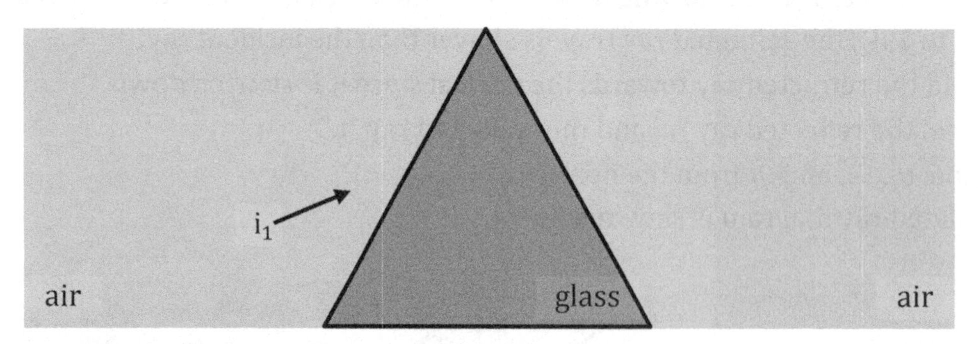

(D) Draw 1 reflection and 1 refraction.

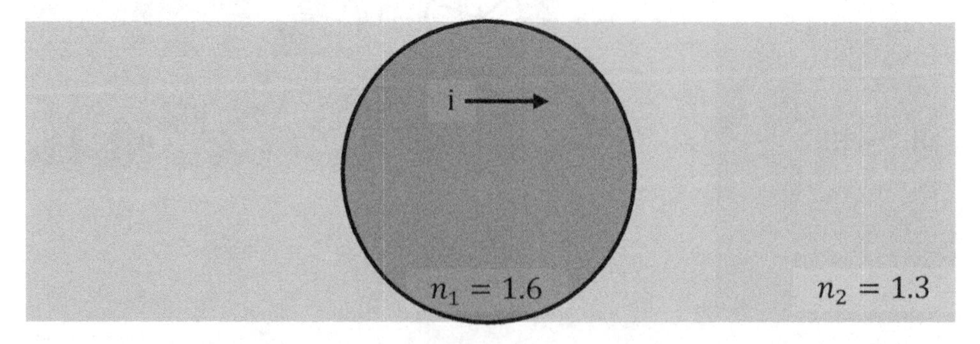

Want help? These problems are fully solved in the back of the book.

22 SNELL'S LAW

Relevant Terminology

Reflection – the ray of light that travels back into the same medium upon striking a surface.
Refraction – the bending of a ray of light associated with its change in speed as it passes from one medium into another medium.
Vacuum – a region of space completely devoid of matter. There is not even air.
Index of refraction – the ratio of the speed of light in vacuum to the speed of light in a medium. A **larger** index of refraction corresponds to a **slower** speed in the medium.
Normal – an imaginary line that is **perpendicular** to the surface.
Incident – a ray of light that strikes the boundary between two different media.
Angle of incidence – the angle between the **incident** ray and the **normal**.
Angle of reflection – the angle between the **reflected** ray and the **normal**.
Angle of refraction – the angle between the **refracted** ray and the **normal**.

Index of Refraction

The **index of refraction** (n) of a medium is the ratio of the speed of light in **vacuum** (c) to the speed of light in the medium (v).

$$n = \frac{c}{v}$$

The speed of light in **vacuum** equals $c = 3.00 \times 10^8$ m/s to three significant figures. The index of refraction for vacuum is exactly 1 (since $\frac{c}{c} = 1$). The speed of light is nearly the same in **air** (only slightly less) as vacuum, so the index of refraction for air is slightly greater than 1. We will round the index of refraction for air to 1 unless a particular problem requires five or more significant figures. For water, the index of refraction equals 1.333, which we will approximate as $\frac{4}{3}$ in order to solve the problems without a calculator.

Medium	Index of Refraction
vacuum	1 (exactly)
air	1 (approximately)
water	$\frac{4}{3}$ (good to 4 significant figures)

The index of refraction (n) of a medium is always **greater than 1** except in vacuum. The reason is that light travels faster in vacuum than any other medium.

Snell's Law

When a ray of light travels through a medium and strikes the boundary between that medium and an adjacent medium:

- Part or all of the ray of light **reflects** back into the same medium.
- Part of the ray may also **refract** into the adjacent medium.

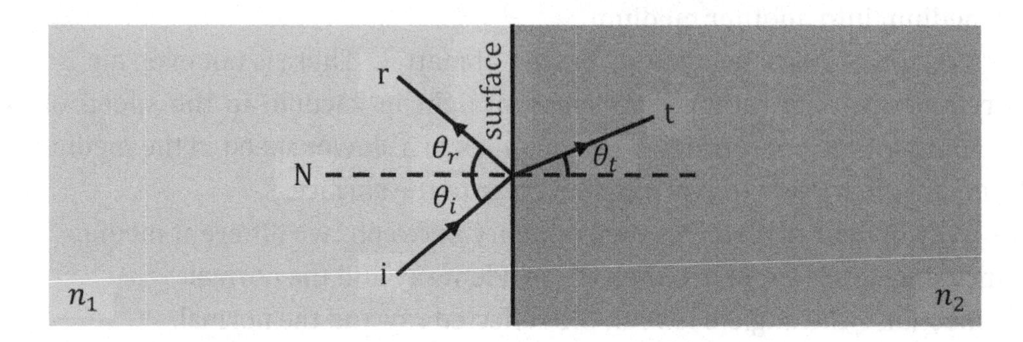

We measure angles relative to the **normal**, which is **perpendicular** to the surface.

- The angle of **incidence** (θ_i) is the angle between the **incident** ray and the **normal**.
- The angle of **reflection** (θ_r) is the angle between the **reflected** ray and the **normal**.
- The angle of **refraction**[1] (θ_t) is the angle between the **refracted** ray and the **normal**.

The **law of reflection** states that the **reflected** angle (θ_r) equals the incident angle (θ_i).

$$\theta_r = \theta_i$$

The **law of refraction** (**Snell's law**) relates the **refracted** angle (θ_t) to the incident angle (θ_i) via trig functions.

$$n_i \sin\theta_i = n_t \sin\theta_t$$

Basic Trig Functions for Special Angles

θ	0°	30°	45°	60°	90°
$\sin\theta$	0	$\frac{1}{2}$	$\frac{\sqrt{2}}{2}$	$\frac{\sqrt{3}}{2}$	1
$\cos\theta$	1	$\frac{\sqrt{3}}{2}$	$\frac{\sqrt{2}}{2}$	$\frac{1}{2}$	0
$\tan\theta$	0	$\frac{\sqrt{3}}{3}$	1	$\sqrt{3}$	undef.

[1] Think "t" for "transmitted," since the subscript "r" is already used for "reflected."

Symbols and Units

Symbol	Name	Units
v	speed of light in a medium	m/s
c	speed of light in vacuum	m/s
i	the incident ray	
r	the reflected ray	
t	the refracted ray	
N	the normal line	
n	index of refraction	unitless
n_i	index of refraction of the incident medium	unitless
n_t	index of refraction of the refracting medium	unitless
θ_i	angle of incidence	° or rad
θ_r	angle of reflection	° or rad
θ_t	angle of refraction	° or rad

Important Distinctions

Pay careful attention to whether a problem asks about reflection or refraction, since these two words only differ by two letters, but the solution is drastically different for the two cases. The law of **reflection** is $\theta_r = \theta_i$, whereas the law of **refraction** (Snell's law) is $n_i \sin\theta_i = n_t \sin\theta_t$.

Note that the **index** of refraction (n_t) and the **angle** of refraction (θ_t) are two different quantities.

Snell's Law Strategy

To solve a problem involving Snell's law, follow these steps:

1. Make a list of the known quantities. Be sure to measure[2] all of the angles from the **normal**. Remember that the normal is **perpendicular** to the surface.
2. Relate the incident angle (θ_i) to the refracted angle (θ_t) via Snell's law.
$$n_i \sin\theta_i = n_t \sin\theta_t$$
3. You should know the following values for **index of refraction**:
   - $n_v = 1$ (exactly) in **vacuum**.
   - $n_a \approx 1$ (approximately) in **air**. For a rare problem where the given values are good to five significant figures, look up the precise value in a textbook.
   - $n_w = \frac{4}{3}$ (to four significant figures) in **water**.

   For a **textbook** problem that gives you a specific material (like zircon), you may need to look up the index of refraction in a table.
4. Beware that the index of refraction for glass varies depending upon the type of glass. So if you solve for the index of refraction for "glass" in one problem, you **can't** use the same value in a different problem because it might be a different type of glass. However, if the type of glass is specified (like flint glass), then you may look up the index of refraction in a table in a textbook.
5. To solve for an angle, first apply algebra to isolate the sine function, and then take the inverse sine of both sides, as illustrated in the example. It may help to review essential trigonometry skills in Volume 1, Chapter 9. If you're not using a calculator, note the handy table on page 252 (there is a tip for memorizing this in Volume 1).
6. Once you know the index of refraction for a medium, you can solve for the speed of light in that medium using the following equation, knowing that the speed of light in **vacuum** is $c = 3.00 \times 10^8$ m/s.
$$n = \frac{c}{v} \quad , \quad v = \frac{c}{n}$$
7. If a problem asks you for the angle of **reflection** (**not** to be confused with refraction), note that $\theta_r = \theta_i$.
8. If you need to convert between degrees and **radians**, note that π rad $= 180°$.
9. If a problem involves **total internal reflection**, see Chapter 23. If a problem involves a **prism** or **dispersion**, see Chapter 24.
10. If a Snell's law problem asks about **wavelength** or **frequency**, see Chapter 20.
11. If a problem involves **Brewster's angle** or **polarization**, see Chapter 30.

[2] Some problems intentionally label a different angle, knowing that some students naively use whichever angle happens to be given. That's certainly fair: The student who remembers how the angle is defined and proceeds to apply this knowledge correctly deserves a better score. It's also important to remember this in lab, where a few students incorrectly measure the angle from the surface rather than from the normal.

Example: A monkey shines a ray of banana-colored light from air to glass. The incident angle equals 45° and the index of refraction of the glass is $\sqrt{2}$.
(A) What is the angle of refraction in the glass?
Identify the known quantities in appropriate units.

- The incident angle is $\theta_i = 45°$.
- The index of refraction of the air (the incident medium) is $n_i \approx 1$.
- The index of refraction of the glass (the refracting medium) is $n_t = \sqrt{2}$.

Apply Snell's law.

$$n_i \sin\theta_i = n_t \sin\theta_t$$
$$(1)\sin 45° = \sqrt{2}\sin\theta_t$$

Recall from trig that $\sin 45° = \frac{\sqrt{2}}{2}$ (or consult the table on page 252).

$$\frac{\sqrt{2}}{2} = \sqrt{2}\sin\theta_t$$

Isolate the sine function. Divide both sides of the equation by $\sqrt{2}$. Note that the $\sqrt{2}$'s cancel.

$$\frac{1}{2} = \sin\theta_t$$

Take the inverse sine of both sides of the equation. Note that $\sin^{-1}[\sin(\theta_t)] = \theta_t$.

$$\theta_t = \sin^{-1}\left(\frac{1}{2}\right) = 30°$$

The refracted angle is $\theta_t = 30°$ because $\sin(30°) = \frac{1}{2}$ (see page 252). A nice feature of Snell's law problems is that the angle **always**[3] lies in Quadrant I ($0° \leq \theta \leq 90°$).

(B) What is the speed of light in the glass?
Use the equation that relates the index of refraction to the speed of light in the medium. You should know that the speed of light in vacuum is $c = 3.00 \times 10^8$ m/s.

$$v_t = \frac{c}{n_t} = \frac{3 \times 10^8}{\sqrt{2}} = \frac{3}{\sqrt{2}}10^8 = \frac{3}{\sqrt{2}}\frac{\sqrt{2}}{\sqrt{2}}10^8 = \frac{3\sqrt{2}}{2}10^8 = 3\sqrt{2}\frac{10^8}{2} = 3\sqrt{2}\frac{10^7 10}{2}$$
$$v_t = 15\sqrt{2} \times 10^7 \text{ m/s}$$

We multiplied by $\frac{\sqrt{2}}{\sqrt{2}}$ in order to **rationalize** the denominator. Note that $\sqrt{2}\sqrt{2} = 2$. Also, we wrote 10^8 as $10^7 \times 10$ in order to simplify $\frac{10^8}{2}$ as $\frac{10^7 10}{2} = 5 \times 10^7$. Note that $3\sqrt{2} \times 5 = 15\sqrt{2}$. The speed of light in the glass is $v_t = 15\sqrt{2} \times 10^7$ m/s, which is the same as $v_t = \frac{3}{\sqrt{2}} \times 10^8$ m/s or $v_t = \frac{3\sqrt{2}}{2} \times 10^8$ m/s. If you use a calculator, this works out to $v_t = 21 \times 10^7$ m/s, which is the same as $v_t = 2.1 \times 10^8$ m/s. It's always a good idea to check that v_t is **less than** 3×10^8 m/s (otherwise, you know you made a mistake).

---

[3] As long as the answer is **real**. If the argument of the inverse sine exceeds one, the answer is imaginary. That's usually a sign that the student made a **mistake**, but it could be due to **total internal reflection** (Chapter 23).

88. A monkey shines a ray of light from banana juice to glass. The incident angle equals 60°, the index of refraction of the banana juice is $\sqrt{3}$, and the index of refraction of the glass is $\frac{3\sqrt{2}}{2}$.

(A) What is the angle of refraction in the banana juice?

(B) What is the speed of light in the banana juice?

(C) What is the speed of light in the glass?

Want help? Check the hints section at the back of the book.

Answers: (A) 45° (B) $\sqrt{3} \times 10^8$ m/s $= 1.7 \times 10^8$ m/s (C) $\sqrt{2} \times 10^8$ m/s $= 1.4 \times 10^8$ m/s

89. A monkey shines a ray of light from one medium to another. The ray of light travels $\sqrt{6} \times 10^8$ m/s in the first medium and $\sqrt{2} \times 10^8$ m/s in the second medium. The incident angle equals 60°.

(A) What is the index of refraction of the first medium?

(B) What is the index of refraction of the second medium?

(C) What is the angle of refraction?

Want help? Check the hints section at the back of the book.

Answers: (A) $\frac{\sqrt{6}}{2} = 1.2$ (B) $\frac{3\sqrt{2}}{2} = 2.1$ (C) 30°

90. A monkey shines a ray of light as illustrated below. The incident angle equals 60°.

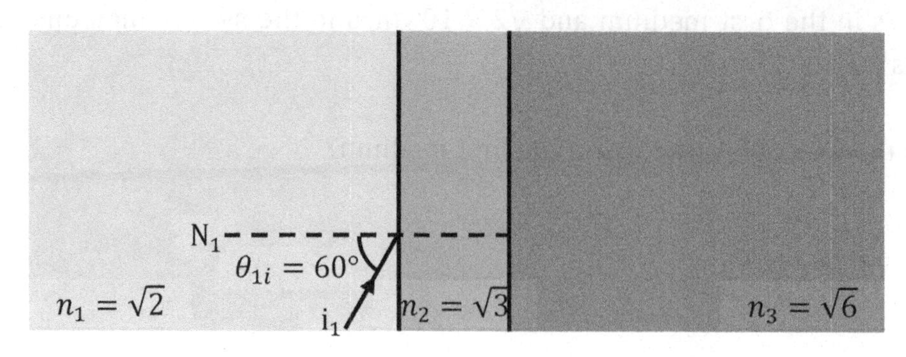

(A) What is the angle of refraction in the third medium?

(B) If the second medium were removed, what would be the angle of refraction in the third medium?

(C) Explain your answers to parts (A) and (B).

Want help? Check the hints section at the back of the book.

Answers: (A) 30° (B) 30°

23 TOTAL INTERNAL REFLECTION

Relevant Terminology

Reflection – the ray of light that travels back into the same medium upon striking a surface.
Refraction – the bending of a ray of light associated with its change in speed as it passes from one medium into another medium.
Vacuum – a region of space completely devoid of matter. There is not even air.
Index of refraction – the ratio of the speed of light in vacuum to the speed of light in a medium. A **larger** index of refraction corresponds to a **slower** speed in the medium.
Normal – an imaginary line that is **perpendicular** to the surface.

Total Internal Reflection

The sequence of diagrams below illustrates the concept of **total internal reflection.**

- Top left: The ray of light traveling from glass to air refracts **away from the normal** because it is **speeding up** (light travels faster in air than glass).
- Top right: When the incident angle is a little **larger**, the refracted ray bends even **farther away** from the normal.
- Bottom left: The incident angle is just large enough that the refracted angle equals 90°. The incident angle (θ_i) is called the **critical angle** (θ_c) for total internal reflection.
- Bottom right: The incident angle exceeds the critical angle ($\theta_i > \theta_c$). There is no refraction at all. There is only reflection. We call this **total internal reflection.**

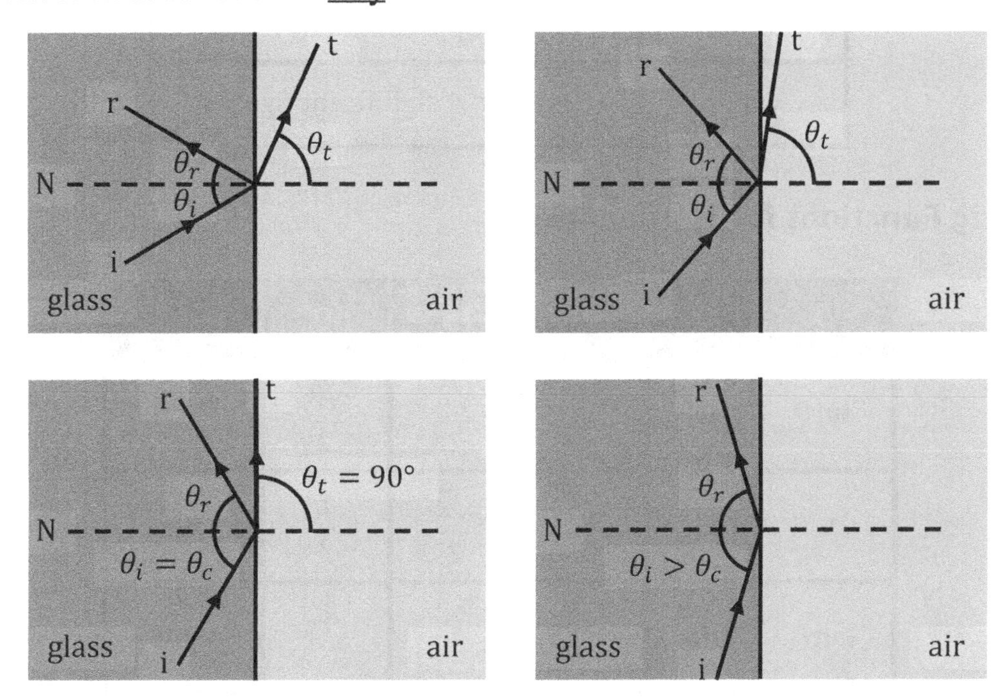

Note that total internal reflection is **only possible** when the refracted ray of light would be **speeding up** (which means that the index of refraction is getting **smaller**: $n_t < n_i$). For example, total internal reflection may occur when a ray of light is traveling from glass to air (since light travels faster in air than glass), but it is **impossible** for a ray of light that is traveling from air to glass (since light travels slower in glass than air).

Equations for Total Internal Reflection

The incident angle (θ_i) equals the **critical angle** (θ_c) for **total internal reflection** to occur when the refracted angle (θ_t) is 90°. Set $\theta_i = \theta_c$ and $\theta_t = 90°$ in Snell's law (Chapter 22).

$$n_i \sin\theta_i = n_t \sin\theta_t$$
$$n_i \sin\theta_c = n_t \sin 90°$$

Recall from trig that $\sin 90° = 1$.

$$n_i \sin\theta_c = n_t$$

The **index of refraction** (n) of a medium is the ratio of the speed of light in **vacuum** (c) to the speed of light in the medium (v).

$$n = \frac{c}{v}$$

Recall that the speed of light in **vacuum** equals $c = 3.00 \times 10^8$ m/s.

Medium	Index of Refraction
vacuum	1 (exactly)
air	1 (approximately)
water	$\frac{4}{3}$ (good to 4 significant figures)

Basic Trig Functions for Special Angles

θ	0°	30°	45°	60°	90°
$\sin\theta$	0	$\frac{1}{2}$	$\frac{\sqrt{2}}{2}$	$\frac{\sqrt{3}}{2}$	1
$\cos\theta$	1	$\frac{\sqrt{3}}{2}$	$\frac{\sqrt{2}}{2}$	$\frac{1}{2}$	0
$\tan\theta$	0	$\frac{\sqrt{3}}{3}$	1	$\sqrt{3}$	undef.

Symbols and Units

Symbol	Name	Units
v	speed of light in a medium	m/s
c	speed of light in vacuum	m/s
i	the incident ray	
r	the reflected ray	
t	the refracted ray	
N	the normal line	
n	index of refraction	unitless
n_i	index of refraction of the incident medium	unitless
n_t	index of refraction of the refracting medium	unitless
θ_i	angle of incidence	° or rad
θ_r	angle of reflection	° or rad
θ_t	angle of refraction	° or rad
θ_c	critical angle for total internal reflection	° or rad

Important Distinction

Whether or not **total internal reflection** is even **possible** depends upon whether the ray of light would be speeding up or slowing down:

- Total internal reflection can occur if the refracted ray would be traveling **faster** than the incident ray, meaning that the index or refraction for the refracted medium is **smaller** than the index of refraction for the incident medium ($n_t < n_i$).
- Total internal reflection is **not possible** if the refracted ray would be traveling **slower** than the incident ray, meaning that the index or refraction for the refracted medium is **larger** than the index of refraction for the incident medium ($n_t > n_i$).

It may help if you can remember that total internal reflection can occur from glass to air, but **can't** occur from air to glass. Light travels faster in air than glass, and air has a smaller index of refraction than glass (it's approximately 1 for air, but ranges from 1.4 to 1.7 for most types of glass).

Total Internal Reflection Strategy

To solve a problem involving total internal reflection, follow these steps:

1. Make a list of the known quantities. Be sure to measure all of the angles from the **normal**. Remember that the normal is **perpendicular** to the surface.
2. Relate the incident angle (θ_i) to the refracted angle (θ_t) via Snell's law.
$$n_i \sin\theta_i = n_t \sin\theta_t$$
3. Set the incident angle (θ_i) equals to the **critical angle** (θ_c) for **total internal reflection** to occur, $\theta_i = \theta_c$, and set the refracted angle (θ_t) equal to ninety degrees, $\theta_t = 90°$.
$$n_i \sin\theta_c = n_t \sin 90°$$
Recall from trig that $\sin 90° = 1$.
$$n_i \sin\theta_c = n_t$$
4. You should know the following values for **index of refraction**:
   - $n_v = 1$ (exactly) in **vacuum**.
   - $n_a \approx 1$ (approximately) in **air**. For a rare problem where the given values are good to five significant figures, look up the precise value in a textbook.
   - $n_w = \frac{4}{3}$ (to four significant figures) in **water**.

   For a **textbook** problem that gives you a specific material (like zircon), you may need to look up the index of refraction in a table.
5. Beware that the index of refraction for glass varies depending upon the type of glass. So if you solve for the index of refraction for "glass" in one problem, you **can't** use the same value in a different problem because it might be a different type of glass. However, if the type of glass is specified (like flint glass), then you may look up the index of refraction in a table in a textbook.
6. To solve for an angle, first apply algebra to isolate the sine function, and then take the inverse sine of both sides, as illustrated in the example. It may help to review essential trigonometry skills in Volume 1, Chapter 9. If you're not using a calculator, note the handy table on page 260 (there is a tip for memorizing this in Volume 1).
7. Once you know the index of refraction for a medium, you can solve for the speed of light in that medium using the following equation, knowing that the speed of light in **vacuum** is $c = 3.00 \times 10^8$ m/s.
$$n = \frac{c}{v} \quad , \quad v = \frac{c}{n}$$
8. If a problem asks you for the angle of **reflection** (**not** to be confused with refraction), note that $\theta_r = \theta_i$.
9. If you need to convert between degrees and **radians**, note that π rad $= 180°$.
10. If a problem involves a **prism** or **dispersion**, see Chapter 24.
11. If a Snell's law problem asks about **wavelength** or **frequency**, see Chapter 20.
12. If a problem involves **Brewster's angle** or **polarization**, see Chapter 30.

Example: A monkey shines a ray of banana-colored light from glass to air. The index of refraction of the glass is $\sqrt{2}$. What is the critical angle for total internal reflection?

Identify the known quantities in appropriate units.

- The index of refraction of the glass (the incident medium) is $n_i = \sqrt{2}$.
- The index of refraction of the air (the would-be refracting medium) is $n_t \approx 1$.
- For the critical angle for total internal reflection, the refracted angle is $\theta_t = 90°$.

Apply Snell's law with $\theta_i = \theta_c$ and $\theta_t = 90°$.

$$n_i \sin\theta_i = n_t \sin\theta_t$$
$$\sqrt{2}\sin\theta_c = (1)\sin 90°$$

Recall from trig that $\sin 90° = 1$ (or consult the table on page 260).

$$\sqrt{2}\sin\theta_c = (1)(1)$$
$$\sqrt{2}\sin\theta_c = 1$$

Isolate the sine function. Divide both sides of the equation by $\sqrt{2}$.

$$\sin\theta_c = \frac{1}{\sqrt{2}}$$

Rationalize the denominator: Multiply the numerator and denominator both by $\sqrt{2}$. Recall that $\sqrt{2}\sqrt{2} = 2$.

$$\sin\theta_c = \frac{1}{\sqrt{2}}\frac{\sqrt{2}}{\sqrt{2}} = \frac{\sqrt{2}}{2}$$

Take the inverse sine of both sides of the equation. Note that $\sin^{-1}[\sin(\theta_c)] = \theta_c$.

$$\theta_c = \sin^{-1}\left(\frac{\sqrt{2}}{2}\right) = 45°$$

The **critical angle** for total internal reflection to occur is $\theta_c = 45°$ because $\sin(45°) = \frac{\sqrt{2}}{2}$ (see page 260).

91. Medium A is adjacent to medium B. Medium A has an index of refraction of $\sqrt{3}$, while medium B has an index of refraction of $\sqrt{6}$.

(A) It is possible for total internal reflection to occur for a ray of light in medium A that is incident upon medium B? If so, what is the critical angle for total internal reflection?

(B) It is possible for total internal reflection to occur for a ray of light in medium B that is incident upon medium A? If so, what is the critical angle for total internal reflection?

Want help? Check the hints section at the back of the book.

Answers: (A) no (B) yes, 45°

24 DISPERSION AND SCATTERING

Relevant Terminology

Dispersion – the separation of a ray of light into its component **colors**.
Photon – a particle of light. A beam of light is made up of photons.
Scattering – the redirection of photons (particles of light) as they collide (or interact) with gas molecules (or other particles of matter).
Wavelength – the **horizontal** distance between two consecutive crests.
Power – the instantaneous rate at which work is done.
Intensity – power per unit area.
Refraction – the bending of a ray of light associated with its change in speed as it passes from one medium into another medium.
Index of refraction – the ratio of the speed of light in vacuum to the speed of light in a medium. A **larger** index of refraction corresponds to a **slower** speed in the medium.
Normal – an imaginary line that is **perpendicular** to the surface.

Dispersion in a Prism

The separation of a ray of light into its component colors is called **dispersion**. White light is made up of red, orange, yellow, green, blue, indigo, and violet colors. The acronym ROY G. BIV can help you remember the order of the colors in decreasing wavelength. One way to see dispersion is to shine a ray of white light through a glass prism, as illustrated below.

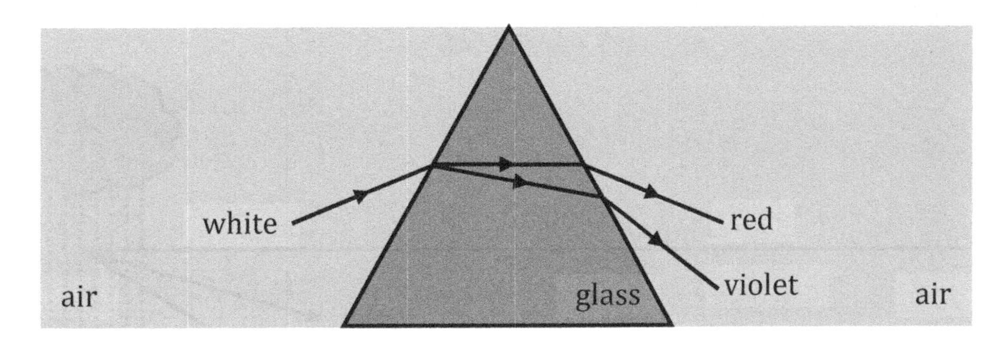

In a dispersive medium, such as glass or water, the different colors travel different speeds in the medium. (In vacuum, however, all of the colors travel the same speed.) Violet travels slowest while red travels fastest in a dispersive medium, and therefore violet refracts (changes direction) the most while red refracts the least. The incident beam of white light enters the prism, and the colors that make up the incident beam of white light emerge from the prism at different angles.

Rainbow Formation

A rainbow is formed when sunlight enters a spherical raindrop. As illustrated below, when a ray of sunlight enters a raindrop, it separates into its component colors (only the two extremes - red and violet - are shown) via **dispersion** when it refracts, reflects at the back, and refracts a second time as it returns back into the air. Since violet slows down the most, it changes direction the most during the refraction. As a result, the violet ray emerges from the raindrop at a shallower angle, while red emerges from the raindrop at a steeper angle.

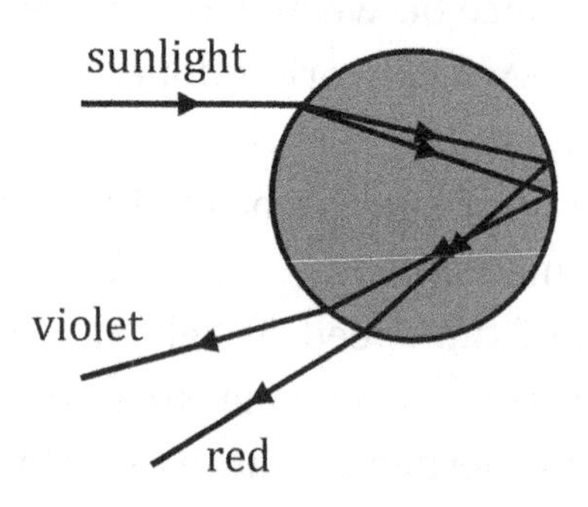

You can see a rainbow when the sun is behind you and it is raining in front of you. You can't see the red and violet rays from the same raindrop because your eye can't be in two places at once. Rather, you see different colors coming from different raindrops. Since red light emerges from the raindrop at a steeper angle, you must look higher to see red light. Similarly, since violet light emerges from the raindrop at a shallower angle, you must look lower to see violet light. This is why red appears at the top and violet appears at the bottom of the primary rainbow.

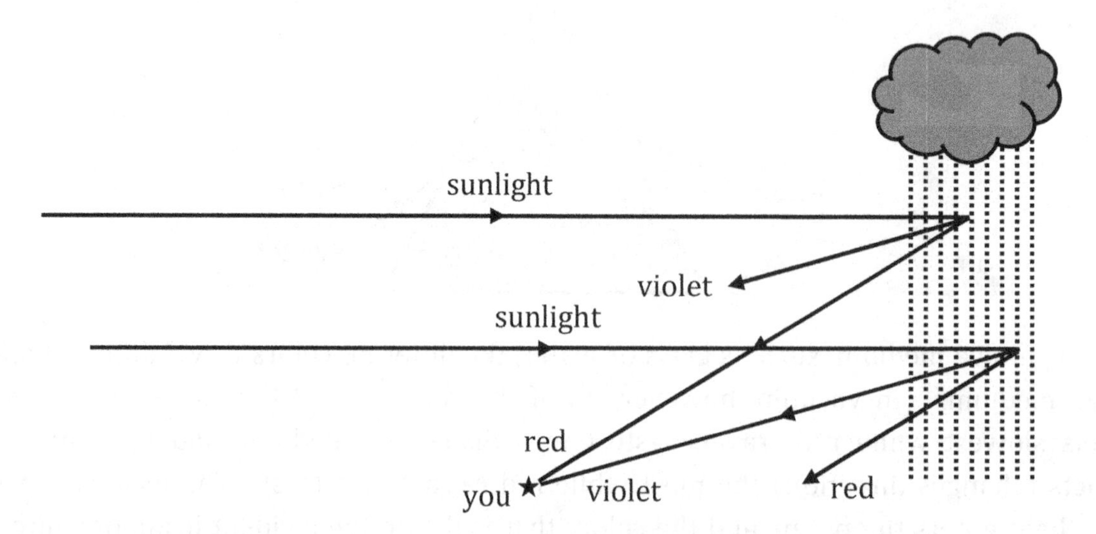

The next time you see a rainbow, check to see if the sun is indeed behind you and if red appears at the top while violet appears at the bottom of the primary rainbow.

Sometimes, if you look closely, you can see a faint secondary rainbow above the primary rainbow. The secondary rainbow forms when a ray of sunlight enters near the bottom of the raindrop and reflects twice inside as shown below. Compare the diagram below with the diagram for the primary rainbow on the previous page. In the secondary, red light emerges at a shallower angle and violet light emerges at a steeper angle, which is opposite to the primary. Therefore, the secondary rainbow has the colors reversed (with violet at the top and red at the bottom) compared to the primary rainbow. The secondary rainbow is also much fainter than the primary rainbow because more light is lost (due to refraction) with the additional reflection inside of the raindrop.

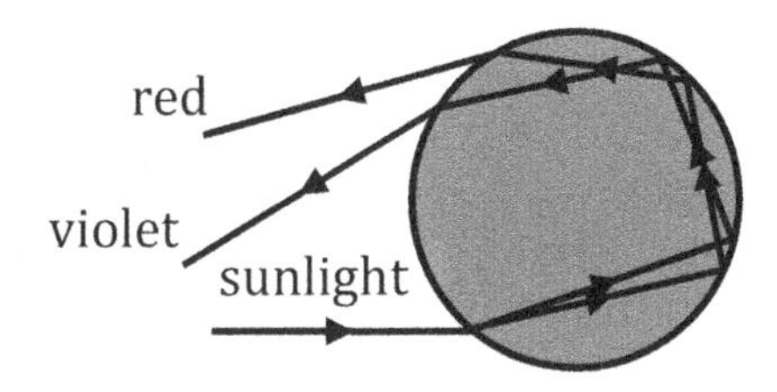

Scattering

When sunlight travels through the atmosphere, the particles of light (called photons) **scatter** off of gas molecules in the atmosphere. Scattering refers to the redirection of the path of the ray of light during the collision (or interaction) of the photon with the gas molecule. The relative intensity of the scattered light is inversely proportional to the fourth power of the wavelength: $I \propto \frac{1}{\lambda^4}$. The symbol $\propto$ translates as "is proportional to." Since wavelength (λ) is in the denominator, **shorter wavelengths** (like violet and blue) **scatter more** on average, whereas **longer wavelengths** (like red and orange) **scatter less** on average.

Why Does the Sky Appear Blue?

The reason that the sky appears blue is due to the scattering of sunlight off of gas molecules in earth's atmosphere. Shorter wavelengths scatter more on average, so violet and blue photons tend to scatter multiple times before reaching earth's surface. Longer wavelengths scatter less on average, so a greater percentage of reds and oranges take a direct route compared to violets and blues. When you look towards the sun (which you shouldn't do if you value your retinas), you see a greater percentage of longer wavelengths, which is why the sun appears red or yellow. When you look anywhere else in the sky during daylight, you see a greater percentage of shorter wavelengths, which is why the sky appears blue.

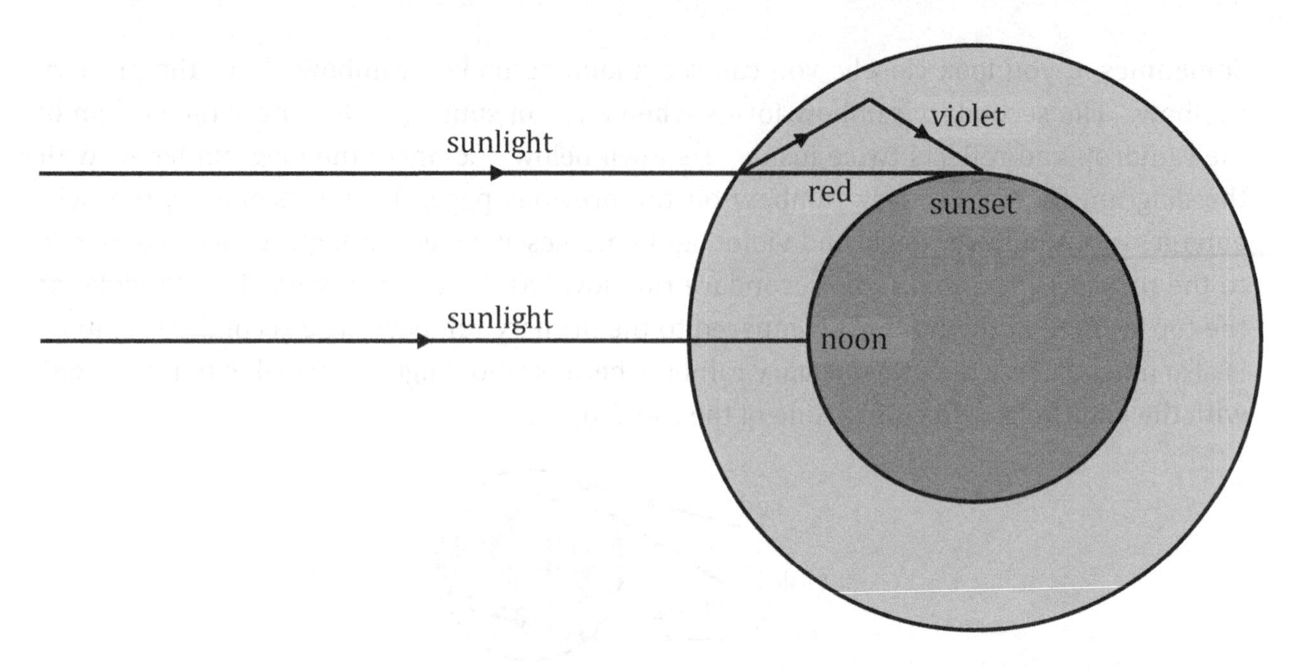

The fact that shorter wavelengths tend to scatter more on average also explains why the sun appears yellow at noon, but red at sunrise or sunset. If you study the diagram above, you will see that the distance that direct sunlight travels **through earth's atmosphere** is longer at sunset and shorter at noon. Therefore, more scattering occurs at sunset than at noon, and the sun appears a longer wavelength (red) at sunset than at noon (yellow). That is, there is more scattering of shorter wavelengths at sunset (or sunrise), so that you see predominantly red light when you view the sun at this time of the day.

Basic Trig Functions for Special Angles

θ	0°	30°	45°	60°	90°
$\sin\theta$	0	$\frac{1}{2}$	$\frac{\sqrt{2}}{2}$	$\frac{\sqrt{3}}{2}$	1
$\cos\theta$	1	$\frac{\sqrt{3}}{2}$	$\frac{\sqrt{2}}{2}$	$\frac{1}{2}$	0
$\tan\theta$	0	$\frac{\sqrt{3}}{3}$	1	$\sqrt{3}$	undef.

Symbols and Units

Symbol	Name	Units
v	speed of light in a medium	m/s
c	speed of light in vacuum	m/s
i	the incident ray	
r	the reflected ray	
t	the refracted ray	
N	the normal line	
n	index of refraction	unitless
n_i	index of refraction of the incident medium	unitless
n_t	index of refraction of the refracting medium	unitless
θ_i	angle of incidence	° or rad
θ_r	angle of reflection	° or rad
θ_t	angle of refraction	° or rad
A	prism angle	° or rad
D	angle of deviation	° or rad

Important Distinction

In everyday conversation, people tend to use the words "scatter" and "disperse" almost interchangeably. However, in physics, scattering and dispersion are two entirely different phenomena:

- **Dispersion** refers to the separation of a beam of light into its component colors during refraction through a medium (such as glass or water) for which the different colors travel different speeds.
- **Scattering** refers to the redirection of a ray of light when photons (particles of light) collide (or interact) with gas molecules (or other particles of matter)

Prism Strategy

To solve a problem involving a **prism**, follow these steps:

1. Draw a ray diagram, following the strategy of Chapter 21. Recall that the **normal** is **perpendicular** to the surface. Also recall that the refracted ray bends towards the normal when it slows down away from the normal when it speeds up. It may help to review Chapter 21, especially Problem 87, part (C), which involves a prism.
2. Relate the incident angle (θ_i) to the refracted angle (θ_t) via **Snell's law** (Chapter 22).
$$n_i \sin\theta_i = n_t \sin\theta_t$$
If the outside medium is **air**, $n_i = 1$.
3. It may help to recall a few principles from geometry:
    - Two angles that form a right angle are **complementary**: $\theta_1 + \theta_2 = 90°$.

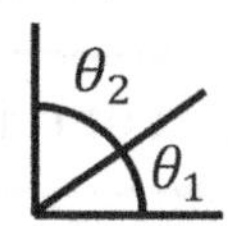

    - Two angles that form a 180° angle are **supplementary**: $\theta_1 + \theta_2 = 180°$.

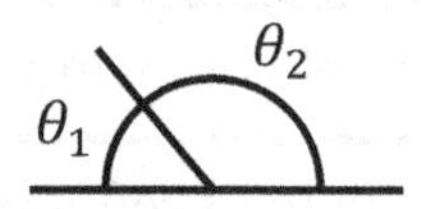

    - The three angles of a **triangle** add up to 180°: $\theta_1 + \theta_2 + \theta_3 = 180°$.
    - The four angles of a **quadrilateral** add up to 360°: $\theta_1 + \theta_2 + \theta_3 + \theta_4 = 360°$.

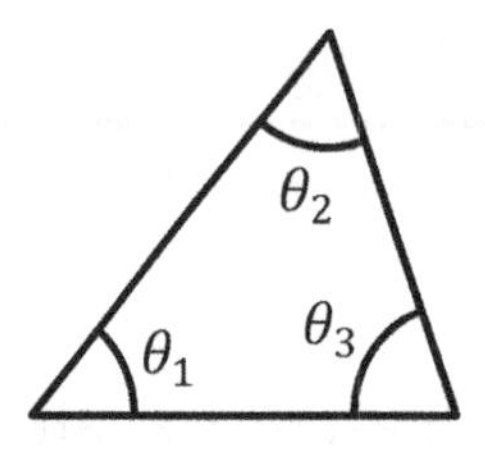

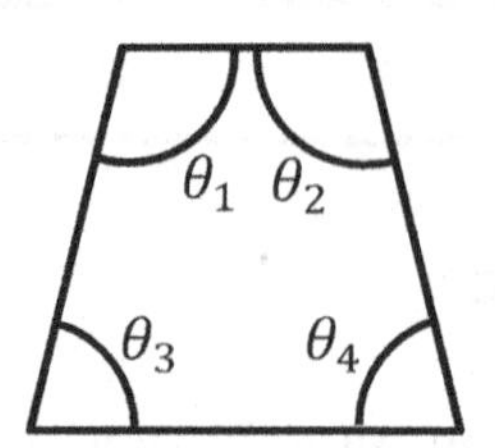

    - **Alternate interior angles** (defined for **parallel** lines as shown below) are equal: $\theta_1 = \theta_2$. **Vertical angles** (right diagram) are also equal: $\theta_3 = \theta_4$.

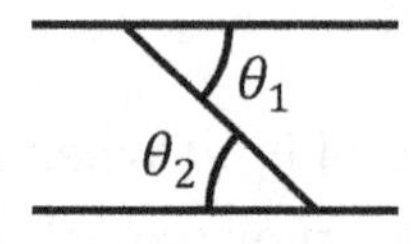

4. Apply algebra to solve for the desired unknown.

Example: Consider the prism illustrated below. Although the triangle itself is neither equilateral nor isosceles,[1] the ray of light passes symmetrically through the prism, meaning that $\theta_{1i} = \theta_{2t}$ and $\theta_{1t} = \theta_{2i}$.

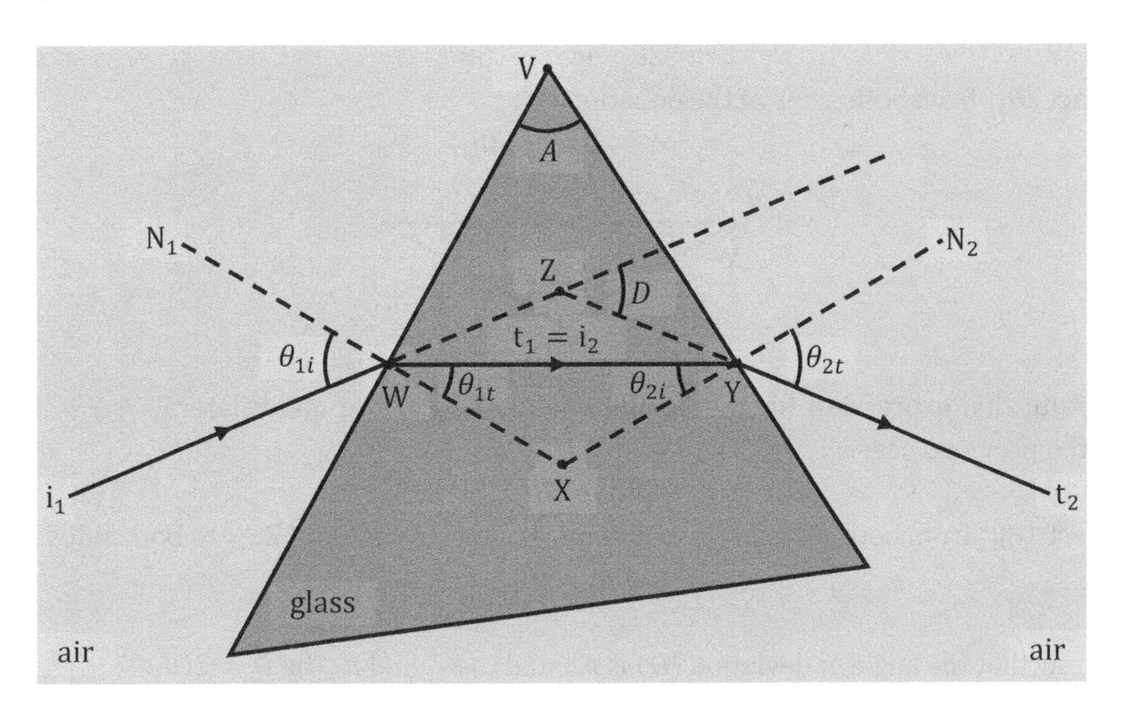

(A) Show that the **prism angle** (A) is related to θ_{1t} by $A = 2\theta_{1t}$.
Consider quadrilateral VWXYV and angle α shown below. Two of the angles are 90° because each **normal** is **perpendicular** to the surface. The four angles of any quadrilateral add up to 360°.

$$A + 90° + 90° + \alpha = 360°$$

Subtract 180° from both sides of the equation.

$$A + \alpha = 180°$$

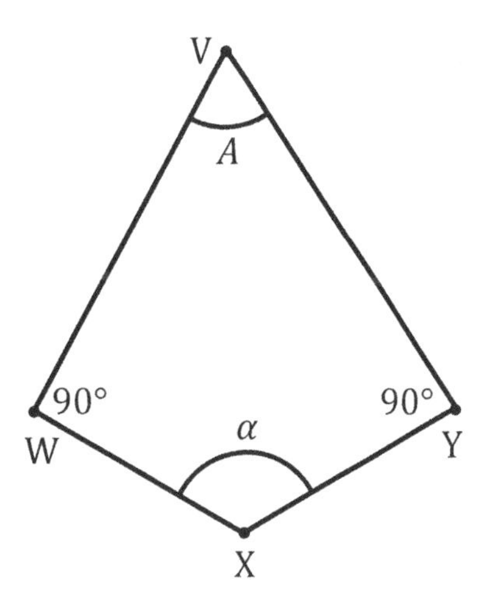

[1] The technical word for this is "**scalene.**" A scalene triangle has **unequal** sides and angles.

Now consider triangle WXYW shown below. The angles of any triangle add up to 180°.

$$\theta_{1t} + \theta_{2i} + \alpha = 180°$$

The problem states that the ray of light passes symmetrically through the prism such that $\theta_{1t} = \theta_{2i}$. Therefore, $\theta_{1t} + \theta_{2i} = \theta_{1t} + \theta_{1t} = 2\theta_{1t}$.

$$2\theta_{1t} + \alpha = 180°$$

Subtract $2\theta_{1t}$ from both sides of the equation.

$$\alpha = 180° - 2\theta_{1t}$$

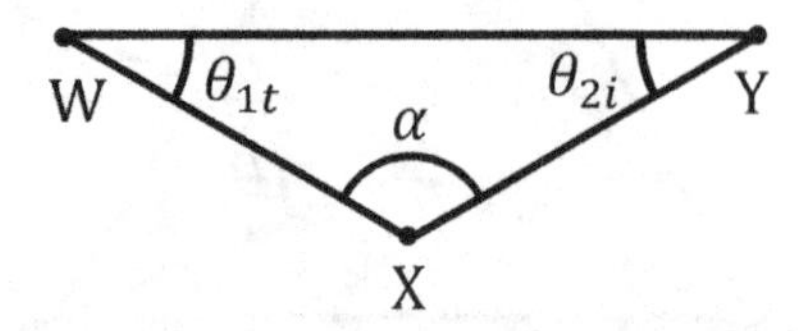

Substitute this expression for α into the equation from the quadrilateral, $A + \alpha = 180°$, from the previous page.

$$A + 180° - 2\theta_{1t} = 180°$$

Subtract 180° from both sides of the equation. It cancels out. Add $2\theta_{1t}$ to both sides.

$$A = 2\theta_{1t}$$

(B) Show that the **angle of deviation** (D) is related to θ_{1i} and θ_{1t} by $D = 2(\theta_{1i} - \theta_{1t})$.

Consider triangle WYZW shown below. The angles of any triangle add up to 180°.

$$2\beta + \gamma = 180°$$

Note that angles D and γ are **supplementary**.

$$D + \gamma = 180°$$

Comparing the previous two equations, it should be clear that:

$$D = 2\beta$$

Angles θ_{1i} and $(\beta + \theta_{1t})$ are **vertical angles**. Recall from geometry that vertical angles are equal. See page 270.

$$\theta_{1i} = \beta + \theta_{1t}$$

Subtract θ_{1t} from both sides of the equation.

$$\beta = \theta_{1i} - \theta_{1t}$$

Plug this expression for β into the previous equation for D.

$$D = 2(\theta_{1i} - \theta_{1t})$$

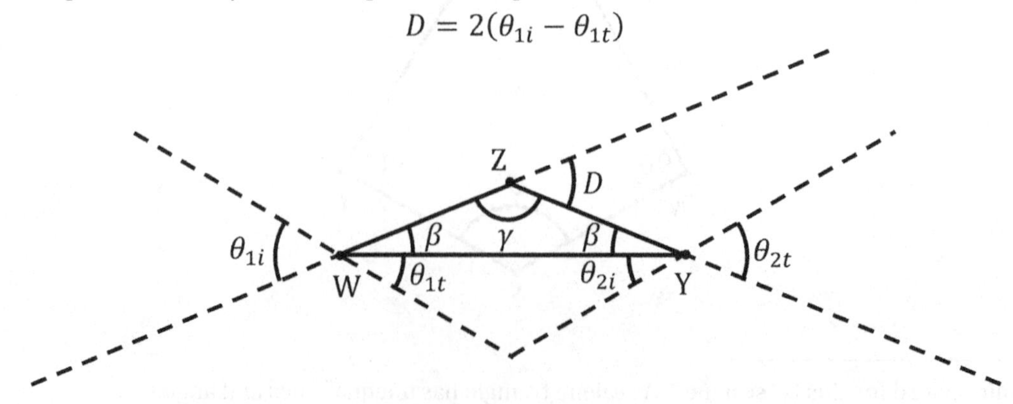

(C) Derive an equation for the index of refraction of the glass in terms of A and D only.

Apply Snell's law (Chapter 22).

$$n_a \sin\theta_{1i} = n_g \sin\theta_{1t}$$

The index of refraction of air is approximately equal to one: $n_a \approx 1$.

$$\sin\theta_{1i} = n_g \sin\theta_{1t}$$

Divide both sides of the equation by $\sin\theta_{1t}$.

$$n_g = \frac{\sin\theta_{1i}}{\sin\theta_{1t}}$$

In part (A), we found that $A = 2\theta_{1t}$. Solve for θ_{1t} to get:

$$\theta_{1t} = \frac{A}{2}$$

In part (B), we found that:

$$D = 2(\theta_{1i} - \theta_{1t})$$

Divide both sides of this equation by 2.

$$\frac{D}{2} = \theta_{1i} - \theta_{1t}$$

Plug $\theta_{1t} = \frac{A}{2}$ into this equation.

$$\frac{D}{2} = \theta_{1i} - \frac{A}{2}$$

Add $\frac{A}{2}$ to both sides of the equation.

$$\theta_{1i} = \frac{D}{2} + \frac{A}{2} = \frac{D+A}{2}$$

Plug $\theta_{1i} = \frac{D+A}{2}$ and $\theta_{1t} = \frac{A}{2}$ into the previous equation for n_g.

$$n_g = \frac{\sin\left(\frac{D+A}{2}\right)}{\sin\left(\frac{A}{2}\right)}$$

92. In the diagram below (which is **not** drawn to scale), the prism angle is $A = 75°$.

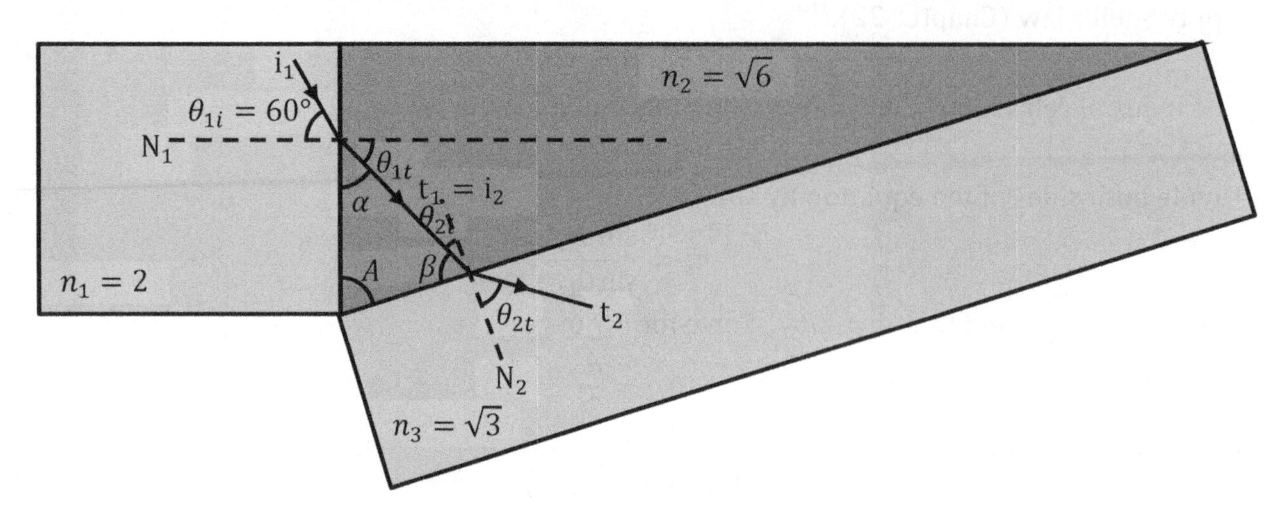

(A) Determine the first angle of refraction, θ_{1t}.

(B) Determine the **second** angle of refraction, θ_{2t}.

(C) What is the angle between the first and second refracted rays?

Want help? Check the hints section at the back of the book.

Answer: 45°, 45°, 15°

25 THIN LENSES

Relevant Terminology

Convex – a lens that bulges outward in the middle. A convex lens is a **converging** lens.
Concave – a lens that curves inward in the middle. A concave lens is a **diverging** lens.
Focus – a point where parallel rays of light would converge after passing through a lens or reflecting from a mirror. A lens has two foci – one on each side.
Optic axis – an imaginary line passing through the center of a lens and through its foci such that a ray of light traveling along this line would experience no refraction.
Principal axis – the same as the **optic axis**.
Object – a source of light rays, either by emitting or by reflecting light. In a lens system, this emitted or reflected light passes through the lens to form an image.
Image – the visual reproduction of an object after the object's light is refracted through a lens (or reflected by a mirror). In lab, a real image is viewed on a **screen**.
Focal length – the distance from the **focus** to the **center** of the lens.
Object distance – the distance from the **object** to the **center** of the lens.
Image distance – the distance from the **image** to the **center** of the lens.
Real image – formed when the output light rays of a lens system actually **converge** at the location of the image. A real image can be viewed on a screen. A real image forms on the opposite side of a lens compared to the object (for a single lens acting by itself).
Virtual image – formed when the output light rays of a lens system appear (after extrapolating backwards) to **diverge** from the location of the image. A virtual image **cannot** be viewed on a screen like a real image can. A virtual image forms on the same side of a lens as the object (for a single lens acting by itself).
Virtual object – when the light rays of a lens system appear (after extrapolating backwards) to originate from the location of the second (or third, fourth, etc.) object in a system with two or more lenses, but where the rays of light do not actually pass through that location.
Upright – an image which extends in the same direction as the object from the optic axis.
Inverted – an image which appears **upside down** compared to the object. Note that left/right are actually interchanged as well as up/down for an inverted image.
Character – whether the image formed is **real** or **virtual**.
Orientation – whether the image formed is **upright** or **inverted**.
Magnification – the ratio of the image height to the object height.
Microscope – an instrument that uses lenses to make small objects appear **larger**.
Telescope – an instrument that uses lenses/mirrors to make distant objects appear **closer**.
Eyepiece – the lens of a microscope or telescope which is nearest to the viewer's **eye**.
Objective – the lens of a microscope or telescope which is nearest to the **object**.

Essential Concepts

A **convex** (or **converging**) lens causes parallel rays of light to **converge** at a **focus** (F) whereas a **concave** (or **diverging**) lens causes parallel rays of light to **diverge** from a **focus.**

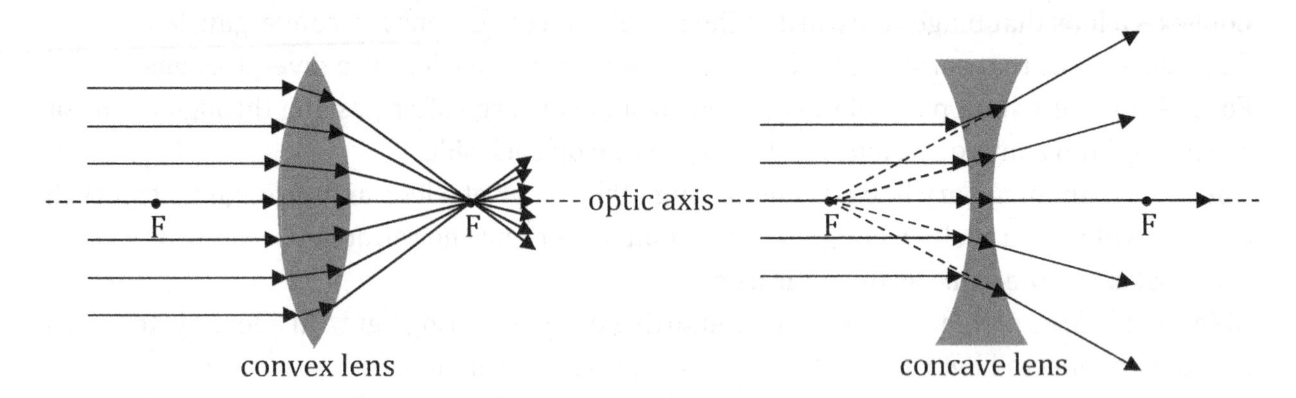

When an **object** (O) is placed before a lens, the output rays of light emanating from the object either converge at a real **image** or appear to diverge from a virtual image. (We will explore this in more detail in the section entitled Ray Diagrams.)

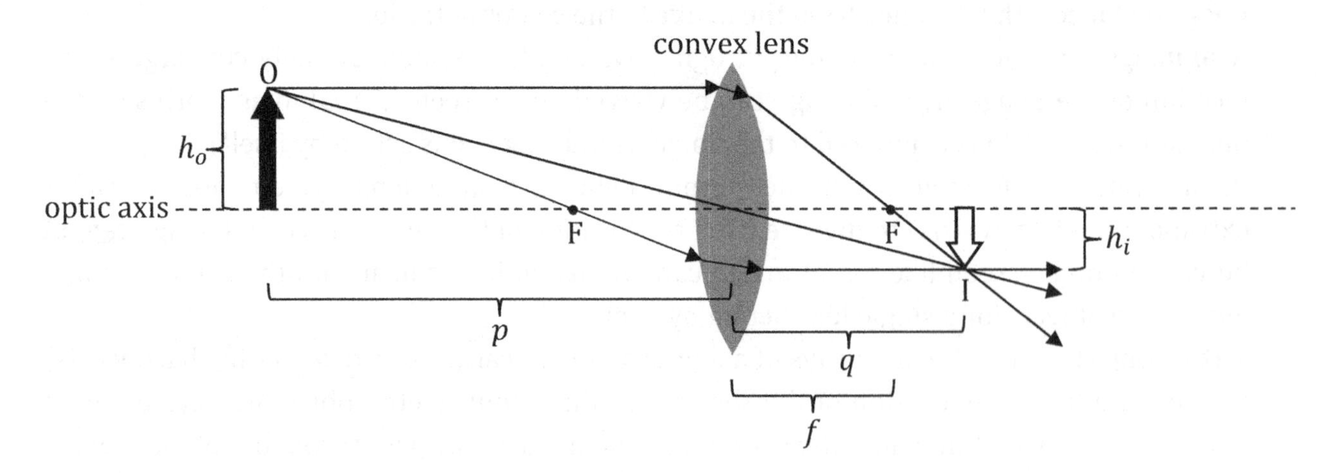

Remember how the following distances are defined:

- The **object distance** (p) extends from the **object** (O) to the **center** of the lens.
- The **image distance** (q) extends from the **image** (I) to the **center** of the lens.
- The **focal length** (f) extends from the **focus** (F) to the **center** of the lens.
- The **object height** (h_o) extends from the **optic axis** to the top of the **object** (O).
- The **image height** (h_i) extends from the **optic axis** to the top of the **image** (I).

It's important to realize that these distances may be **negative.** Study the sign conventions on the following page.

Thin Lens Equations

The **object distance** (p), **image distance** (q), and **focal length** (f) are related to one another through the following equation.[1]

$$\frac{1}{p} + \frac{1}{q} = \frac{1}{f}$$

The **magnification** (M) equals the ratio of the **image height** (h_i) to the **object height** (h_o), which turns out to be the same as $-\frac{q}{p}$.

$$M = \frac{h_i}{h_o} = -\frac{q}{p}$$

Sign Conventions

Symbol	Name	Sign Convention
f	focal length	positive for a **convex** lens negative for a **concave** lens
p	object distance	positive for the first object (in a multi-lens system, $p_2 < 0$ if $q_1 > L$)
q	image distance	positive if the image and object are on **opposite** sides of the lens (the image is **real**) negative if the image and object are on the **same** side of the lens (the image is **virtual**)
h_o	object height	positive for the first object (in a multi-lens system, h_{2o} has the same sign as h_{1i})
h_i	image height	positive if the image and object extend in the **same direction** from the optic axis negative if the image appears upside down compared to the object
M	magnification	positive if image is **upright**, negative if image is **inverted**

[1] The notation for thin lenses varies wildly among textbooks and instructors. Some use d_o and d_i or s_o and s_i for the object and image distances, though that results in extra subscripts when solving a problem with multiple lenses. Another practice is to use o and i for the object and image distances, which is simple and easy to remember, but the o may look like a zero and the i may look like a 1 (and in advanced optics, in the context of complex numbers, it would get even more confusing with imaginary numbers). The point is that if you're taking a class or reading another book, there is a good chance that the symbols are different from the p and q that we're using in this book, in which case you'll need to know which symbols correspond to p and q.

Multiple Lenses

When there are two or more lenses, let the image of the first lens serve as the object of the second lens, and apply the thin lens equations to each lens separately.

$$\frac{1}{p_1} + \frac{1}{q_1} = \frac{1}{f_1} \quad , \quad \frac{1}{p_2} + \frac{1}{q_2} = \frac{1}{f_2}$$

$$h_{2o} = h_{1i} \quad , \quad p_2 = L - q_1$$

$$M_1 = \frac{h_{1i}}{h_{1o}} = -\frac{q_1}{p_1} \quad , \quad M_2 = \frac{h_{2i}}{h_{2o}} = -\frac{q_2}{p_2}$$

$$M = M_1 M_2 = \frac{h_{2i}}{h_{1o}}$$

- The height of the object for the second lens equals the height of the image for the first lens: $h_{2o} = h_{1i}$.
- The object distance for the second lens is related to the image distance for the first lens by the distance between the lenses: $p_2 = L - q_1$.
- The overall magnification equals the ratio of the final image height to the original object height, and also equals the product of the magnifications for the individual lenses: $M = M_1 M_2 = \frac{h_{2i}}{h_{1o}}$.

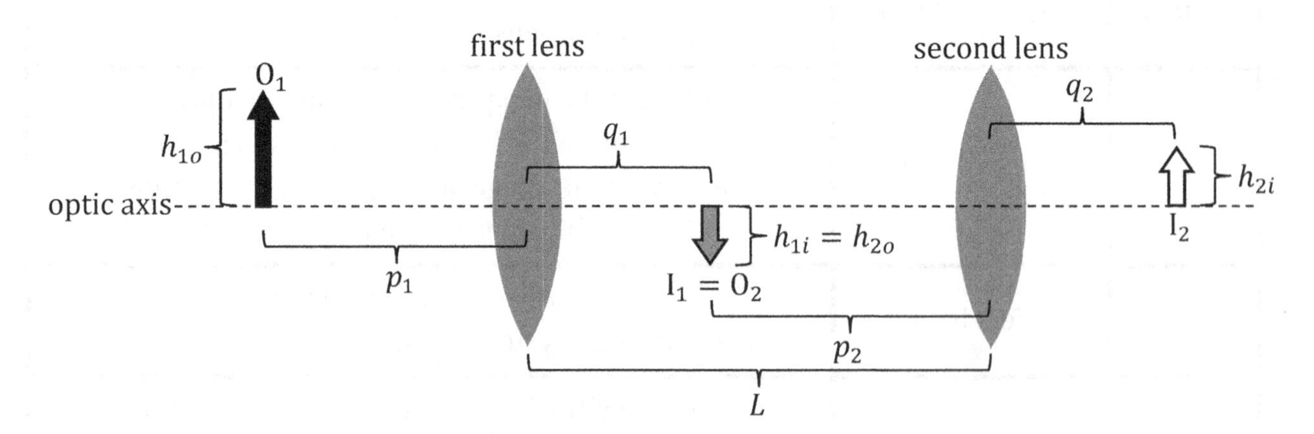

Lens Makers' Equation

For a thin lens with spherical surfaces on both sides, the **focal length** (f) is related to the **index of refraction** (n) of the lens and the **radii** (R_L and R_R) of the two spherical surfaces by the lens makers' equation (assuming that the lens is in air or vacuum). When the original object is drawn to the left of a lens, R is **positive** if the center of curvature lies to the **right** of the lens and **negative** if the center of curvature lies to the **left** of the lens. For example, for a double convex lens like the one above, $R_L > 0$, $R_R < 0$, and $\frac{1}{R_L} - \frac{1}{R_R} > 0$.

$$\frac{1}{f} = (n-1)\left(\frac{1}{R_L} - \frac{1}{R_R}\right)$$

Ray Diagrams

The image formed by a lens can be located by drawing a ray diagram as follows:

1. Draw one ray from the top of the object, **parallel** to the optic axis to the lens, and through the appropriate **focus**.
    - For a **convex** lens, the output ray converges through the focus on the **opposite** side of the lens compared to the object.
    - For a **concave** lens, the output ray diverges from the focus on the **same** side of the lens as the object. Use a dashed line to connect the focus to the point where the ray meets the lens, and use a solid line to make the actual ray of light diverge from the focus.
2. Draw another ray from the top of the object straight through the **center** of the lens, virtually undeflected (this ray experiences negligible refraction).
3. Draw the last ray from the top of the object towards the **unused focus** (that is, whichever focus was **not** used by ray 1), and then **parallel** to the optic axis. How you interpret this rule depends on the nature of the lens.
    - For a **convex** lens, the output ray passes through the near focus (on the same side of the lens as the object), and then runs parallel to the optic axis when it reaches the lens.
    - For a **concave** lens, the output ray heads towards the far focus (on the opposite side of the lens as the object), but doesn't actually get there. When this ray reaches the lens, it then runs parallel to the optic axis. Draw a dashed line for the extension of the ray from the lens to the focus that it was heading towards, and use a solid line for the rest.

The **image** forms where the three output rays intersect. Be sure to look at the **output** rays (the rays leaving the lens), and **not** the input rays (the rays entering the lens). The input rays intersect at the object, and you're **not** looking for the object. You're looking for the image, so you must look at the output rays. **Tip**: If the output rays are **diverging**, you must **extrapolate backwards** to the **other side** of the lens (as shown in the examples).

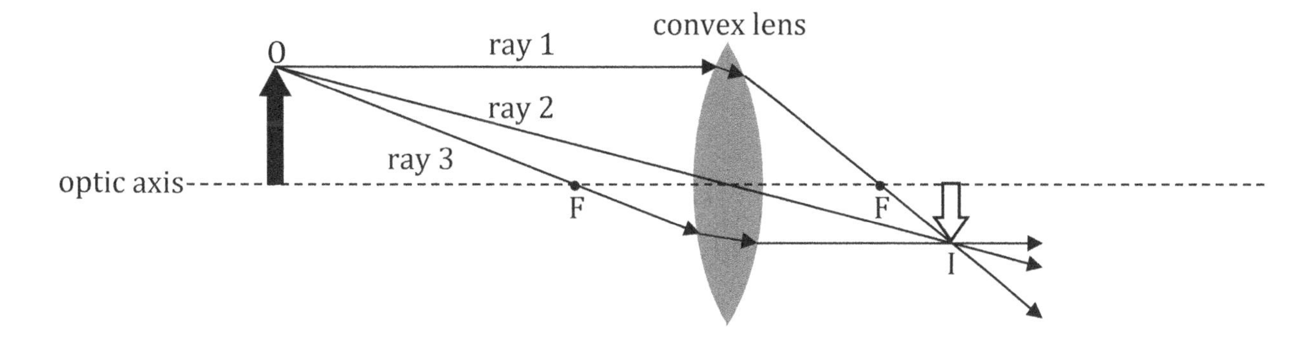

Apparent Depth

When an object is submerged in a liquid, such as a penny at the bottom of a cup of water, the object appears to be closer to the surface than it really is due to refraction. The **apparent depth** (the image distance, q) is related to the **actual depth** (the object distance, p) by the following formula, where n_1 is the index of refraction of the liquid and n_2 is the index of refraction of the substance above the liquid (which is usually air).

$$q = -\frac{n_2}{n_1}p$$

Magnifiers, Microscopes, and Telescopes

A **simple magnifier** consists of a single **convex** (or converging) lens. The magnification is greatest when the lens is placed at the **near point** of the eye, which is approximately 25 cm for humans with normal eyesight. (When an object is closer to the eye than the near point, the eye is unable to focus on it.) The eye is most relaxed when the image appears far away, which corresponds to placing the object at the focus of the lens.[2] If you place a simple magnifier at the optimal position (maximizing the magnification of the lens and the comfort of your eye), the object will be at the focus and the distance between the lens and your eye will be about 25 cm. In this case, the **angular magnification** (M_a) is given by the following formula, where the 0.25 m is a distance in meters.

$$M_a = \frac{\theta}{\theta_0} = \frac{0.25 \text{ m}}{f}$$

A **compound microscope** consists of two convex lenses. The **objective** lens has a short focal length (f_o) – less than a centimeter – while the **eyepiece** has a greater focal length (f_e). The **overall magnification** of a compound microscope is given by the following formula, where L is the distance between the lenses.

$$M = -\frac{L}{f_o}\left(\frac{0.25 \text{ m}}{f_e}\right)$$

A **refracting** telescope consists of two lenses, where the **objective** lens has a greater focal length than the **eyepiece**. The distance between the lenses equals the sum of the focal lengths: $f_o + f_e = L$. The **angular magnification** (M_a) of a telescope is negative (meaning that the image appears inverted) if both lenses are convex:[3]

$$M_a = \frac{\theta}{\theta_0} = -\frac{f_o}{f_e}$$

[2] As q approaches ∞, $\frac{1}{q}$ approaches zero and p approaches f according to $\frac{1}{p} + \frac{1}{q} = \frac{1}{f}$.

[3] A Galilean telescope uses a **concave** eyepiece, such that $f_e < 0$ and $M_a > 0$, yielding an **upright** image.

Symbols and Units

Symbol	Name	Units
F	focus	
O	object	
I	image	
p	object distance (object to lens center)	m
q	image distance (image to lens center)	m
f	focal length (focus to lens center)	m
h_o	object height	m
h_i	image height	m
M	magnification	unitless
M_a	angular magnification	unitless
L	distance between the centers of two lenses	m
R	radius of curvature	m
n	index of refraction	unitless
θ	angular measure when viewed through an instrument	° or rad
θ_0	angular measure (from near point) when viewed with no lens	° or rad

Important Distinctions

It's important to distinguish between the **object** and the **image** that it forms. Study the distinction in the figure on page 276, where p is the object distance and q is the image distance. Also note the distinction between a **virtual image** and a **virtual object**. The more common term is "virtual **image**," as it can be formed by a single lens (and is always formed by a single concave lens). The term "virtual **object**" comes up rarely: First, you need a multi-lens system, and secondly $L - q_1$ must turn out to be negative for the term to even apply. For any single lens system (even concave), the **object** will be **real**, not virtual (whereas the **image** may be virtual – the sign of q will determine this).

Thin Lens Strategy

To solve a problem involving one or more thin lenses, follow these steps:

1. To draw a **ray diagram** for a single lens, draw the three rays described on page 279, as illustrated in the examples.
2. Make a list of the known quantities and identify the desired unknown. For a lens:
   - The **object distance** (p) extends from the **object** to the **center** of the lens.
   - The **image distance** (q) extends from the **image** to the **center** of the lens.
   - The **focal length** (f) extends from the **focus** to the **center** of the lens.
   - The **object height** (h_o) extends from the **optic axis** to the top of the **object**.
   - The **image height** (h_i) extends from the **optic axis** to the top of the **image**.
3. Study the sign conventions on page 277. Note that problems give you numerical values without any signs, while you must put the signs in yourself. For example, you must use a **negative sign** for the **focal length** (f) of a **concave** (or diverging) lens. An incorrect **sign** often results in a drastically incorrect answer.
4. For a single lens, use the following equations.
$$\frac{1}{p}+\frac{1}{q}=\frac{1}{f} \quad , \quad M=\frac{h_i}{h_o}=-\frac{q}{p} \quad , \quad \frac{1}{f}=(n-1)\left(\frac{1}{R_L}-\frac{1}{R_R}\right)$$
Tip: A common way to solve for the image height is to use $\frac{h_i}{h_o}=-\frac{q}{p}$.
5. For a system of lenses, use the following equations. Basically, you treat the image of the first lens as the object for the second lens. See the diagram on page 278.
$$\frac{1}{p_1}+\frac{1}{q_1}=\frac{1}{f_1} \quad , \quad \frac{1}{p_2}+\frac{1}{q_2}=\frac{1}{f_2}$$
$$h_{2o}=h_{1i} \quad , \quad p_2=L-q_1$$
$$M_1=\frac{h_{1i}}{h_{1o}}=-\frac{q_1}{p_1} \quad , \quad M_2=\frac{h_{2i}}{h_{2o}}=-\frac{q_2}{p_2}$$
$$M=M_1M_2=\frac{h_{2i}}{h_{1o}}$$
6. For an object submerged in a liquid, the **apparent depth** (q) is related to the **actual depth** (p) by (where n_1 is for the liquid and n_2 is for the substance above the liquid):
$$q=-\frac{n_2}{n_1}p$$
7. To find the **angular magnification** (M) or **overall magnification** (M_a) of one of the following optical instruments, choose the appropriate formula.
   - For a simple magnifier: $M_a=\frac{\theta}{\theta_0}=\frac{0.25\text{ m}}{f}$.
   - For a compound microscope: $M=-\frac{L}{f_o}\left(\frac{0.25\text{ m}}{f_e}\right)$.
   - For a refracting telescope: $M_a=\frac{\theta}{\theta_0}=-\frac{f_o}{f_e}$, $f_o+f_e=L$.
8. If a problem involves a **spherical mirror**, see Chapter 26.

Example: A 5.0-mm tall monkey figurine is placed 40 cm before a concave lens which has foci that are 60 cm from the center of the lens.
(A) Where does the image form?
Identify the given information. **Note**: It's okay to leave the distances in the units given (as long as p, q, and f have the same units, and as long as h_o and h_i have the same units).

- The object distance is $p = 40$ cm. The first object distance is always positive.
- The focal length is $f = -60$ cm. Focal length is **negative** for a **concave** lens.
- The object height is $h_o = 5.0$ mm. The first object height is always positive. **Note**: It's okay for h_o to have different units (as long as it the same units as h_i).

Solve for the **image distance** (q).

$$\frac{1}{p} + \frac{1}{q} = \frac{1}{f}$$

$$\frac{1}{40} + \frac{1}{q} = \frac{1}{-60}$$

Subtract $\frac{1}{40}$ from both sides of the equation.

$$\frac{1}{q} = -\frac{1}{60} - \frac{1}{40}$$

Multiply $\frac{1}{60}$ by $\frac{2}{2}$ and multiply $\frac{1}{40}$ by $\frac{3}{3}$ in order to make a **common denominator**.

$$\frac{1}{q} = -\frac{2}{120} - \frac{3}{120}$$

$$\frac{1}{q} = -\frac{5}{120}$$

Cross multiply.

$$-5q = 120$$

Divide both sides of the equation by -5.

$$q = -\frac{120}{5} = -24 \text{ cm}$$

The image forms at $q = -24$ cm, which means that the image is 24 cm from the lens, where the **minus sign** indicates that the image forms on the **same side** of the lens as the object.

(B) What is the image height?
Combine the two magnification equations together. That is:

$$\frac{h_i}{h_o} = -\frac{q}{p}$$

Multiply both sides of the equation by h_o. Be careful with the minus signs.

$$h_i = -\frac{q}{p} h_o = -\frac{(-24)}{40}(5) = 3.0 \text{ mm}$$

The image height is $h_i = 3.0$ mm. Note that the minus signs and the centimeters (cm) **canceled** out. The image height is in millimeters (mm) because h_o was put in millimeters.

(C) What is the magnification of the image?
Use one of the formulas for magnification.

$$M = \frac{h_i}{h_o} = \frac{3}{5} \times = 0.6 \times$$

The magnification is $M = \frac{3}{5} \times = 0.6 \times$. The image appears 60% as tall as the object.

(D) What is the orientation of the image?
The word "**orientation**" means that the question is asking if the image is upright or inverted. Review the sign conventions on page 277. Examine the **sign** of the **magnification**: Since $M > 0$, the image is **upright**.

(E) What is the character of the image?
The word "**character**" means that the question is asking if the image is real or virtual. Review the sign conventions on page 277. Examine the **sign** of the **image distance**: Since $q < 0$, the image is **virtual**.

Example: Prove that a concave lens acting by itself can't form a real image.
Review the sign conventions on page 277. An image is **real** if q is **positive**. Solve for $\frac{1}{q}$.

$$\frac{1}{q} = \frac{1}{f} - \frac{1}{p}$$

The focal length (f) is **negative** for a **concave** lens, such that $\frac{1}{f}$ must be negative. The first object distance (p) is always **positive**, such that $-\frac{1}{p}$ must be negative. Since $\frac{1}{f}$ is negative and $-\frac{1}{p}$ is also negative, $\frac{1}{f} - \frac{1}{p}$ must be negative and $\frac{1}{q}$ must be negative. Therefore, a **concave** lens acting by itself can only produce a **virtual** image ($q < 0$).

Example: Prove that a concave lens acting by itself can't form an inverted image.
Review the sign conventions on page 277. An image is **inverted** if h_i is **negative**. Solve for h_i in the magnification equations.

$$h_i = -\frac{q}{p} h_o$$

The first object distance (p) is always **positive**, the first object height (h_o) is always **positive**, and we just showed in the previous example that the image distance (q) must be **negative** for a concave lens acting by itself. Since $p > 0$, $h_o > 0$, and $q < 0$, the combination $-\frac{q}{p} h_o$ must be positive (the two minus signs make a plus sign). Therefore, a **concave** lens acting by itself can only produce an **upright** image ($h_i > 0$).

Example: A 20-mm tall banana figurine is placed 60 cm before a lens system. The first lens is convex and has foci that are 40 cm from its center. The second lens is also convex, but has foci that are 30 cm from its center. The centers of the lenses are 200 cm apart.
(A) Where does the final image form?
Identify the given information. **Note**: It's okay to leave the distances in the units given (as long as the p's, the q's, the f's, and L have the same units, and as long as the h's have the same units).

- The first object distance is $p_1 = 60$ cm. The first object distance is always positive.
- The focal lengths are $f_1 = 40$ cm and $f_2 = 30$ cm. Focal length is **positive** for a **convex** lens. (Note: For a concave lens, focal length would be negative.)
- The distance between the centers of the lenses is $L = 200$ cm.
- The first object height is $h_{1o} = 20$ mm. The first object height is always positive. **Note**: It's okay for h_{1o} to have different units (as long as all the h's are consistent).

Solve for the first **image distance** (q_1).

$$\frac{1}{p_1} + \frac{1}{q_1} = \frac{1}{f_1}$$

$$\frac{1}{60} + \frac{1}{q_1} = \frac{1}{40}$$

Subtract $\frac{1}{60}$ from both sides of the equation.

$$\frac{1}{q_1} = \frac{1}{40} - \frac{1}{60}$$

Multiply $\frac{1}{40}$ by $\frac{3}{3}$ and multiply $\frac{1}{60}$ by $\frac{2}{2}$ in order to make a **common denominator**.

$$\frac{1}{q_1} = \frac{3}{120} - \frac{2}{120}$$

$$\frac{1}{q_1} = \frac{1}{120}$$

Take the **reciprocal** of both sides of the equation.[4]

$$q_1 = 120 \text{ cm}$$

This is <u>not</u> the <u>final</u> image. Treat the image of the first lens as the object for the second lens. Since the first image is 120 cm past (since q_1 is positive) the first lens, and since the lenses are 200 cm apart, the object distance for the second lens is 80 cm. (See the diagram on page 278.) If you struggle to reason this out, just use the following handy equation.

$$p_2 = L - q_1 = 200 - 120 = 80 \text{ cm}$$

[4] Be careful <u>not</u> to do this until there is only **one** term on both sides of the equation. It wouldn't help to do this back when the equation was $\frac{1}{q_1} = \frac{1}{40} - \frac{1}{60}$ because you would get the nested expression $q_1 = \frac{1}{\frac{1}{40} - \frac{1}{60}}$. (You most certainly do <u>not</u> get $40 - 60$, which you can easily verify gives the **wrong** answer for q_1. Recall from algebra that $\frac{1}{\frac{1}{a}+\frac{1}{b}}$ does <u>not</u> simplify to $a + b$. You need to find a common denominator before you find the reciprocal.)

Now that we know the second object distance, use $p_2 = 80$ cm and $f_2 = 30$ cm to find the final image distance.

$$\frac{1}{p_2} + \frac{1}{q_2} = \frac{1}{f_2}$$

$$\frac{1}{80} + \frac{1}{q_2} = \frac{1}{30}$$

Subtract $\frac{1}{80}$ from both sides of the equation.

$$\frac{1}{q_2} = \frac{1}{30} - \frac{1}{80}$$

Multiply $\frac{1}{30}$ by $\frac{8}{8}$ and multiply $\frac{1}{80}$ by $\frac{3}{3}$ in order to make a **common denominator**.

$$\frac{1}{q_2} = \frac{8}{240} - \frac{3}{240}$$

$$\frac{1}{q_2} = \frac{5}{240}$$

Cross multiply.

$$5q_2 = 240$$

Divide both sides of the equation by 5.

$$q_2 = \frac{240}{5} = 48 \text{ cm}$$

The **final** image forms at $q_2 = 48$ cm, which means that the image is 48 cm past (since q_2 is positive, it's on the opposite side of the lens compared to the object) the second lens.

(B) What is the overall magnification of the system?

First find the magnification of each lens individually, and then multiply them together.

$$M_1 = -\frac{q_1}{p_1} = -\frac{120}{60} = -2 \times$$

$$M_2 = -\frac{q_2}{p_2} = -\frac{48}{80} = -0.6 \times$$

$$M = M_1 M_2 = (-2)(-0.6) = 1.2 \times$$

The **overall** magnification is $M = 1.2 \times$. The two minus signs make a plus sign.

(C) What is the final image height?

Now use the other equation for the overall magnification.

$$M = \frac{h_{2i}}{h_{1o}}$$

Multiply both sides of the equation by h_{1o}. Recall that $h_{1o} = 20$ mm.

$$h_{2i} = Mh_{1o} = (1.2)(20) = 24 \text{ mm}$$

The final image height is $h_{2i} = 24$ mm.

(D) What is the orientation of the final image?
The word "**orientation**" means that the question is asking if the final image is upright or inverted. Review the sign conventions on page 277. Examine the **sign** of the overall **magnification**: Since $M > 0$, the final image is **upright**.

(E) What is the character of the final image?
The word "**character**" means that the question is asking if the final image is real or virtual. Review the sign conventions on page 277. Examine the **sign** of the second **image distance**: Since $q_2 > 0$, the final image is **real**.

Example: Draw a ray diagram for a convex lens where the object distance is three times the focal length.

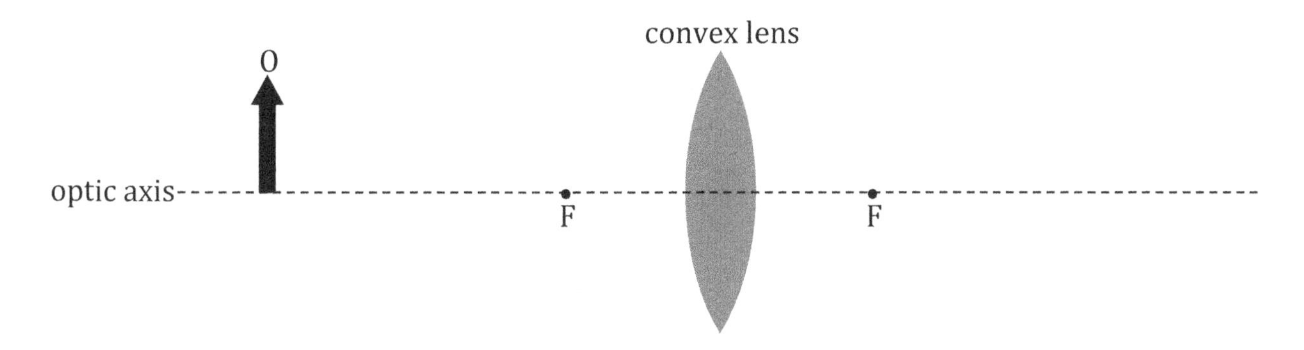

Begin by drawing a convex lens. (Page 276 shows how convex and concave lenses look.) Draw a line through the center of the lens that is perpendicular to the lens to serve as the **optic axis**. Draw and label **two** foci (F) the same distance from each side of the lens. Draw and label an object (O) such that the object distance (from the object to the center of the lens) is three times the focal length (from the focus to the center of the lens), as specified in the problem.

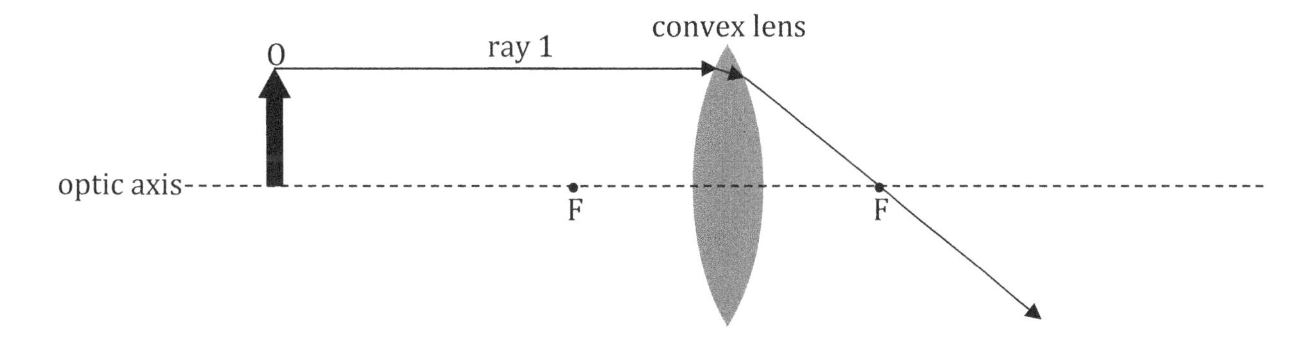

Draw the first ray from the top of the object **parallel** to the optic axis until it reaches the lens. Since the lens is **convex**, the refracted ray will converge to the **far focus**.

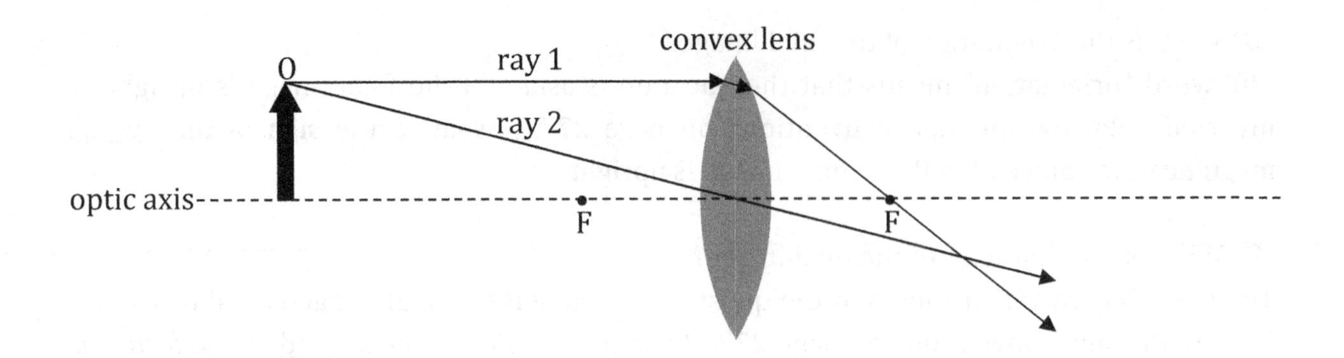

Draw the second ray straight through the center of the lens (it is virtually undeflected).

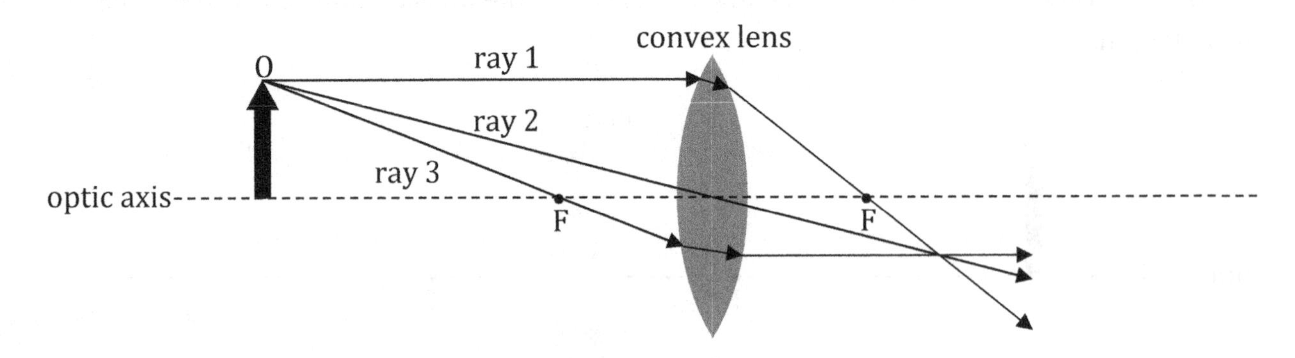

Draw the third ray from the top of the object through the **near focus** (since that focus hasn't yet been used) until it reaches the lens. Since the lens is convex, the refracted ray will then travel **parallel** to the optic axis.

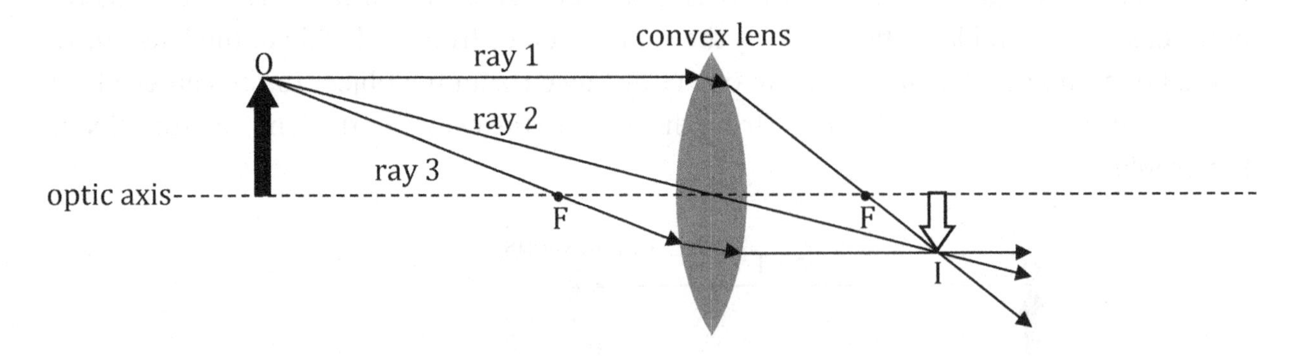

Draw and label the image where the three **output** rays intersect. (The output rays – the rays leaving the lens – intersect at the image. Be careful not to look at the input rays – the rays entering the lens – which intersect at the object.)

It would be wise to review the directions from page 279, and compare them closely with how we applied them to this example.

Example: Draw a ray diagram for a concave lens where the object distance is three times the focal length.

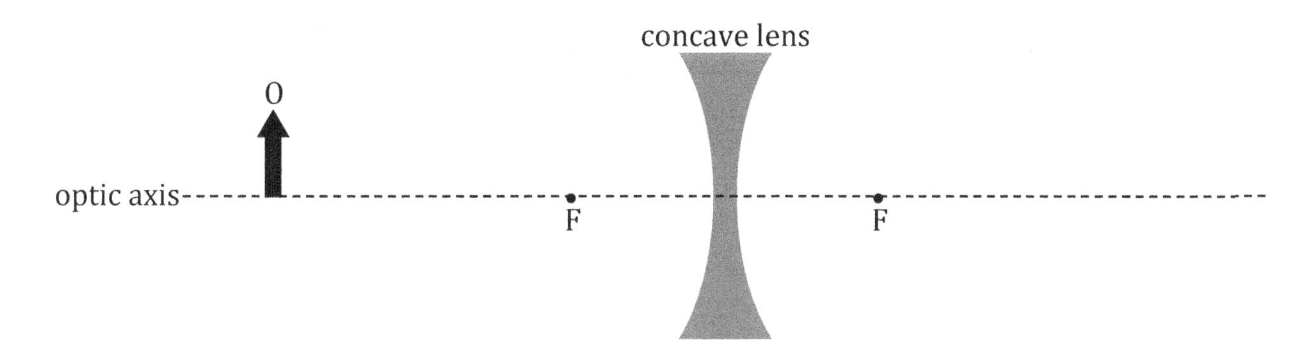

Begin by drawing a concave lens. (Page 276 shows how convex and concave lenses look.) Draw a line through the center of the lens that is perpendicular to the lens to serve as the **optic axis**. Draw and label **two** foci (F) the same distance from each side of the lens. Draw and label an object (O) such that the object distance (from the object to the center of the lens) is three times the focal length (from the focus to the center of the lens), as specified in the problem.

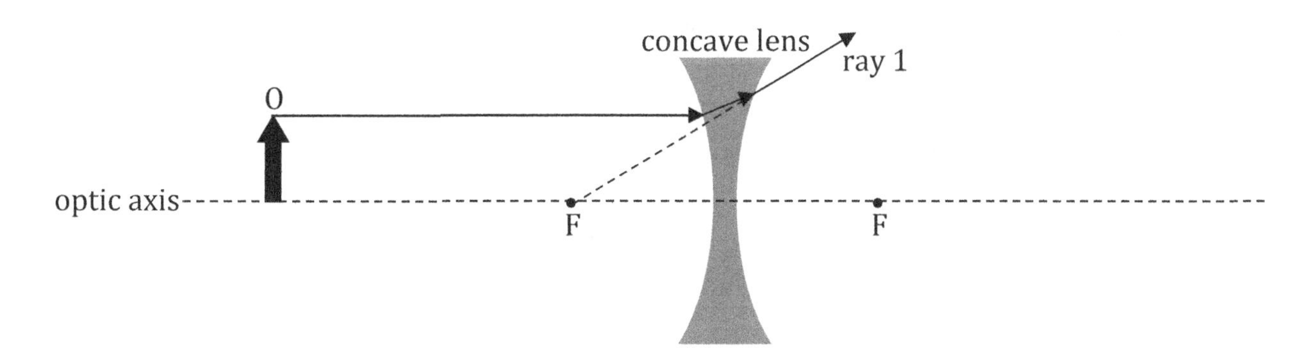

Draw the first ray from the top of the object **parallel** to the optic axis until it reaches the lens. Since the lens is **concave**, the refracted ray will diverge from the **near focus**. Use a dashed line to show that this ray diverges from the near focus.

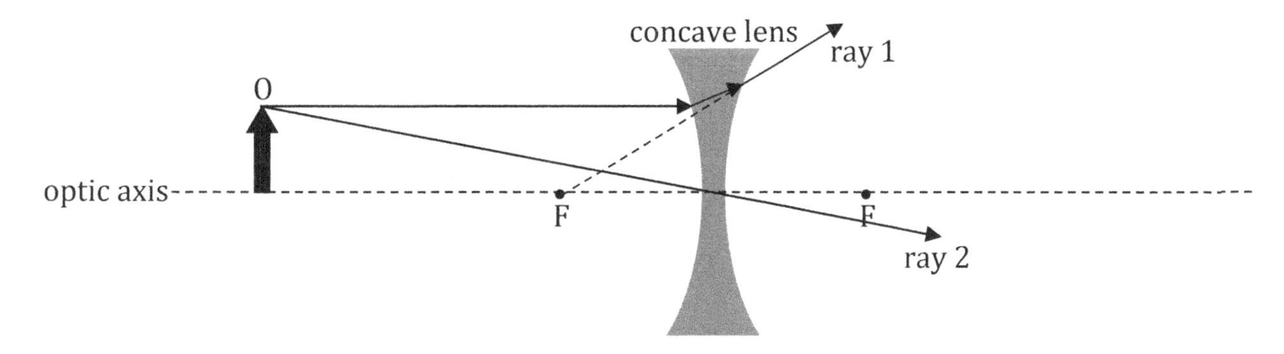

Draw the second ray straight through the center of the lens (it is virtually undeflected).

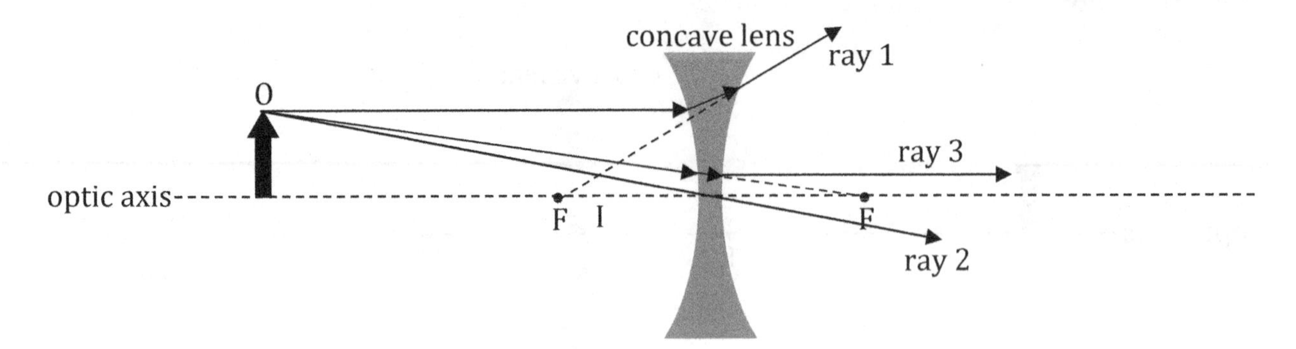

Draw the third ray from the top of the object towards the <u>**far**</u> **focus** (since that focus hasn't yet been used) until it reaches the lens. Since the lens is concave, the refracted ray will then travel **parallel** to the optic axis before it actually reaches the far focus.

Where is the image? The image forms where the three output rays intersect. Study the ray diagram above. Where do the three **output** (<u>**not**</u> input) rays intersect? (Look at the three rays that leave the lens.) They <u>**do**</u> intersect. Since they don't intersect to the right of the lens, you must **extrapolate backwards to the left of the lens**. We added a horizontal dashed line below, extrapolating ray 3 to the left. Rays 1, 2, and 3 intersect to the <u>**left**</u> of the lens. That's where the image is located. Look closely to see how all three output rays extrapolate back to the top of the image in the diagram below.

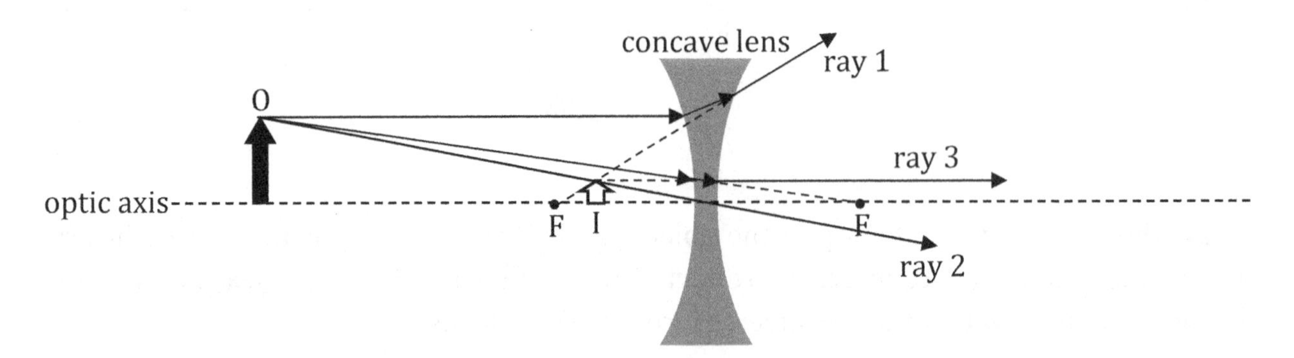

Note that this is a **virtual** image.

It would be wise to review the directions from page 279, and compare them closely with how we applied them to this example. It is also instructive to compare and contrast this example with the previous example.

93. An 8.0-mm tall monkey figurine is placed 200 cm before a convex lens which has foci that are 40 cm from the center of the lens.

(A) Where does the image form?

(B) What is the image height?

(C) What is the magnification of the image?

(D) What is the orientation of the image?

(E) What is the character of the image?

Want help? Check the hints section at the back of the book.

Answers: (A) 50 cm (B) -2.0 mm (C) $-\frac{1}{4}\times$ (D) inverted (E) real

94. A 24-mm tall monkey figurine is placed 192 cm before a concave lens which has foci that are 64 cm from the center of the lens.

(A) Where does the image form?

(B) What is the image height?

(C) What is the magnification of the image?

(D) What is the orientation of the image?

(E) What is the character of the image?

Want help? Check the hints section at the back of the book.

Answers: (A) −48 cm (B) 6.0 mm (C) $\frac{1}{4}\times$ (D) upright (E) virtual

95. A 16-mm tall banana figurine is placed 30 cm before a convex lens which has foci that are 90 cm from the center of the lens.

(A) Where does the image form?

(B) What is the image height?

(C) What is the magnification of the image?

(D) What is the orientation of the image?

(E) What is the character of the image?

Want help? Check the hints section at the back of the book.

Answers: (A) −45 cm (B) 24 mm (C) 1.5× (D) upright (E) virtual

96. A 36-mm tall banana figurine is placed 48 cm before a lens system. The first lens is concave and has foci that are 24 cm from its center. The second lens is convex and has foci that are 32 cm from its center. The centers of the lenses are 80 cm apart.

(A) Where does the final image form?

(B) What is the overall magnification of the system?

(C) What is the final image height?

(D) What is the orientation of the final image?

(E) What is the character of the final image?

Want help? Check the hints section at the back of the book.

Answers: (A) 48 cm (B) $-\frac{1}{6}\times$ (C) -6.0 mm (D) inverted (E) real

97. (A) Prove that a convex lens acting by itself can't form an image that is both real and upright.

(B) Prove that a convex lens acting by itself can't form an image that is both virtual and inverted.

Want help? Check the hints section at the back of the book.

98. Draw a ray diagram for a convex lens where the object distance is twice the focal length.

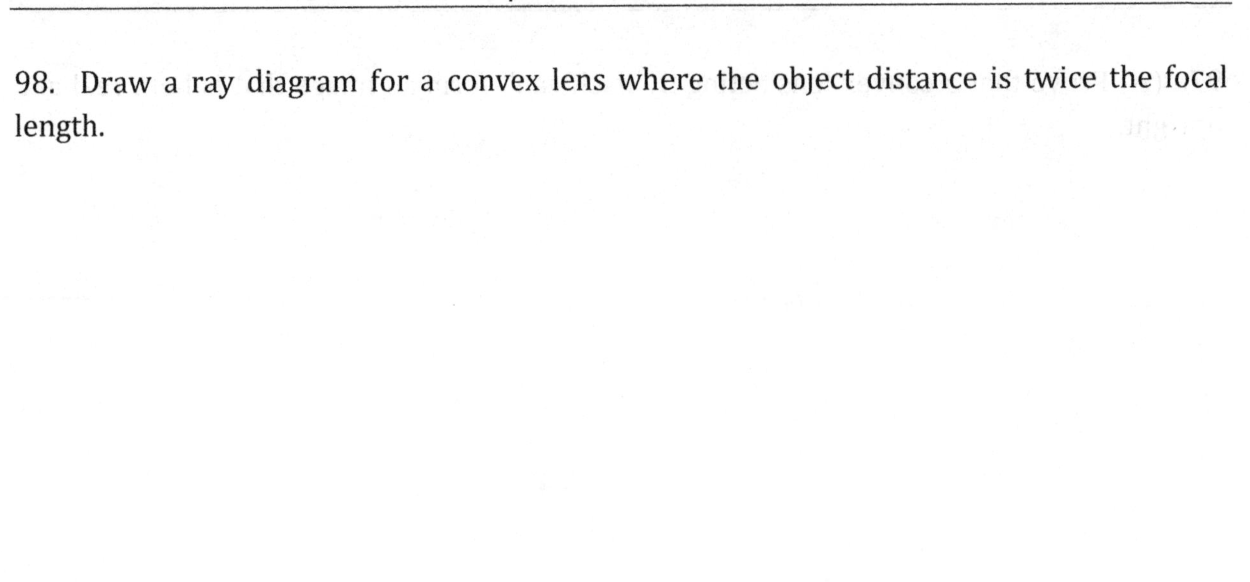

Want help? Check the hints section at the back of the book.

99. Draw a ray diagram for a convex lens where the object distance is one-half of the focal length.

Want help? Check the hints section at the back of the book.

100. Draw a ray diagram for a concave lens where the object distance is one-half of the focal length.

Want help? Check the hints section at the back of the book.

26 SPHERICAL MIRRORS

Relevant Terminology

Convex – a mirror that bulges outward (towards the object) in the middle. A convex mirror is a **diverging** mirror (whereas a convex lens is a converging lens).
Concave – a mirror that curves away from the object in the middle. A concave mirror is a **converging** mirror (whereas a concave lens is a diverging lens).
Focus – a point where parallel rays of light would converge after passing through a lens or reflecting from a mirror. A mirror has only one focus, whereas a lens has two foci.
Optic axis – an imaginary line passing through the center of a mirror and through its focus such that a ray of light traveling along this line would experience a reflected angle of 0°.
Principal axis – the same as the **optic axis**.
Object – a source of light rays, either by emitting or by reflecting light. For a spherical mirror, this emitted or reflected light then reflects off the mirror to form an image.
Image – the visual reproduction of an object after the object's light is reflected by a mirror (or refracted through a lens). In lab, a real image is viewed on a **screen**.
Focal length – the distance from the **focus** to where the mirror intersects the optic axis.
Object distance – the distance from the **object** to where the mirror intersects the optic axis.
Image distance – the distance from the **image** to where the mirror intersects the optic axis.
Real image – formed when the output light rays of a mirror actually **converge** at the location of the image. A real image can be viewed on a screen. A real image forms in front of a mirror.
Virtual image – formed when the output light rays of a mirror appear (after extrapolating backwards) to **diverge** from the location of the image. A virtual image <u>cannot</u> be viewed on a screen like a real image can. A virtual image forms behind a mirror.
Upright – an image which extends in the same direction as the object from the optic axis.
Inverted – an image which appears **upside down** compared to the object. Note that left/right are actually interchanged as well as up/down for an inverted image.[1]
Character – whether the image formed is **real** or **virtual**.
Orientation – whether the image formed is **upright** or **inverted**.
Magnification – the ratio of the image height to the object height.

---

[1] It is interesting to note that, although words appear backwards when you hold a sheet of paper up to a plane mirror, a plane mirror does <u>**not**</u> invert left/right or up/down. To see this, stand before a mirror mounted on an east or west facing wall and raise your **north** hand: Your reflection will also raise its **north** hand. What is inverted is your **perception**: From your experience with looking at other people, you expect to see another person's left hand on your right side when they face you, but it's the opposite when you view your reflection in a mirror. Here's the real test of left/right inversion: Lie down on a countertop in front of a mirror, such that your feet are on the right side of the counter and your head is on the left side of the counter. You surely won't see your reflection's feet on the left side of the mirror. (However, note that this chapter is about spherical mirrors, not plane mirrors. As we will see, the image of a spherical mirror may be inverted, and when it is, it is truly inverted both left/right <u>and</u> up/down.)

Essential Concepts

A **concave** (or **converging**) mirror causes parallel rays of light to **converge** at a **focus** (F) whereas a **convex** (or **diverging**) mirror causes parallel rays of light to **diverge** from a **focus.**

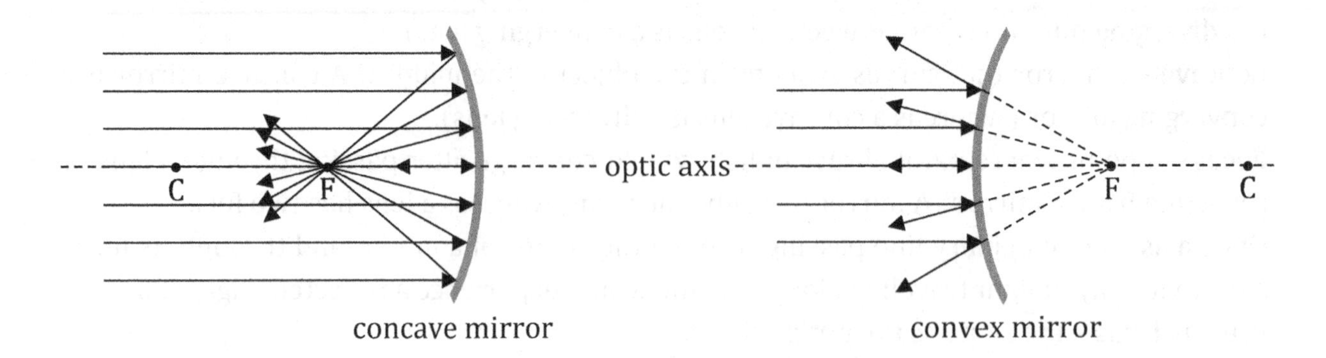

When an **object** (O) is placed before a mirror, the output rays of light emanating from the object either converge at a real **image** or appear to diverge from a virtual image. (We will explore this in more detail in the section entitled Ray Diagrams.)

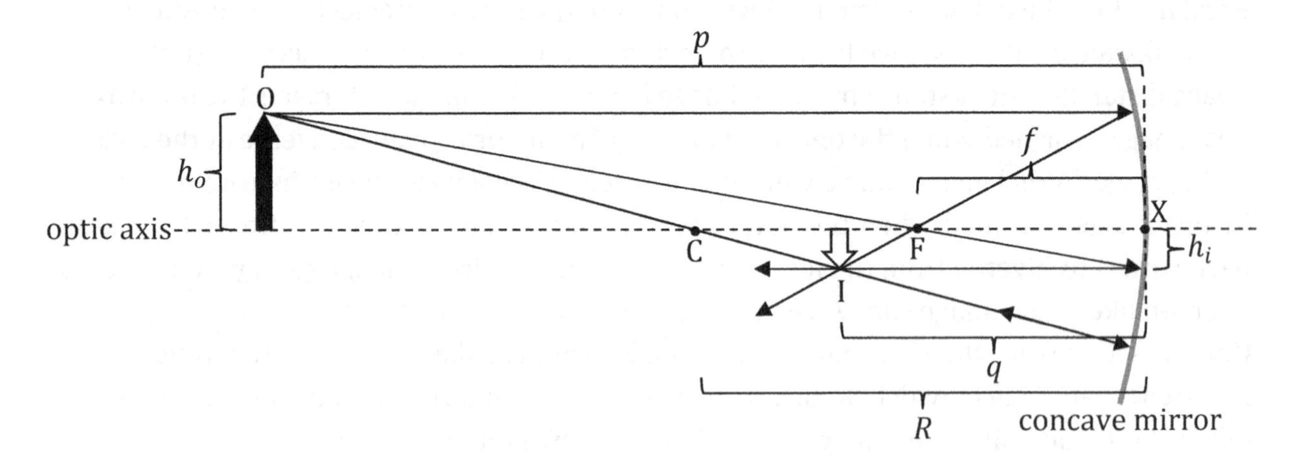

Remember how the following distances are defined. Note that point X marks the spot where the **mirror's surface** (its left edge in the diagram above) intersects the **optic axis.**

- The **object distance** (p) extends from the **object** (O) to point X.
- The **image distance** (q) extends from the **image** (I) to point X.
- The **focal length** (f) extends from the **focus** (F) to point X.
- The **radius of curvature** (R) extends from the center of curvature (C) to point X.
- The **object height** (h_o) extends from the **optic axis** to the top of the **object** (O).
- The **image height** (h_i) extends from the **optic axis** to the top of the **image** (I).

It's important to realize that these distances may be **negative**. Study the sign conventions on the following page.

Spherical Mirror Equations

The **object distance** (p), **image distance** (q), and **focal length** (f) are related to one another through the following equation.

$$\frac{1}{p}+\frac{1}{q}=\frac{1}{f}$$

The **radius of curvature** (R) is twice the **focal length** (f).

$$R = 2f$$

The **magnification** (M) equals the ratio of the **image height** (h_i) to the **object height** (h_o), which turns out to be the same as $-\frac{q}{p}$.

$$M = \frac{h_i}{h_o} = -\frac{q}{p}$$

Sign Conventions

Symbol	Name	Sign Convention
f	focal length	positive for a **concave** mirror negative for a **convex** mirror
R	radius of curvature	positive for a **concave** mirror negative for a **convex** mirror
p	object distance	positive for the first object (in a multi-mirror/lens system, $p_2 < 0$ if $q_1 > L$)
q	image distance	positive if the image and object are on the **same** side of the mirror (the image is **real**) negative if the image and object are on **opposite** sides of the mirror (the image is **virtual**)
h_o	object height	positive for the first object (in a multi-mirror system, h_{2o} has the same sign as h_{1i})
h_i	image height	positive if the image and object extend in the **same direction** from the optic axis negative if the image appears upside down compared to the object
M	magnification	positive if image is **upright**, negative if image is **inverted**

Ray Diagrams

The image formed by a spherical mirror can be found by drawing a ray diagram as follows:

1. Draw one ray from the top of the object, **parallel** to the optic axis to the front surface of the mirror, and through the **focus**. Unlike a lens, a mirror only has one focus.
    - For a **concave** mirror, the output ray converges through the focus on the **same** side of the mirror as the object.
    - For a **convex** mirror, the output ray diverges from the focus on the **opposite** side of the mirror as the object. Use a dashed line to connect the focus to the point where the ray meets the mirror, and use a solid line to make the actual ray of light diverge from the focus.
2. Draw another ray from the top of the object through the **center** of curvature of the lens (point C, which is twice as far from the surface of the mirror as the focus, F). This ray is perpendicular to the mirror and thus reflects back on itself.[2]
3. Draw the last ray from the top of the object through the **focus**, and then **parallel** to the optic axis. Remember that a mirror only has one focus. This same point F is used twice – for rays 1 and 3 both.
    - For a **concave** mirror, the output ray passes through the focus (on the same side of the mirror as the object), and then runs parallel to the optic axis.
    - For a **convex** mirror, the output ray heads towards the focus (on the opposite side of the mirror as the object), but doesn't actually get there. When this ray reaches the mirror, it then runs parallel to the optic axis. Draw a dashed line to connect the focus to the point where the rays meets the mirror, and use a solid line for the rest.

The **image** forms where the three output rays intersect. Be sure to look at the **output** rays (the reflected rays leaving the mirror), and **not** the input rays (the rays incident upon the mirror). The input rays intersect at the object, and you're **not** looking for the object. You're looking for the image, so you must look at the output rays. **Tip**: If the output rays are **diverging**, you must **extrapolate backwards** (as shown in the examples).

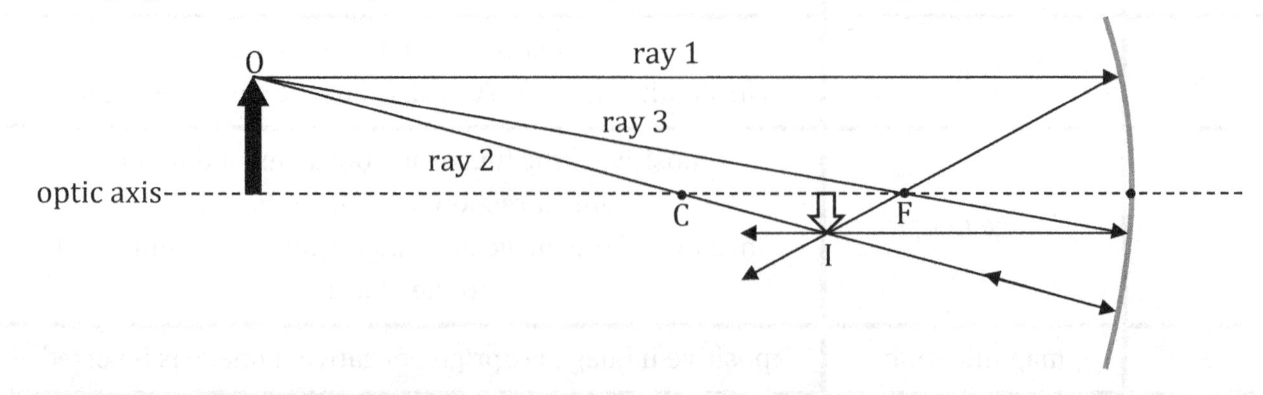

[2] A few books instead draw a ray through point X (see the previous page) and apply the law of reflection to it. A ray reflecting from point X is a convenient alternative when the object is located near the center of curvature.

Symbols and Units

Symbol	Name	Units
F	focus	
C	center of curvature	
O	object	
I	image	
p	object distance (object to surface of mirror)	m
q	image distance (image to surface of mirror)	m
f	focal length (focus to surface of mirror)	m
R	radius of curvature (twice the focal length)	m
h_o	object height	m
h_i	image height	m
M	magnification	unitless

Important Distinctions

Note that convex and concave mirrors have a much different effect than the corresponding lenses:

- A **concave** mirror is a **converging** mirror, whereas a **concave** lens is a **diverging** lens.
- A **convex** mirror is a **diverging** mirror, whereas a **convex** lens is a **converging** lens.

Concave mirrors and **convex** lenses have **positive** focal lengths because they cause parallel rays of light to **converge** at the focus, whereas **convex** mirrors and **concave** lenses have **negative** focal lengths because they cause parallel rays of light to **diverge** from the focus.

For a mirror, the image distance is **positive** if the object and image are on the **same** side of the mirror, whereas for a lens, the image distance is **positive** if the object and image are on **opposite** sides of the lens.

Note that the point C represents the center of curvature of a spherical mirror. Imagine extending the arc to draw a complete circle: Point C lies at the center of that circle. If you're used to drawing C for a thin lens (Chapter 25), be careful not confuse C with point X in the context of spherical mirrors (find C and X on the diagram on page 300).

Spherical Mirror Strategy

To solve a problem involving a spherical mirror, follow these steps:

1. To draw a **ray diagram** for a spherical mirror, draw the three rays described on page 302, as illustrated in the examples.
2. Make a list of the known quantities and identify the desired unknown. For a spherical mirror:
   - The **object distance** (p) extends from the **object** to point X (where the surface of the mirror intersects the optic axis, as shown on page 300).
   - The **image distance** (q) extends from the **image** to point X.
   - The **focal length** (f) extends from the **focus** to point X.
   - The **object height** (h_o) extends from the **optic axis** to the top of the **object**.
   - The **image height** (h_i) extends from the **optic axis** to the top of the **image**.
3. Study the sign conventions on page 301. Note that problems give you numerical values without any signs, while you must put the signs in yourself. For example, you must use a **negative sign** for the **focal length** (f) or **radius of curvature** (R) of a **convex** (or diverging) mirror.
4. For a single spherical mirror, use the following equations.

   $$\frac{1}{p} + \frac{1}{q} = \frac{1}{f} \quad , \quad M = \frac{h_i}{h_o} = -\frac{q}{p} \quad , \quad R = 2f$$

   Tip: A common way to solve for the image height is to use $\frac{h_i}{h_o} = -\frac{q}{p}$.
5. For a system of mirrors or lenses (or a combination of both), use the strategy for a multi-lens system from Chapter 25. The only differences are:
   - The equation $R = 2f$ also applies to a spherical mirror. You may need this equation in addition to the thin lens equations. (Note that all of the thin lens equations from the multi-lens strategy do apply to mirrors.)
   - The sign conventions are different for spherical mirrors than for lenses.

Example: A 9.0-mm tall gorilla figurine is placed 18 cm before a convex mirror for which the center of curvature is 72 cm from the surface of the mirror.

(A) Where does the image form?

Identify the given information. **Note**: It's okay to leave the distances in the units given (as long as p, q, f, and R have the same units, and as long as h_o and h_i have the same units).

- The object distance is $p = 18$ cm. The first object distance is always positive.
- The radius of curvature is $R = -72$ cm. The radius of curvature and the focal length are both **negative** for a **convex** mirror.
- The object height is $h_o = 9.0$ mm. The first object height is always positive. **Note**: It's okay for h_o to have different units (as long as it the same units as h_i).

We need to find the focal length before we can determine where the image forms. The radius of curvature is twice the focal length.

$$R = 2f$$

$$f = \frac{R}{2} = -\frac{72}{2} = -36 \text{ cm}$$

The focal length is $f = -36$ cm. Now we can solve for the **image distance** (q).

$$\frac{1}{p} + \frac{1}{q} = \frac{1}{f}$$

$$\frac{1}{18} + \frac{1}{q} = \frac{1}{-36}$$

Subtract $\frac{1}{18}$ from both sides of the equation.

$$\frac{1}{q} = -\frac{1}{36} - \frac{1}{18}$$

Multiply $\frac{1}{18}$ by $\frac{2}{2}$ in order to make a **common denominator**.

$$\frac{1}{q} = -\frac{1}{36} - \frac{2}{36}$$

$$\frac{1}{q} = -\frac{3}{36}$$

Take the **reciprocal** of both sides of the equation.

$$q = -\frac{36}{3} = -12 \text{ cm}$$

The image forms at $q = -12$ cm, which means that the image is 12 cm from the surface of the mirror, where the **minus sign** indicates that the image forms **behind the mirror** compared to the object.

(B) What is the image height?

Combine the two magnification equations together. That is:

$$\frac{h_i}{h_o} = -\frac{q}{p}$$

Multiply both sides of the equation by h_o. Be careful with the minus signs.

$$h_i = -\frac{q}{p} h_o = -\frac{(-12)}{18}(9) = 6.0 \text{ mm}$$

The image height is $h_i = 6.0$ mm. Note that the minus signs and the centimeters (cm) **canceled** out. The image height is in millimeters (mm) because h_o was put in millimeters.

(C) What is the magnification of the image?
Use one of the formulas for magnification.

$$M = \frac{h_i}{h_o} = \frac{6}{9} = \frac{2}{3} \times$$

The magnification is $M = \frac{2}{3} \times$. If you use a calculator, $M = 0.67 \times$.

(D) What is the orientation of the image?
The word "**orientation**" means that the question is asking if the image is upright or inverted. Review the sign conventions on page 301. Examine the **sign** of the **magnification**: Since $M > 0$, the image is upright.

(E) What is the character of the image?
The word "**character**" means that the question is asking if the image is real or virtual. Review the sign conventions on page 301. Examine the **sign** of the **image distance**: Since $q < 0$, the image is **virtual**.

Example: Draw a ray diagram for a concave mirror where the object is twice as far from the mirror as the center of curvature is.

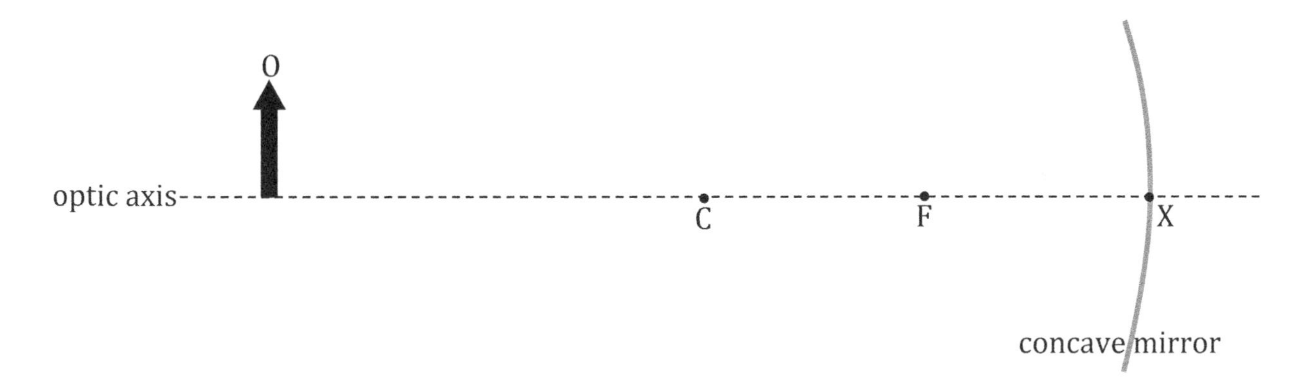

Begin by drawing a spherical mirror with the concave surface facing the object. (Page 300 shows how convex and concave mirrors look.) Draw the center of curvature (C), and draw a line through the center of curvature to serve as the **optic axis**. Draw and label a focus (F) **halfway** (since $R = 2f$) between the center of curvature and the point where the surface of the mirror meets the optic axis (point X). Draw and label an object (O) such that the object distance (from the object to point X) is twice the radius of curvature (from the center of curvature to point X), as specified in the problem.

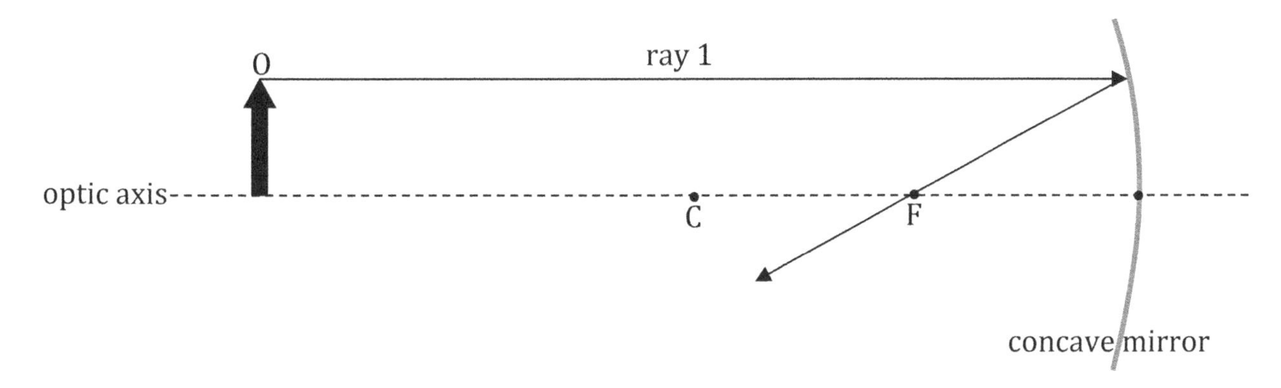

Draw the first ray from the top of the object **parallel** to the optic axis until it reaches the mirror. The reflected ray will pass through the **focus**.

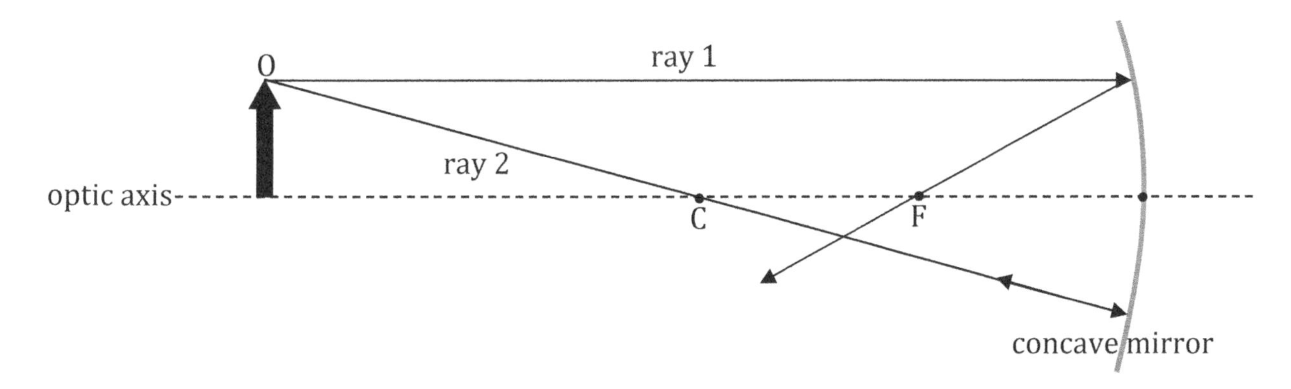

Draw the second ray through the center of curvature. Since this ray is perpendicular to the mirror (because this ray travels along a radius, and a radius is perpendicular to the surface of a sphere), it reflects back on itself.

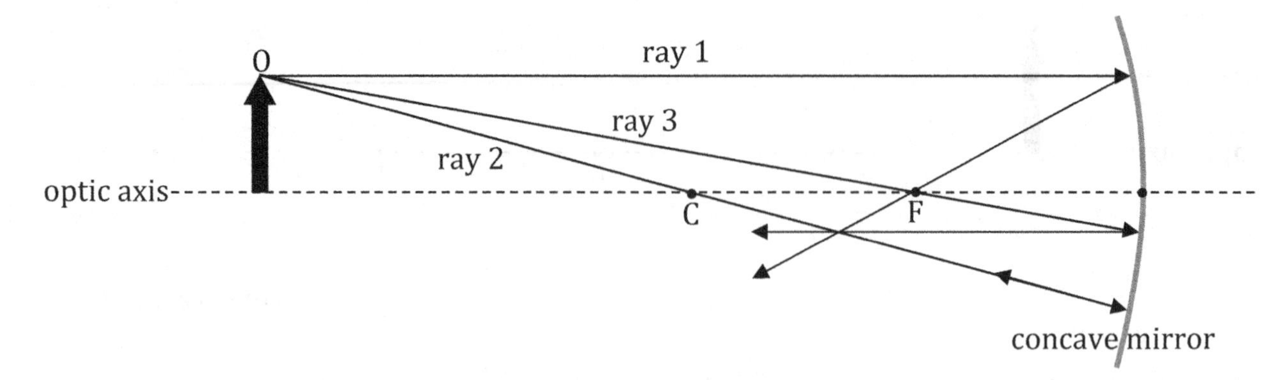

Draw the third ray from the top of the object through the **focus** until it reaches the mirror. The reflected ray will then travel **parallel** to the optic axis.

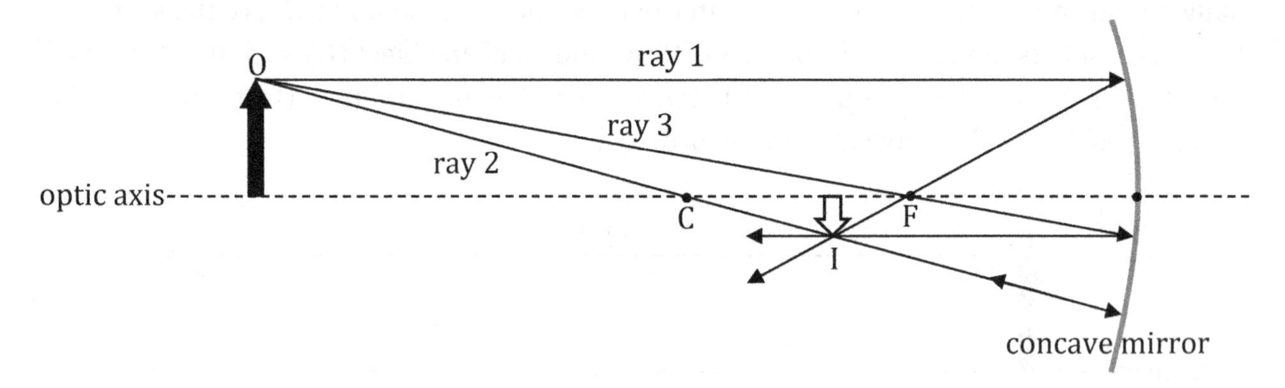

Draw and label the image where the three **output** rays intersect. (The output rays – the reflected rays leaving the mirror – intersect at the image. Be careful not to look at the input rays – the rays incident upon the mirror – which intersect at the object.)

It would be wise to review the directions from page 302, and compare them closely with how we applied them to this example.

Example: Draw a ray diagram for a convex mirror where the object is just as far from the mirror as the center of curvature is.

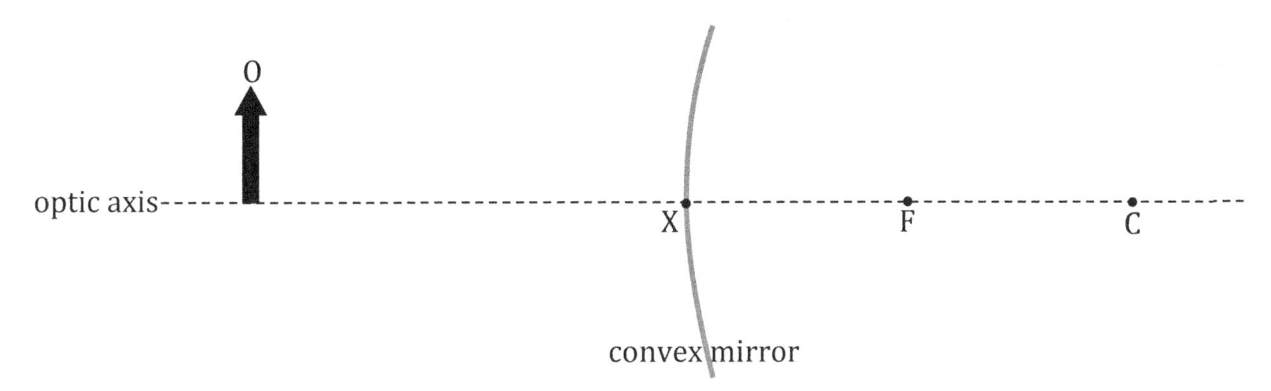

Begin by drawing a spherical mirror with the convex surface facing the object. (Page 300 shows how convex and concave mirrors look.) Draw the center of curvature (C) **behind** the convex mirror, and draw a line through the center of curvature to serve as the **optic axis**. Draw and label a focus (F) **halfway** (since $R = 2f$) between the center of curvature and the point where the surface of the mirror meets the optic axis (point X). Draw and label an object (O) such that the object distance (from the object to point X) is the same distance as the radius of curvature (from the center of curvature to point X), as specified in the problem.

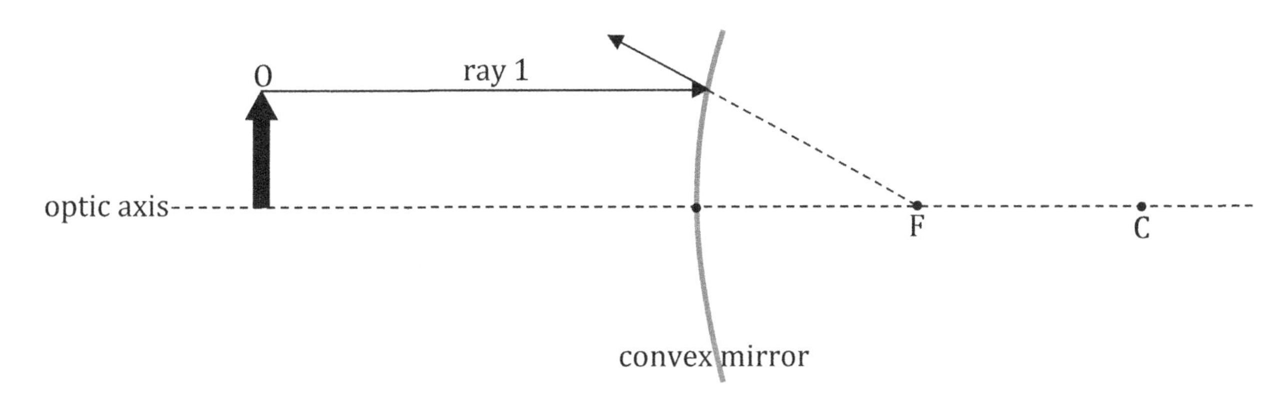

Draw the first ray from the top of the object **parallel** to the optic axis until it reaches the mirror. The reflected ray will diverge from the **focus**. Use a dashed line to show that this ray diverges from the focus.

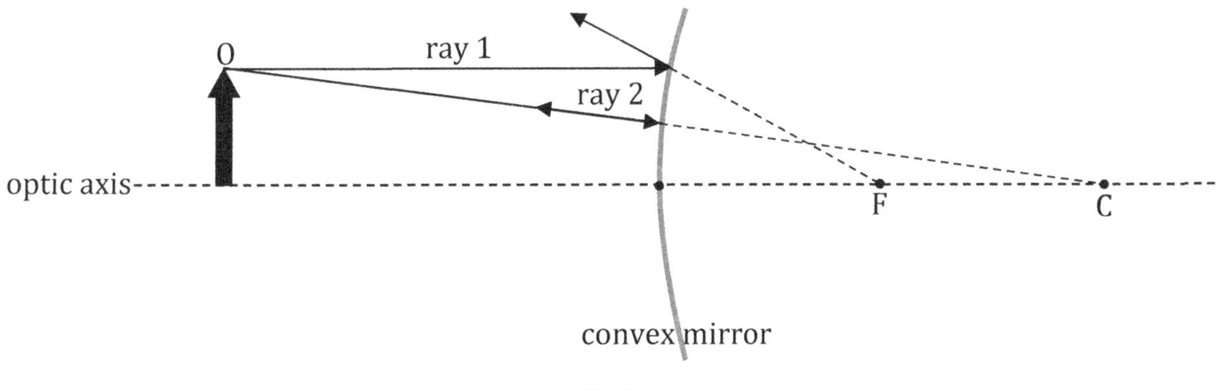

Draw the second ray towards the center of curvature. Since this ray is perpendicular to the mirror (because this ray travels along a radius, and a radius is perpendicular to the surface of a sphere), it reflects back on itself when it reaches the mirror. Use a dashed line to show that this ray is heading towards the center of curvature.

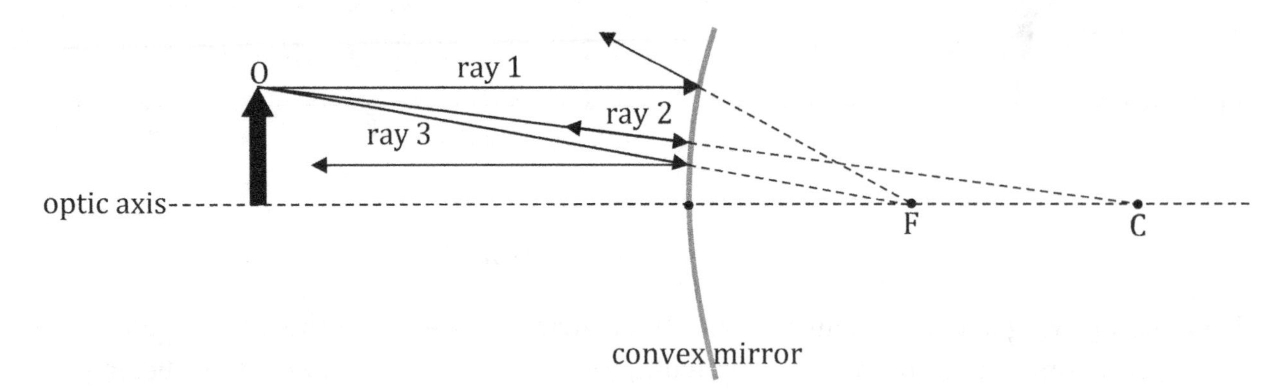

Draw the third ray from the top of the object towards the **focus** until it reaches the mirror. The reflected ray will then travel **parallel** to the optic axis. Use a dashed line to show that this ray is heading towards the focus.

Where is the image? The image forms where the three output rays intersect. Study the ray diagram above. Where do the three **output** (**not** input) rays intersect? (Look at the three reflected rays leaving the mirror.) They **do** intersect. Since they don't intersect in front of the mirror, you must **extrapolate backwards to the back of the mirror**. We added a horizontal dashed line below, extrapolating ray 3 to the right. Rays 1, 2, and 3 intersect **behind** the mirror. That's where the image is located. Look closely to see how all three output rays extrapolate back to the top of the image in the diagram below.

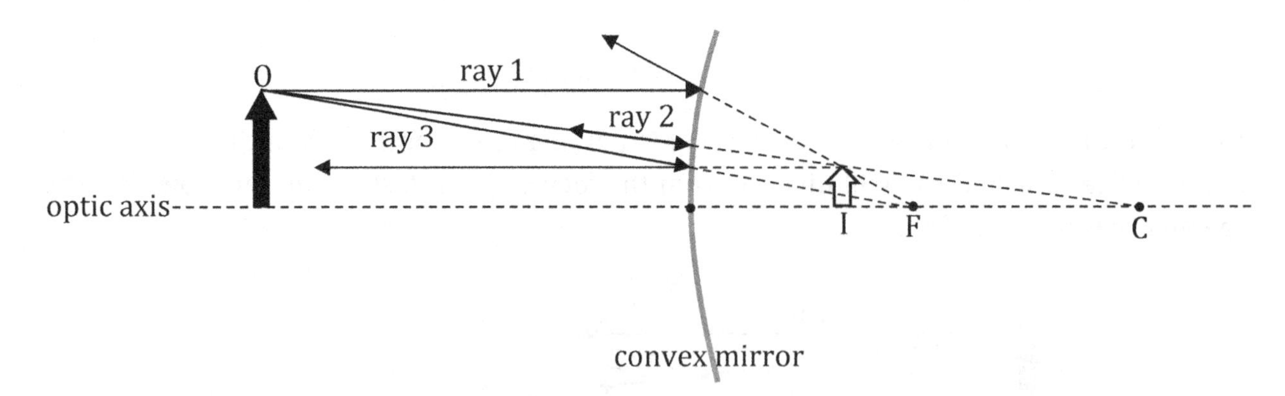

It would be wise to review the directions from page 302, and compare them closely with how we applied them to this example. It is also instructive to compare and contrast this example with the previous example.

101. A 5.0-mm tall chimpanzee figurine is placed 21 cm before a concave mirror for which the center of curvature is 28 cm from the surface of the mirror.

(A) Where does the image form?

(B) What is the image height?

(C) What is the magnification of the image?

(D) What is the orientation of the image?

(E) What is the character of the image?

Want help? Check the hints section at the back of the book.

Answers: (A) 42 cm (B) -10 mm (C) $-2.0 \times$ (D) inverted (E) real

102. A 15-mm tall orangutan figurine is placed 80 cm before a convex mirror for which the center of curvature is 40 cm from the surface of the mirror.

(A) Where does the image form?

(B) What is the image height?

(C) What is the magnification of the image?

(D) What is the orientation of the image?

(E) What is the character of the image?

Want help? Check the hints section at the back of the book.

Answers: (A) -16 cm (B) 3.0 mm (C) $\frac{1}{5}\times$ (D) upright (E) virtual

103. A 7.0-mm tall lemur figurine is placed 27 cm before a concave mirror for which the center of curvature is 108 cm from the surface of the mirror.

(A) Where does the image form?

(B) What is the image height?

(C) What is the magnification of the image?

(D) What is the orientation of the image?

(E) What is the character of the image?

Want help? Check the hints section at the back of the book.

Answers: (A) −54 cm (B) 14 mm (C) 2.0× (D) upright (E) virtual

104. Draw a ray diagram for a concave mirror where the object distance is three times the focal length.

Want help? Check the hints section at the back of the book.

105. Draw a ray diagram for a concave mirror where the object is halfway between the mirror and the focus.

Want help? Check the hints section at the back of the book.

106. Draw a ray diagram for a convex mirror where the object is just as far from the mirror as the focus is.

Want help? Check the hints section at the back of the book.

27 SINGLE-SLIT DIFFRACTION

Relevant Terminology

Diffraction – the bending of light around an obstacle, such as when light passes through a narrow slit.
Wavelength – the **horizontal** distance between two consecutive crests.
Monochromatic – light with a single, precise wavelength. For light in the visible spectrum, monochromatic light is light consisting of a single **color**.
Fringe – a bright or dark band appearing in a diffraction or interference pattern.

Essential Concepts

When light passes through a narrow slit, it produces a diffraction pattern. When viewed on a screen, the diffraction pattern consists of a series of bright and dark bands (or **fringes**) similar to the diagram below (except that the "gray" bands represents "light" and the "white" space represents "dark").

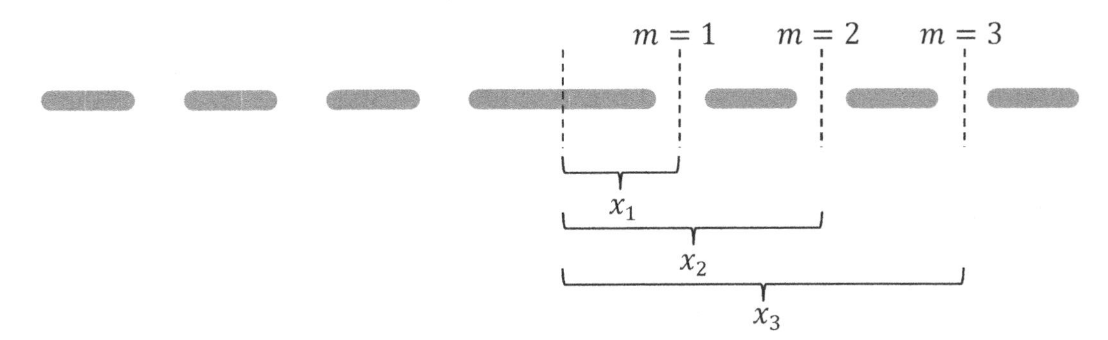

We label the dark fringes (the "white" spots above) with the index m: The value $m = 1$ corresponds to the first dark fringe (from the central bright fringe), the value $m = 2$ corresponds to the second dark fringe, and so on. The distance x_m is measured from the center of the central bright fringe to the center of the mth dark fringe, as shown above and below.

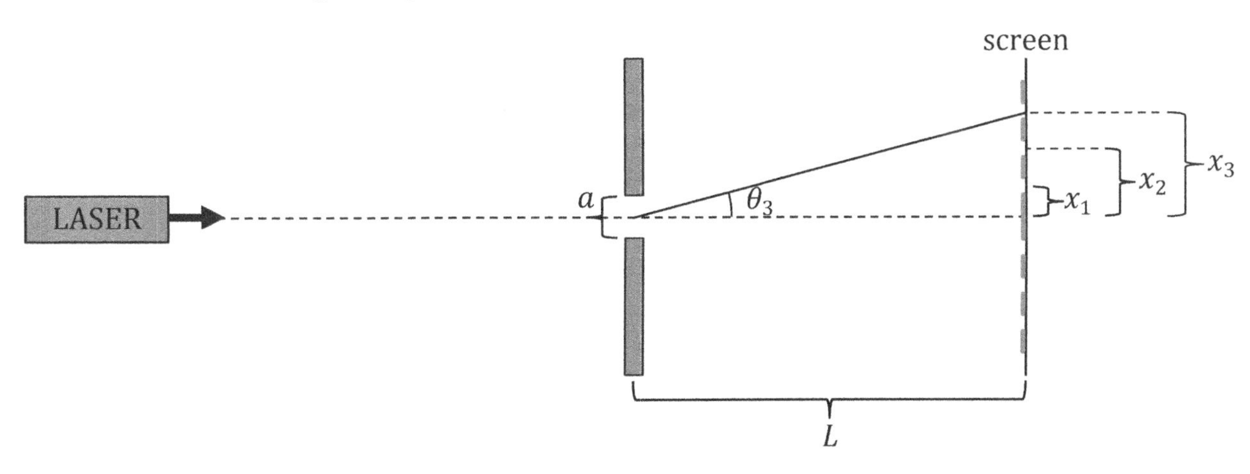

Equations for Single-slit Diffraction

The center of the mth **dark** fringe (measured from the center of the slit relative to the optic axis) corresponds to the angle θ_m (see the diagram on the previous page). In the following equation (which is the condition for destructive interference – that is, the location of the dark fringes), λ represents **wavelength** and a is the **slit width**.

$$a \sin \theta_m = m\lambda \quad (m = \pm 1, \pm 2, \pm 3 \dots)$$

From the diagram on the previous page, θ_m can also be related to x_m (the distance from the optic axis to the center of the mth **dark** fringe – remember that the "dark" fringes are the "white" spaces in the diagrams in this book, whereas the "bright" fringes are the "gray" bands) and the distance L (measured along the optic axis from the slit to the screen).

$$\tan \theta_m = \frac{x_m}{L}$$

Important Distinctions

Diffraction and refraction both involve the bending of light, but in different ways.

- **Diffraction** involves the bending of light around an obstacle, such as when light passes through a narrow slit.
- **Refraction** (see Chapters 21-22) involves the bending of light as it travels from one medium into a second medium.

Metric Prefixes and the Angstrom

One **Angstrom** (Å) equals 10^{-10} m, which equates to 0.1 nm.

Prefix	Name	Power of 10
c	centi	10^{-2}
m	milli	10^{-3}
μ	micro	10^{-6}
n	nano	10^{-9}

Symbols and Units

Symbol	Name	Units
m	a nonzero integer corresponding to m^{th} dark fringe from the optic axis	unitless
λ	wavelength	m
θ_m	the angle of the center of the mth dark fringe from the slit relative to the optic axis	° or rad
x_m	the distance from the optic axis to the center of the mth dark fringe	m
a	the width of the slit	m
L	the distance along the optic axis from the slit to the screen	m

Single-slit Diffraction Strategy

To solve a problem involving single-slit diffraction, follow these steps:

1. Make a list of the known quantities. Identify the desired unknown symbol.
2. If a problem gives you a picture of the diffraction pattern or a diagram of the experimental setup, compare with the diagrams on page 317 to help identify any information that may be given in the picture.
3. The following equations apply to the centers of the dark fringes in a single-slit diffraction pattern, where m is an integer corresponding the mth **dark** fringe from the optic axis (which passes through the center of the central bright fringe). Note that λ represents **wavelength** and a represents the **slit width**. See the diagrams on page 317, which show how θ_m, x_m, and L are defined.
$$a \sin\theta_m = m\lambda \quad (m = \pm 1, \pm 2, \pm 3 \ldots)$$
$$\tan\theta_m = \frac{x_m}{L}$$
4. If a problem involves a diffraction **grating**, see Chapter 28.
5. If a problem involves **two or three slits**, see Chapter 29.

Example: A monkey shines a laser beam with a wavelength of 500 nm through a slit with a width of 0.020 mm. A diffraction pattern forms on a screen that is 300 cm from the slit. How far is the 20th dark fringe from the optic axis?

Identify the given information in consistent units (nm, mm, and cm are all inconsistent). We'll put all of the distances in meters. Also, identify the desired unknown symbol.

- The wavelength is $\lambda = 5.00 \times 10^{-7}$ m (since 1 nm $= 10^{-9}$ m and $500 = 5.00 \times 10^{2}$).
- The slit width is $a = 2.0 \times 10^{-5}$ m (since 1 mm $= 10^{-3}$ m and $0.020 = 2.0 \times 10^{-2}$).
- The distance from the slit to the screen is $L = 3.00$ m (since 1 cm $= 10^{-2}$ m).
- The question specifies the 20th dark fringe from the optic axis: $m = 20$.
- We are looking for x_{20}.

We can't yet use the equation for x_{20} because we don't yet know θ_{20}. Therefore, we must solve for θ_{20} in the other equation.

$$a \sin\theta_m = m\lambda$$

Set $m = 20$, as specified in the problem.

$$a \sin\theta_{20} = 20\lambda$$

Divide both sides of the equation by a.

$$\sin\theta_{20} = \frac{20\lambda}{a} = \frac{20(5 \times 10^{-7})}{2 \times 10^{-5}} = 50 \times 10^{-2} = 0.50$$

Note that $\frac{10^{-7}}{10^{-5}} = 10^{-7-(-5)} = 10^{-7+5} = 10^{-2}$ according to the rule $\frac{x^m}{x^n} = x^{m-n}$. Take the inverse sine of both sides of the equation.

$$\theta_{20} = \sin^{-1}(0.50)$$

The angle is $\theta_{20} = 30°$ because $\sin(30°) = \frac{1}{2} = 0.50$. Now use the equation that has x_{20}. Note that the "m" in x_m is a **subscript** (it's **not** x times m).

$$\tan\theta_m = \frac{x_m}{L}$$

Set $m = 20$ in this equation also.

$$\tan\theta_{20} = \frac{x_{20}}{L}$$

Multiply both sides of the equation by L.

$$x_{20} = L\tan\theta_{20} = (3)\tan 30° = (3)\frac{\sqrt{3}}{3} = \sqrt{3}\text{ m}$$

The center of the 20th dark fringe appears $x_{20} = \sqrt{3}$ m from the optic axis.[1] If you use a calculator, $x_{20} = 1.7$ m.

---

[1] The intensity of the bright spots diminishes significantly as you get further from the optic axis, so that in practice it may be difficult or impossible to see the 20th fringe on a screen when it is 30° offline, as in this example. Once again, we chose numbers to make the math easy to follow without a calculator. Problem 107 will show you a "trick" for working with much smaller angles without a calculator (which allows you to work with realistic numbers to very good approximation). However, a diffraction grating (Chapter 28) can produce clear fringes at large angles, like 30° (though in that case, the fringes also tend to be more spread out, so that it wouldn't likely be the 20th fringe located at 30°).

107. A monkey shines a laser beam through a slit with a width of 0.0080 mm. A diffraction pattern forms on a screen that is 150 cm from the slit. The center of the 10th dark fringe appears $\frac{\sqrt{3}}{2}$ m from the optic axis. What is the wavelength of the laser light? (**Note**: It is **not** necessary to make the small-angle approximation in order to solve this problem.)

108. A monkey shines a laser beam through a slit with a width of 0.026 mm. A diffraction pattern forms on a screen that is 80 cm from the slit. The center of the 3rd dark fringe appears 6.0 cm from the optic axis. What is the wavelength of the laser light? **Note**: If you're not using a calculator, note that this problem can be solved by making the following small angle approximations. For sufficiently small angles (much smaller than 30°), $\sin\theta \approx \theta$ and $\tan\theta \approx \theta$ provided that θ is expressed in **radians**.[2]

Want help? Check the hints section at the back of the book.

Answers: 400 nm, 650 nm

[2] One way to see this is to draw a triangle with small θ on the unit circle. You should see that for a small angle, $x \approx 1$ and that y is approximately equal to the arc length. For the unit circle, $R = 1$ such that the arc length simply equals $s = R\theta = \theta$. For small angles, $y \approx \theta$ and you get $\sin\theta = y \approx \theta$ and $\tan\theta = \frac{y}{x} \approx \frac{\theta}{1} = \theta$. Another way to see it is with a calculator. For example, for $\theta = 30°$, first convert to radians to get $\theta = \frac{\pi}{6}$ rad, which is approximately $\theta = 0.5236$ rad, and compare $\theta = 0.5236$ rad with $\sin 30° = \sin\left(\frac{\pi}{6} \text{ rad}\right) = 0.5000$ and $\tan 30° = \tan\left(\frac{\pi}{6} \text{ rad}\right) = 0.5774$. You can see that tangent is noticeably off for $\theta = 30°$, but when θ is very **small** (which is the case in many single-slit diffraction problems), the approximation is much better.

109. A monkey shines red laser light through a narrow slit and views the diffraction pattern on a screen.

(A) What will happen to the diffraction pattern if the monkey replaces the slit with a narrower slit?

(B) What will happen to the diffraction pattern if the monkey replaces the laser with a laser that emits green light?

(C) What will happen to the diffraction pattern if the monkey moves the screen farther from the slit?

Want help? This problem is fully explained in the back of the book.

28 DIFFRACTION GRATING

Relevant Terminology

Diffraction – the bending of light around an obstacle, such as when light passes through a narrow slit.
Wavelength – the **horizontal** distance between two consecutive crests.
Grating – closely spaced, thin parallel lines machined onto a transparent material.
Fringe – a bright or dark band appearing in a diffraction or interference pattern.

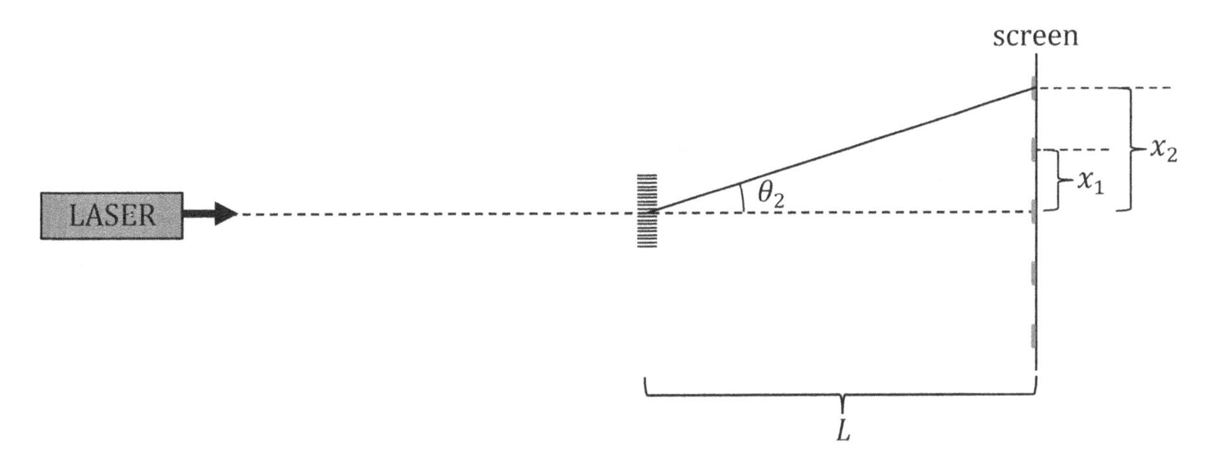

Diffraction Grating Equations

A diffraction grating typically consists of hundreds or thousands of thin, closely spaced parallel lines machined onto a piece of glass or plastic. The **grating spacing** (d) is the distance between neighboring lines, while the **grating constant** $\left(\frac{1}{d}\right)$ indicates the number of lines per unit length. The center of the mth **bright** spot (measured from the center of the grating relative to the optic axis) corresponds to the angle θ_m. In the following equation (which is the condition for constructive interference – that is, the location of the bright spots), λ represents **wavelength**.

$$d \sin \theta_m = m\lambda \quad (m = 0, \pm 1, \pm 2, \pm 3 \ldots)$$

From the diagram above, θ_m can also be related to x_m (the distance from the optic axis to the center of the mth **bright** spot) and the distance L (measured along the optic axis from the grating to the screen).

$$\tan \theta_m = \frac{x_m}{L}$$

Symbols and Units

Symbol	Name	Units
m	an integer corresponding to m^{th} bright fringe from the optic axis	unitless
λ	wavelength	m
θ_m	the angle of the center of the mth bright fringe from the grating relative to the optic axis	° or rad
x_m	the distance from the optic axis to the center of the mth bright fringe	m
d	grating spacing	m
$\frac{1}{d}$	grating constant	$\frac{1}{\text{m}}$
L	the distance along the optic axis from the grating to the screen	m

Important Distinctions

You can tell the grating spacing and grating constant apart by looking at the **units**:

- The **grating spacing** (d) is a distance in m, cm, or mm, for example.
- The **grating constant** $\left(\frac{1}{d}\right)$ is an inverse distance in lines/m, lines/cm, or lines/mm.

The diffraction **grating** equation, $d \sin\theta_m = m\lambda$, provides the location of **bright** spots, whereas the similar **single-slit** diffraction equation (Chapter 27), $a \sin\theta_m = m\lambda$, provides the location of **dark** spots.

Metric Prefixes and the Angstrom

One **Angstrom** (Å) equals 10^{-10} m, which equates to 0.1 nm.

Prefix	Name	Power of 10
c	centi	10^{-2}
m	milli	10^{-3}
µ	micro	10^{-6}
n	nano	10^{-9}

Diffraction Grating Strategy

To solve a problem involving a diffraction **grating**, follow these steps:

1. Make a list of the known quantities. Identify the desired unknown symbol.
2. Note that a number in lines/m, lines/cm, or lines/mm is the **grating constant** $\left(\frac{1}{d}\right)$. Take the **reciprocal** of this number to find the **grating spacing** (d).
3. The following equations apply to the centers of the bright spots in a diffraction grating pattern, where m is an integer corresponding the mth **bright** fringe from the optic axis (which passes through the center of the central bright fringe). The symbol λ represents **wavelength**. See the diagram on page 323, which shows how θ_m, x_m, and L are defined.
$$d \sin \theta_m = m\lambda \quad (m = 0, \pm 1, \pm 2, \pm 3 \ldots)$$
$$\tan \theta_m = \frac{x_m}{L}$$
4. To determine the maximum value of m (call it m_{max}), set $\theta_{max} = 90°$. The total number of observable fringes[1] (or bright spots) equals $2m_{max} + 1$.

Example: A monkey shines a laser beam with a wavelength of 600 nm through a grating that has 4000 lines/cm. How many fringes can be observed?

Identify the given information in consistent units (nm and cm are inconsistent).

- The wavelength is $\lambda = 6.00 \times 10^{-7}$ m (since 1 nm $= 10^{-9}$ m and $600 = 6.00 \times 10^2$).
- The grating constant[2] is $\frac{1}{d} = 400{,}000 \frac{\text{lines}}{\text{m}}$ (since $4000 \frac{\text{lines}}{\text{cm}} = 4000 \frac{\text{lines}}{\text{cm}} \times \frac{100 \text{ cm}}{1 \text{ m}}$).

The **grating spacing** is the **reciprocal** of the **grating constant**: $d = \frac{1}{400{,}000}$ m $= 0.0000025$ m $= 2.5 \times 10^{-6}$ m. Set $\theta_{max} = 90°$ in the equation for the diffraction grating to determine the maximum value of m (which we will call m_{max}).

$$d \sin \theta_m = m\lambda$$
$$d \sin \theta_{max} = m_{max}\lambda$$
$$d \sin 90° = m_{max}\lambda$$

Recall from trig that $\sin 90° = 1$. Divide both sides of the equation by λ.

$$m_{max} = \frac{d}{\lambda} = \frac{2.5 \times 10^{-6}}{6 \times 10^{-7}} = \frac{25}{6} = 4$$

Note that $m_{max} = 4$ (and **not** 4.17) because m must be an integer. The total number of observable bright fringes is $2m_{max} + 1 = 2(4) + 1 = 8 + 1 = 9$.

---

[1] The +1 is for the central bright spot, while the 2 represents that other (that is, besides the central spot) bright spots form on both sides of the optic axis.

[2] To convert from cm to m, we would divide by 100, but here we're converting from $\frac{1}{\text{cm}}$ to $\frac{1}{\text{m}}$, so we **multiply** by 100. Alternatively, first take the reciprocal of 4000 lines/cm to find that the **grating spacing** is $d = \frac{1}{4000}$ cm $= 0.00025$ cm, and then divide by 100 to get $d = \frac{1}{400{,}000}$ m $= 0.0000025$ m, which is what we ultimately need.

110. A monkey shines a laser beam through a grating that has 5000 lines/cm. The second bright fringe from the optic axis is measured at an angle of 30° relative to the optic axis.[3] What is the wavelength of the laser light?

111. A monkey shines a laser beam with a wavelength of 450 nm through a grating that has 8000 lines/cm. How many fringes can be observed?

Want help? Check the hints section at the back of the book.

Answers: 500 nm, 5

[3] Unlike the footnote on page 320 regarding **single-slit** diffraction, it is common to get large angles with a diffraction **grating**. By machining lines on a piece of glass or plastic, we can create much smaller values of d for a grating than typical values of a for a single slit. As a result, a very visible bright spot can be produced by a **grating** at large angles like 30° or more, whereas the visible spots produced by single-slit diffraction tend to be much smaller.

29 DOUBLE-SLIT INTERFERENCE

Relevant Terminology

Interference – the superposition of two waves of the same frequency, reinforcing or canceling (in part or in whole) one another depending upon their relative phase angles. **Constructive** interference results in a **bright** spot (**maximum** intensity), while **destructive** interference results in a dark spot (**minimum** intensity, which is **zero**).
Wavelength – the **horizontal** distance between two consecutive crests.
Monochromatic – light with a single, precise wavelength. For light in the visible spectrum, monochromatic light is light consisting of a single **color**.
Fringe – a bright or dark band appearing in a diffraction or interference pattern.

Essential Concepts

When light passes through two or more closely spaced narrow slits, it produces an interference pattern. When viewed on a screen, the interference pattern consists of a series of bright and dark bands (or **fringes**) similar to the diagram below (except that the "gray" bands represents "light" and the "white" space represents "dark").[1]

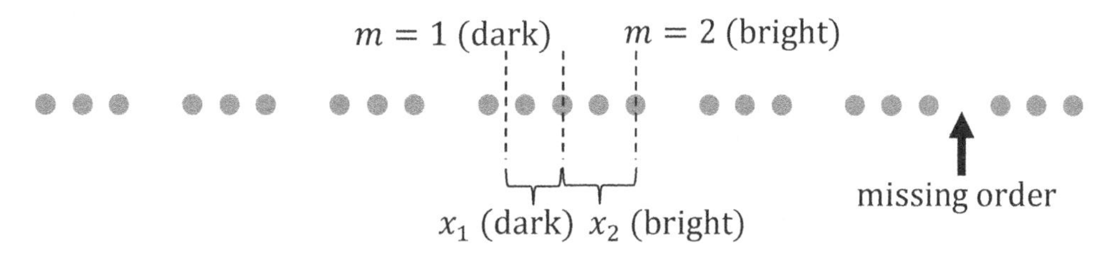

We label the bright and dark fringes with the index m. The distance x_m is measured from the optic axis to the center of the mth bright or $(m + 1)^{th}$ dark fringe, as shown above and below.

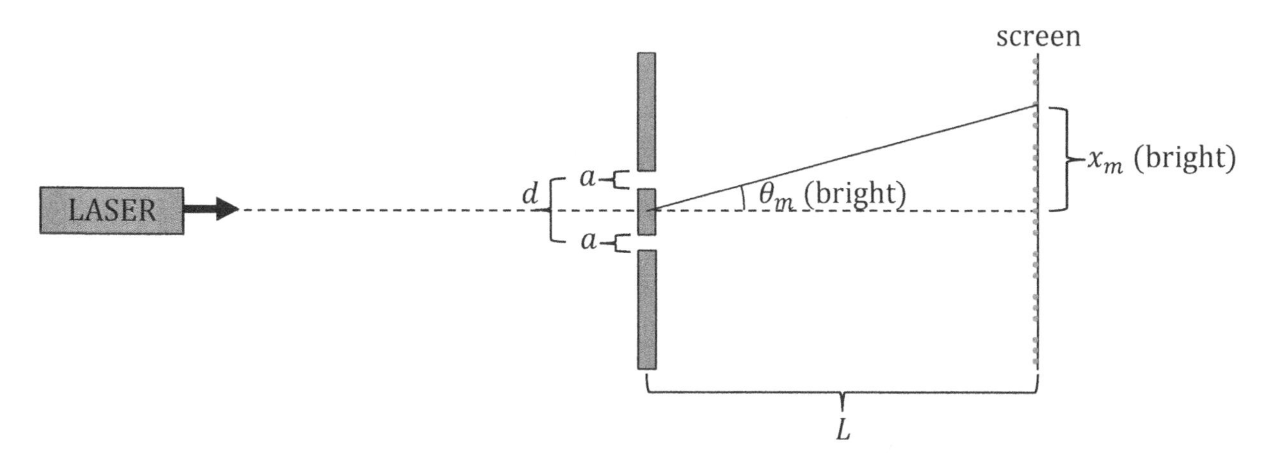

[1] Does $m = 1$ seem "wrong" to you for the "second" dark spot? If so, see the note at the bottom of page 328.

Note the "**missing orders**" in the diagram on the previous page. A "**missing order**" occurs when there are two or more slits. The location of the missing orders coincide with the dark fringes from the corresponding single-slit diffraction pattern (Chapter 27). That is, the **slit width** (a) determines the locations of the missing orders, while the **slit spacing** (d) – which is the separation between the centers of the slits – determines the interference pattern (the locations of the bright and dark fringes). This is easiest to see by comparing the patterns for 1, 2, 3, and 4 slits shown below.

- The effect of the second slit is to break up the fringes of the single-slit diffraction pattern into several smaller fringes.
- The effect of the third slit is to add one weak bright fringe between each pair of main bright fringes.
- The effect of the fourth slit is to add two weak bright fringes between each pair of main bright fringes.
- For N slits, there are $N - 2$ weak bright fringes between each pair of main fringes.

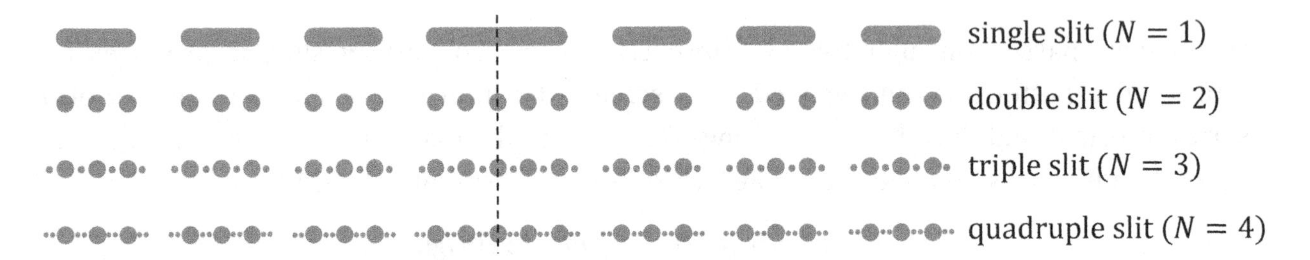

Equations for Multiple-slit Interference

The center of the mth fringe measured from the optic axis corresponds to the angle θ_m. In the following equations, λ represents **wavelength** and d is the **slit spacing** (the distance between the centers of neighboring slits).

$$d \sin\theta_m = m\lambda \quad (m = 0, \pm 1, \pm 2, \pm 3 \ldots) \quad \text{bright fringes (constructive interference)}$$

$$d \sin\theta_m = \left(m + \frac{1}{2}\right)\lambda \quad (0, \pm 1, \pm 2, \pm 3 \ldots) \quad \text{dark fringes (destructive interference)}$$

The counting is different for bright and dark fringes: For **bright** fringes, $m = 0$ gives the central bright fringe (on the optic axis), $m = 1$ gives the first bright fringe from the optic axis, $m = 2$ gives the second bright fringe, etc. For **dark** fringes, $m = 0$ gives the first dark fringe from the optic axis, $m = 1$ gives the second dark fringe, $m = 2$ gives the third dark fringe, etc. From the diagram on the previous page, θ_m can also be related to x_m (from the optic axis to the mth fringe) and the distance L (from the slit to the screen).

$$\tan\theta_m = \frac{x_m}{L}$$

The **missing orders** are determined by the single-slit diffraction equation (Chapter 27).

$$a \sin\theta_n = n\lambda \quad (n = \pm 1, \pm 2, \pm 3 \ldots)$$

Symbols and Units

Symbol	Name	Units
m	an integer corresponding to the mth bright or the $(m+1)^{\text{th}}$ dark fringe from the optic axis	unitless
n	a nonzero integer corresponding to the nth **missing order**	unitless
N	the number of slits	unitless
λ	wavelength	m
θ_m	the angle of the center of the mth bright or the $(m+1)^{\text{th}}$ dark fringe relative to the optic axis	° or rad
x_m	the distance from the optic axis to the center of the mth bright or the $(m+1)^{\text{th}}$ dark fringe	m
d	the distance between the centers of neighboring slits	m
a	the width of each slit	m
L	the distance along the optic axis from the slits to the screen	m

Metric Prefixes and the Angstrom

One **Angstrom** (Å) equals 10^{-10} m, which equates to 0.1 nm.

Prefix	Name	Power of 10
c	centi	10^{-2}
m	milli	10^{-3}
µ	micro	10^{-6}
n	nano	10^{-9}

Multiple-slit Interference Strategy

To solve a problem where light passes through two or more slits (but <u>**not**</u> a grating), follow these steps:

1. Make a list of the known quantities. Identify the desired unknown symbol.
2. If a problem gives you a picture of the interference pattern or a diagram of the experimental setup, compare with the diagrams on pages 327-328 to help identify any information that may be given in the picture.
3. The following equations apply to the bright or dark fringes appearing on an interference pattern formed by two or more slits, where m is an integer correspond-ding the mth bright or the $(m+1)^{th}$ dark fringe from the optic axis (which passes through the center of the central bright fringe). Note that λ represents **wavelength** and d represents the **slit spacing** (the distance between the centers of neighboring slits). See the diagrams on page 327, which show how θ_m, x_m, and L are defined.

$$d \sin\theta_m = m\lambda \quad (m = 0, \pm 1, \pm 2, \pm 3 \ldots) \quad \text{bright fringes (constructive interference)}$$

$$d \sin\theta_m = \left(m + \frac{1}{2}\right)\lambda \quad (0, \pm 1, \pm 2, \pm 3 \ldots) \quad \text{dark fringes (destructive interference)}$$

$$\tan\theta_m = \frac{x_m}{L}$$

4. For **bright** fringes, $m = 0$ gives the central bright fringe (on the optic axis), $m = 1$ gives the first bright fringe from the optic axis, $m = 2$ gives the second bright fringe, etc. For **dark** fringes, $m = 0$ gives the first dark fringe from the optic axis, $m = 1$ gives the second dark fringe, $m = 3$ gives the third dark fringe, etc.
5. To find the location of the nth **missing order** on an interference pattern formed by two or more slits, apply the single-slit diffraction formula (Chapter 27), where a is the **slit width**.

$$a \sin\theta_n = n\lambda \quad (n = \pm 1, \pm 2, \pm 3 \ldots)$$

6. If a problem involves a diffraction **grating**, see Chapter 28.
7. If a problem involves **single-slit** diffraction, see Chapter 27.

Important Distinctions

The symbol d (called the **slit spacing**) is the distance between the centers of neighboring slits, whereas the symbol a (called the **slit width**) is the width of each slit.

- d determines the **interference** pattern, meaning that it determines the locations of the bright and dark fringes.
- a determines the **diffraction** pattern, meaning that it determines the locations of the **missing orders** in the interference pattern.

Note that $d \sin\theta_m = m\lambda$ gives the mth **bright** fringe from the optic axis, but $d \sin\theta_m = \left(m + \frac{1}{2}\right)\lambda$ gives the $(m+1)^{th}$ **dark** fringe from the optic axis. The counting is different.

Example: A monkey shines a laser beam with a wavelength of 400 nm through a pair of slits with a spacing of 0.080 mm and a width of 0.020 mm. An interference pattern forms on a screen that is 60 cm from the slit.

(A) How far is the 3rd bright fringe from the optic axis?
Identify the given information in consistent units (nm, mm, and cm are all inconsistent). We'll put all of the distances in meters. Also, identify the desired unknown symbol.

- The wavelength is $\lambda = 4.00 \times 10^{-7}$ m (since 1 nm $= 10^{-9}$ m and $400 = 4.00 \times 10^{2}$).
- The **slit spacing** is $d = 8.0 \times 10^{-5}$ m (since 1 mm $= 10^{-3}$ m and $0.080 = 8.0 \times 10^{-2}$).
- The **slit width** is $a = 2.0 \times 10^{-5}$ m (since 1 mm $= 10^{-3}$ m and $0.020 = 2.0 \times 10^{-2}$).
- The distance from the slits to the screen is $L = 0.60$ m (since 1 cm $= 10^{-2}$ m).
- The question specifies the 3rd bright fringe from the optic axis: $m = 3$.
- We are looking for x_3.

To solve this problem without a calculator, we will make the small angle approximation (see Problem 108 and footnote 2 in Chapter 27): $\sin\theta_m \approx \tan\theta_m$. Since this question specified a **bright** fringe, $d\sin\theta_m = m\lambda$. Isolate the sine function to get $\sin\theta_m = \frac{m\lambda}{d}$. The other equation for double-slit interference is $\tan\theta_m = \frac{x_m}{L}$. For small angles, $\sin\theta_m \approx \tan\theta_m$, which means that $\frac{m\lambda}{d} \approx \frac{x_m}{L}$. Multiply both sides by L to get $x_m \approx \frac{m\lambda L}{d}$.

$$x_3 \approx \frac{3\lambda L}{d} = \frac{(3)(4\times 10^{-7})(0.6)}{8\times 10^{-5}} = 0.0090 \text{ m} \quad (3^{\text{rd}} \text{ bright fringe})$$

The center of the 3rd bright fringe appears $x_3 \approx 0.0090$ m $= 9.0$ mm from the optic axis.

(B) How far is the 3rd dark fringe from the optic axis?
Note that this question specified the 3rd **dark** fringe, whereas part (A) specified the 3rd **bright** fringe. The difference is that we must use the equation $d\sin\theta_m = \left(m + \frac{1}{2}\right)\lambda$. Divide both sides of the equation by d to get $\sin\theta_m = \left(m + \frac{1}{2}\right)\frac{\lambda}{d}$. The equation $\tan\theta_m = \frac{x_m}{L}$ still applies. For small angles, $\sin\theta_m \approx \tan\theta_m$, which means that $\left(m + \frac{1}{2}\right)\frac{\lambda}{d} \approx \frac{x_m}{L}$. Multiply both sides by L to get $x_m \approx \left(m + \frac{1}{2}\right)\frac{\lambda L}{d}$. Note that $m = 2$ (__not__ 3) for the third **dark** fringe (see the explanation on the bottom of page 328).

$$x_2 \approx \left(2 + \frac{1}{2}\right)\frac{\lambda L}{d} = \left(\frac{5}{2}\right)\frac{(4\times 10^{-7})(0.6)}{(8\times 10^{-5})} = 0.0075 \text{ m} \quad (3^{\text{rd}} \text{ dark fringe})$$

The center of the 3rd bright fringe appears $x_2 \approx 0.0075$ m $= 7.5$ mm from the optic axis.

Note: At the center of the interference pattern lies the central bright fringe (the 0th bright fringe). One spot to the side is the 1st dark fringe, then the 1st bright fringe, then the 2nd dark fringe, then the 2nd bright fringe, then the 3rd dark fringe, then the 3rd bright fringe, etc. Thus, $x_2 \approx 7.5$ mm for the 3rd **dark** fringe is less than $x_3 \approx 9.0$ mm for the 3rd **bright** fringe.

112. A monkey shines a laser beam with a wavelength of 600 nm through multiple closely spaced slits. An interference pattern forms on a screen that is 80 cm from the slits. The interference pattern is shown below. **Note**: If you're not using a calculator, this problem can be solved by making the following small angle approximations. For sufficiently small angles (much smaller than 30°), $\sin\theta \approx \theta$ and $\tan\theta \approx \theta$ provided that θ is expressed in **radians.**[2]

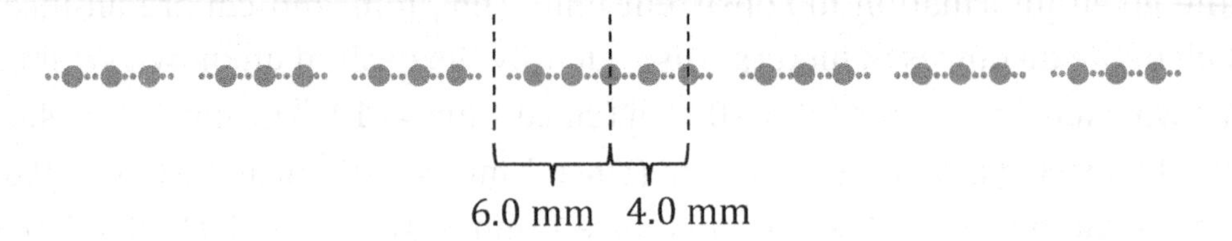

(A) How many slits are there? Explain your answer.

(B) What is the distance between the centers of neighboring slits?

(C) What is the width of each slit?

Want help? Check the hints section at the back of the book.

Answers: (A) 4 (B) 0.24 mm (C) 0.080 mm

[2] See Problem 108 and footnote 2 in Chapter 27.

30 POLARIZATION

Relevant Terminology

Polarization – when the electric fields of all of the waves in a beam of light oscillate in a common direction.
Transverse – a wave for which the amplitude of oscillation is perpendicular to the direction that the wave propagates.
Propagation – the transmission of a wave (often, with the sense of spreading out, like the ripples of water wave – but not necessarily).
Power – the instantaneous rate at which work is done.
Intensity – power per unit area.
Reflection – the ray of light that travels back into the same medium upon striking a surface.
Refraction – the bending of a ray of light associated with its change in speed as it passes from one medium into another medium.
Vacuum – a region of space completely devoid of matter. There is not even air.
Index of refraction – the ratio of the speed of light in vacuum to the speed of light in a medium. A **larger** index of refraction corresponds to a **slower** speed in the medium.
Normal – an imaginary line that is **perpendicular** to the surface.
Birefringence – the splitting of a beam of light in a particular type of crystal (such as calcite) into two different refracted rays which travel at different speeds in the crystal.
Scattering – the redirection of photons (particles of light) as they collide (or interact) with gas molecules (or other particles of matter).

Essential Concepts

Recall from Chapter 20 that light is a transverse electromagnetic wave with its electric field (and the same for its magnetic field) oscillating in a direction that is perpendicular to the direction of propagation. In the diagram below, a ray of light travels along the $+z$-axis while the electric field oscillates back and forth along the $+x$- and $-x$-axes.

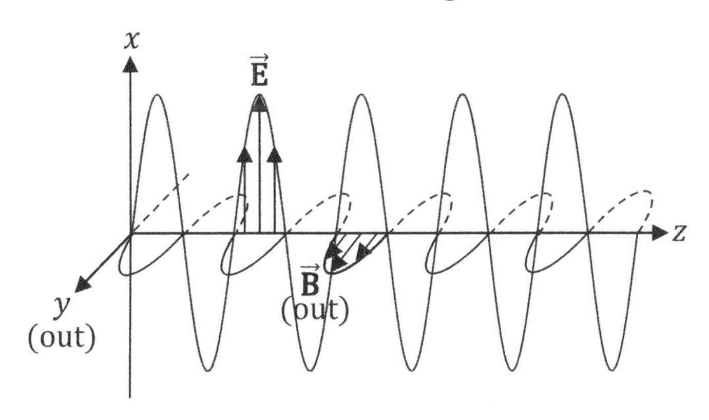

If the beam of light in the previous diagram is **unpolarized**, there would also be electric fields oscillating along the $+y$- and $-y$-axes, along the line $y = x$, along the line $y = 2x$, along the line $y = \frac{x}{2}$, and along every other line (there are an infinite number) between the x- and y-axes. However, if the beam of light in the previous diagram is linearly **polarized**, **all** of the photons in the beam would have their electric fields oscillating back and forth along the $+x$- and $-x$-axes (and no other directions).

Ordinary room light is typically **unpolarized**.

There are four common ways for unpolarized light to become (at least partially) **polarized**.

- Light can become polarized by **reflection**. The polarization (of the reflected ray) is complete when the refracted ray makes a 90° angle with the reflected ray. When the refracted and reflected rays are perpendicular, the incident angle is called **Brewster's angle**. When you view sunlight that reflects off a surface (such as a pond), it becomes partially polarized. Sunglasses are designed with a vertical transmission axis in order to help reduce glare from sunlight reflected off of a horizontal surface (which makes the intensity equal zero according to **Malus's law**, $I = I_0 \cos^2 \theta$, since $\cos 90° = 0$).
- Light becomes polarized when it refracts through a **birefringent** crystal (such as calcite). In this case, the incident ray splits into two refracted rays that travel at different speeds in the crystal.
- Light can become polarized through **scattering** (see Chapter 24). Sunlight that scatters in earth's atmosphere before reaching earth's surface is partially polarized.
- In the laboratory,[1] a material called a **polarizer** can be used to polarize light. A polarizer allows a light wave to pass through it if the direction of oscillation of its electric field matches the alignment of the polarizer, and blocks a light wave if the alignment of the polarizer is perpendicular to the direction of oscillation.
  - When the incident light is **unpolarized**, the intensity (I) of light passing through the polarizer is reduced by a factor of two compared to the intensity (I_0) of light entering the polarizer: $I = \frac{I_0}{2}$.
  - When the incident light is **already polarized** to begin with, the intensity (I) of light passing through the polarizer is reduced compared to the intensity (I_0) of light entering the polarizer according to **Malus's law**: $I = I_0 \cos^2 \theta$.

[1] At least one common optics kit includes both polarizers and retarders that look very similar, but which serve different purposes, so it's worth taking a moment to read the label carefully (a good habit in general).

Polarization Equations

When light passes through a polarizer, the equation depends on whether the light entering the polarizer is unpolarized or already polarized. In either case, I_0 represents the intensity of light entering the polarizer and I represents the intensity of light transmitted through the polarizer.

- If the light entering the polarizer is **unpolarized**, the intensity is reduced by half.

$$I = \frac{I_0}{2}$$

- If the light entering the polarizer is **already polarized**, the intensity depends on the angle (θ) between the light's initial polarization (which way the electric field oscillates before entering the polarizer) and the light's final polarization (which way the electric field oscillates after passing through the polarizer - determined by the orientation of the polarizer).

$$I = I_0 \cos^2 \theta$$

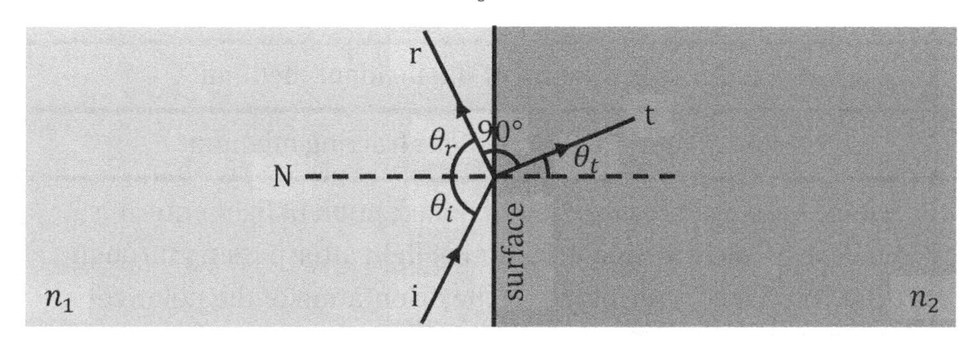

When light is completely polarized by **reflection**, the refracted ray is perpendicular to the reflected ray as illustrated above. In this case, the incident angle (θ_i) is called **Brewster's angle** ($\theta_B = \theta_i$). From the law of reflection, $\theta_r = \theta_i$, the reflected angle also equals Brewster's angle: $\theta_r = \theta_i = \theta_B$. The **refracted** angle (θ_t), however, is different. The refracted angle can be found by applying **Snell's law** (Chapter 22).

$$n_i \sin \theta_B = n_t \sin \theta_t$$

In the diagram above, note that $\theta_r + 90° + \theta_t = 180°$. Subtract 90° from both sides of the equation to get $\theta_r + \theta_t = 90°$. Since $\theta_r = \theta_i = \theta_B$, we can write $\theta_B + \theta_t = 90°$. Subtract θ_B from both sides of the equation to get $\theta_t = 90° - \theta_B$. Plug this into Snell's law.

$$n_i \sin \theta_B = n_t \sin(90° - \theta_B)$$

Apply the trig identity $\sin(x + y) = \sin x \cos y + \sin y \cos x$ with $x = 90°$ and $y = -\theta_B$.

$$n_i \sin \theta_B = n_t \sin 90° \cos(-\theta_B) + n_t \sin(-\theta_B) \cos 90°$$

Recall from trig that $\sin 90° = 1$, $\cos 90° = 0$, $\cos(-x) = \cos(x)$, and $\sin(-x) = -\sin(x)$.

$$n_i \sin \theta_B = n_t(1) \cos(\theta_B) - n_t \sin(\theta_B)\,(0)$$

$$n_i \sin \theta_B = n_t \cos \theta_B$$

Recall from trig that $\frac{\sin x}{\cos x} = \tan x$. Divide both sides of the previous equation by $\cos \theta_B$.

$$n_i \tan \theta_B = n_t$$

Symbols and Units

Symbol	Name	Units
I	intensity after passing through a polarizer	$\frac{\text{W}}{\text{m}^2}$
I_0	intensity before passing through a polarizer	$\frac{\text{W}}{\text{m}^2}$
i	the incident ray	
r	the reflected ray	
t	the refracted ray	
N	the normal line	
n	index of refraction	unitless
n_i	index of refraction of the incident medium	unitless
n_t	index of refraction of the refracting medium	unitless
θ	angle between the direction of polarization of light entering a polarizer and the repolarization of light after passing through the polarizer (determined by the orientation of the polarizer)	° or rad
θ_i	angle of incidence	° or rad
θ_r	angle of reflection	° or rad
θ_t	angle of refraction	° or rad
θ_B	Brewster's angle	° or rad

Common Values of the Index of Refraction

Medium	Index of Refraction
vacuum	1 (exactly)
air	1 (approximately)
water	$\frac{4}{3}$ (good to 4 significant figures)

Polarization Strategy

How you solve a problem involving polarization depends on which kind of problem it is:

- If a problem involves a polarizer, the equation depends on whether the light entering the polarizer is unpolarized or if it is already polarized to begin with. The symbol I_0 represents the intensity of light entering the polarizer and the symbol I represents the intensity of light transmitted through the polarizer.
  - If the light entering the polarizer is **unpolarized**, the intensity of light after passing through the polarizer is reduced by half.
  $$I = \frac{I_0}{2}$$
  - If the light entering the polarizer is **already polarized**, the intensity depends on the angle (θ) between the light's initial polarization and the orientation of the polarizer. The symbol θ equals how much the angle of polarization **changes** after passing through the polarizer.
  $$I = I_0 \cos^2 \theta$$
- A **reflected** ray is **completely** polarized when the reflected ray is perpendicular to the refracted ray. In this case, the incident angle is called Brewster's angle (θ_B) and Snell's law (Chapter 22) reduces to the following equation (as shown on page 335).
  $$n_i \tan \theta_B = n_t$$
  If you need to determine the angle of **reflection**, it is the same as the incident angle: $\theta_r = \theta_i = \theta_B$. If you need to determine the angle of **refraction**, apply geometry, knowing that the refracted ray is **perpendicular** to the reflected ray when the incident angle equals Brewster's angle.
- For a **birefringent** crystal (such as **calcite**), apply Snell's law (Chapter 22).

Basic Trig Functions for Special Angles

θ	0°	30°	45°	60°	90°
$\sin\theta$	0	$\frac{1}{2}$	$\frac{\sqrt{2}}{2}$	$\frac{\sqrt{3}}{2}$	1
$\cos\theta$	1	$\frac{\sqrt{3}}{2}$	$\frac{\sqrt{2}}{2}$	$\frac{1}{2}$	0
$\tan\theta$	0	$\frac{\sqrt{3}}{3}$	1	$\sqrt{3}$	undef.

Example: A ray of light travels from air to glass. The index of refraction of the glass is $\sqrt{3}$. What angle of incidence would make the reflected ray completely polarized?

Identify the known quantities in appropriate units.

- The index of refraction of the air (the incident medium) is $n_i \approx 1$.
- The index of refraction of the glass (the refracting medium) is $n_t = \sqrt{3}$.

Since the reflected ray is **completely polarized**, we may apply the simplified version of Snell's law involving **Brewster's angle** (see the bottom of page 335).

$$n_i \tan\theta_B = n_t$$
$$(1)\tan\theta_B = \sqrt{3}$$
$$\tan\theta_B = \sqrt{3}$$

Take the inverse tangent of both sides of the equation. Note that $\tan^{-1}[\tan(\theta_B)] = \theta_B$.

$$\theta_B = \tan^{-1}\left(\sqrt{3}\right) = 60°$$

The incident angle (**Brewster's angle**) is $\theta_B = 60°$ because $\tan(60°) = \sqrt{3}$ (see page 337).

Example: A ray of unpolarized light with an intensity of $800\,\frac{\text{W}}{\text{m}^2}$ is incident upon a pair of polarizers, as illustrated below. What is the intensity of the light after passing through both polarizers?

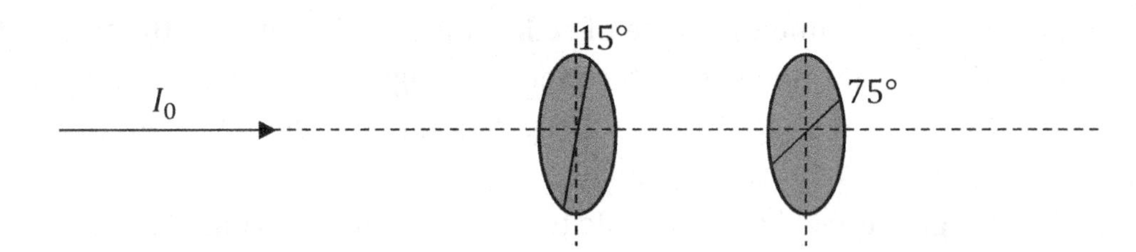

The initial intensity is $I_0 = 800\,\frac{\text{W}}{\text{m}^2}$. Since the light is initially **unpolarized**, the first polarizer simply reduces the intensity by **half**:

$$I_1 = \frac{I_0}{2} = \frac{800}{2} = 400\,\frac{\text{W}}{\text{m}^2}$$

After passing through the first polarizer, the ray of light becomes polarized along the orientation of the first polarizer. The second polarizer then reduces the intensity according to Malus's law, where $\theta = 75° - 15° = 60°$ is the angle between the two polarizers.

$$I_2 = I_1 \cos^2\theta = (400)\cos^2 60° = (400)\left(\frac{1}{2}\right)^2 = (400)\left(\frac{1}{4}\right) = 100\,\frac{\text{W}}{\text{m}^2}$$

The final intensity is $I_2 = 100\,\frac{\text{W}}{\text{m}^2}$.

113. A ray of light travels from one medium with an index of refraction of $\sqrt{2}$ to another medium with an index of refraction of $\sqrt{6}$. If the reflected ray is completely polarized, what are the angles of incidence, reflection, and refraction?

Want help? Check the hints section at the back of the book.

Answers: 60°, 60°, 30°

114. A ray of unpolarized light with an intensity of $960 \frac{\text{W}}{\text{m}^2}$ is incident upon three polarizers, as illustrated below.

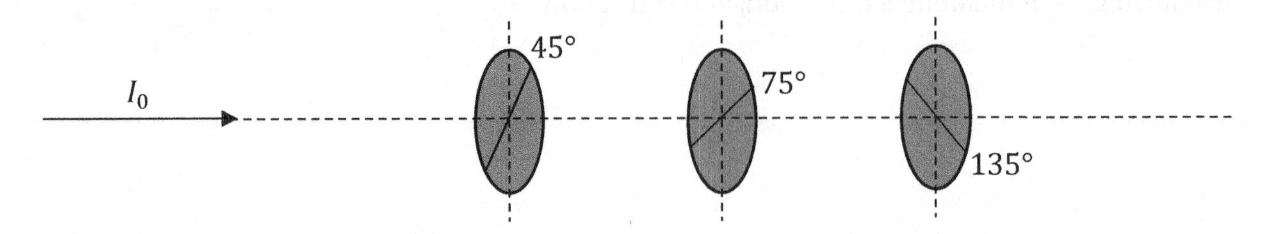

(A) What is the intensity of the light after passing through all three polarizers?

(B) If the middle polarizer is removed, what would be the intensity of the light after passing through both of the remaining two polarizers?

Want help? Check the hints section at the back of the book.

Answers: $90 \frac{\text{W}}{\text{m}^2}$, 0

HINTS, INTERMEDIATE ANSWERS, AND EXPLANATIONS

How to Use This Section Effectively

Think of hints and intermediate answers as training wheels. They help you proceed with your solution. When you stray from the right path, the hints help you get back on track. The answers also help to build your confidence.

However, if you want to succeed in a physics course, you must eventually learn to rely less and less on the hints and intermediate answers. Make your best effort to solve the problem on your own before checking for hints, answers, and explanations. When you need a hint, try to find just the hint that you need to get over your current hurdle. Refrain from reading additional hints until you get further into the solution.

When you make a mistake, think about what you did wrong and what you should have done differently. Try to learn from your mistake so that you don't repeat the mistake in other solutions.

It's natural for students to check hints and intermediate answers repeatedly in the early chapters. However, at some stage, you would like to be able to consult this section less frequently. When you can solve more problems without help, you know that you're really beginning to master physics.

Would You Prefer to See Full Solutions?

Full solutions are like a security blanket: Having them makes students feel better. But full solutions are also dangerous: Too many students rely too heavily on the full solutions, or simply read through the solutions instead of working through the solutions on their own. Students who struggle through their solutions and improve their solutions only as needed tend to earn better grades in physics (though comparing solutions **after** solving a problem is always helpful).

It's a challenge to get just the right amount of help. In the ideal case, you would think your way through every solution on your own, seek just the help you need to make continued progress with your solution, and wait until you've solved the problem as best you can before consulting full solutions or reading every explanation.

With this in mind, full solutions to all problems are contained in a separate book. This workbook contains hints, intermediate answers, explanations, and several directions to help walk you through the steps of every solution, which should be enough to help most students figure out how to solve all of the problems. However, if you need the security of seeing **full solutions to all problems**, look for the book 100 Instructive Calculus-based Physics Examples with ISBN 978-1-941691-21-2. The solution to every problem in this workbook can be found in that book.

How to Cover up Hints that You Don't Want to See too Soon

There is a simple and effective way to cover up hints and answers that you don't want to see too soon:

- Fold a blank sheet of paper in half and place it in the hints and answers section. This will also help you bookmark this handy section of the book.
- Place the folded sheet of paper just below your current reading position. The folded sheet of paper will block out the text below.
- When you want to see the next hint or intermediate answer, just drop the folded sheet of paper down slowly, just enough to reveal the next line.
- This way, you won't reveal more hints or answers than you need.

You learn more when you force yourself to struggle through the problem. Consult the hints and answers when you really need them, but try it yourself first. After you read a hint, try it out and think it through as best you can before consulting another hint. Similarly, when checking intermediate answers to build confidence, try not to see the next answer before you have a chance to try it on your own first.

Chapter 1: Sine Waves

1. First read the values of x_{max} (the vertical coordinate of a crest), x_{min} (the vertical coordinate of a trough), t_{1m} (the horizontal coordinate of the first crest), t_{2m} (the horizontal coordinate of the second crest), and x_0 (the initial vertical coordinate) directly from the graph.

- You should get $x_{max} = 2.0$ m, $x_{min} = -6.0$ m, $t_{1m} \approx 0.3$ s, $t_{2m} \approx 2.3$ s, and $x_0 = 0$.

(A) First use the equation $x_e = \frac{x_{max}+x_{min}}{2}$.

- You should get $x_e = -2.0$ m. Note the **minus** sign.
- Now use the equation $A = x_{max} - x_e$.
- The amplitude is $A = 4.0$ m. Note that the two minus signs make a plus sign.

(B) Use the equation $T = t_{2m} - t_{1m}$.

- The period is $T = 2.0$ s. (It's actually better to subtract 0.3 s from 8.3 s and divide by 4 cycles: Using 4 cycles helps to reduce interpolation error.)

(C) We already found this in part (A).

- The equilibrium position is $x_e = -2.0$ m. It's **negative**.

(D) Use the equation $f = \frac{1}{T}$. Recall that we found the period in part (B).

- The frequency is $f = \frac{1}{2}$ Hz $= 0.50$ Hz.

(E) Read off the vertical coordinate when $t = 5.0$ s.

- The banana's position is $x = -4.0$ m when $t = 5.0$ s.

(F) First determine the correct Quadrant using the method described on pages 10-11.

- Label the angles 0°, 90°, 180°, 270°, and 360° on the sine wave, as discussed on pages 10-11. See the diagram below.

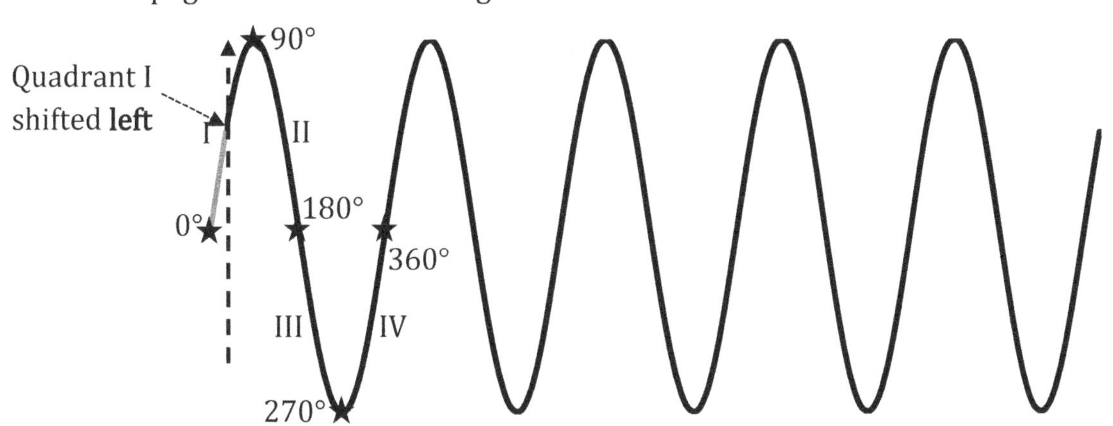

- This sine wave begins in Quadrant I, between 0° and 90°.
- Use the equation $\varphi = \sin^{-1}\left(\frac{x_0 - x_e}{A}\right)$. Plug in numbers.
- You should get $\varphi = \sin^{-1}\left(\frac{1}{2}\right)$. Note that $0 - (-2) = 0 + 2 = +2$ and $\frac{2}{4} = \frac{1}{2}$.
- The phase angle is $\varphi = 30°$ in Quadrant I.
- Convert the phase angle to **radians**. Multiply by $\frac{\pi \text{ rad}}{180°}$.
- The phase angle is $\varphi = \frac{\pi}{6}$ rad.

---

2. First compare the given equation with numbers, $x = 3\sin\left(\pi t - \frac{5\pi}{6}\right) + 2$, to the general equation for a sine wave in symbols, $x = A\sin(\omega_0 t + \varphi) + x_e$, in order to identify the symbols A, ω_0, φ, and x_e.

- You should get $A = 3.0$ m, $\omega_0 = \pi$ rad/s, $\varphi = -\frac{5\pi}{6}$ rad, and $x_e = 2.0$ m.

(A) The amplitude is $A = 3.0$ m. This is found simply by comparing the two equations.

(B) The equilibrium position is $x_e = 2.0$ m. This is also found from the comparison.

(C) The angular frequency is $\omega_0 = \pi$ rad/s. This is also found from the comparison.

(D) Use the equation $x_{min} = x_e - A$.

- The minimum value is $x_{min} = -1.0$ m. Note the **minus** sign.

(E) Use the equation $\omega_0 = \frac{2\pi}{T}$.

- Multiply both sides of the equation by T.
- You should get $\omega_0 T = 2\pi$.
- Divide both sides of the equation by 2π.
- You should get $T = \frac{2\pi}{\omega_0}$. Plug in $\omega_0 = \pi$ rad/s.
- The period is $T = 2.0$ s.

(F) Convert the phase angle, $\varphi = -\frac{5\pi}{6}$ rad, from radians to degrees, using $180° = \pi$ rad.

- The phase angle in degrees is $\varphi = -150°$. Note that it is **negative**. It would also be correct to write $\varphi = +210°$, since adding $360°$ to an angle yields an equivalent angle. However, it would be **incorrect** to write $+150°$ (that has the wrong sign).

(G) Plug $t = \frac{5}{3}$ s into the given equation. Solve for x.

- If you use a calculator, make sure that it is in **radians** mode.
- Leave the phase angle in **radians**. Use $\varphi = -\frac{5\pi}{6}$ rad. (**Don't** use $\varphi = -150°$.)
- Subtract fractions with a **common denominator**.
- Note that $\frac{5\pi}{3} = \frac{10\pi}{6}$. (Multiply the numerator and denominator both by 2.)
- You should get $x = 3\sin\left(\frac{5\pi}{6}\right) + 2$. Note that $\frac{10\pi}{6} - \frac{5\pi}{6} = \frac{5\pi}{6}$.
- Note that $\sin\left(\frac{5\pi}{6}\right) = \frac{1}{2}$ if your calculator is in **radians** mode. (Alternatively, you could convert $\frac{5\pi}{6}$ rad to $150°$ to see that $\sin 150° = \frac{1}{2}$ in degrees mode.)
- The banana's position is $x = 3.5$ m when $t = \frac{5}{3}$ s.

(H) Plug $x = 0.5$ m into the given equation. Solve for t.

- First apply algebra to isolate the sine function.
- You should get $-\frac{1}{2} = \sin\left(\pi t - \frac{5\pi}{6}\right)$. Note that $0.5 - 2 = -1.5$ and $-\frac{1.5}{3} = -\frac{1}{2}$.
- Take the inverse sine of both sides of the equation.
- You should get $\sin^{-1}\left(-\frac{1}{2}\right) = \pi t - \frac{5\pi}{6}$.
- There are two possible answers for the inverse sine. The reference angle is $\frac{\pi}{6}$ rad (corresponding to $30°$) because $\sin\left(\frac{\pi}{6}\right) = \frac{1}{2}$, but $\frac{\pi}{6}$ rad is **not** one the two answers for $\sin^{-1}\left(-\frac{1}{2}\right)$. The correct answers to $\sin^{-1}\left(-\frac{1}{2}\right)$ lie in Quadrants III and IV because the sine function is negative in Quadrants III and IV. If you need a review of trig essentials, review Chapter 9 in Volume 1 of this series.
- In Quadrant III, add π radians (corresponding to $180°$) to the reference angle to get $\frac{7\pi}{6}$ rad (corresponding to $210°$). Note that $\pi + \frac{\pi}{6} = \frac{6\pi}{6} + \frac{\pi}{6} = \frac{7\pi}{6}$.
- In Quadrant IV, subtract the reference angle from 2π radians (corresponding to $360°$) to get $\frac{11\pi}{6}$ rad (corresponding to $330°$). Note that $2\pi - \frac{\pi}{6} = \frac{12\pi}{6} - \frac{\pi}{6} = \frac{11\pi}{6}$.
- Plug both possible angles, $\frac{7\pi}{6}$ rad and $\frac{11\pi}{6}$ rad, into the previous equation for time, which was $\sin^{-1}\left(-\frac{1}{2}\right) = \pi t - \frac{5\pi}{6}$.
- You should get $\frac{7\pi}{6} = \pi t - \frac{5\pi}{6}$ and $\frac{11\pi}{6} = \pi t - \frac{5\pi}{6}$.
- Solve for time in both equations separately using algebra. Note that π cancels.
- You should get $t = 2.0$ s and $t = \frac{8}{3}$ s. However, we're **not** finished yet.

- There are more solutions than the two times, $t = 2.0$ s and $t = \frac{8}{3}$ s, that we have found thus far. The reason is that the sine wave repeats itself. Every period, the sine wave will return to the same position.
- Recall from part (E) that the period is $T = 2.0$ s.
- This means we can add or subtract multiples of 2.0 s to our previous times and also obtain valid answers.
- Let's begin by finding the smallest nonnegative times. Subtract one period ($T = 2.0$ s) from our previous times ($t = 2.0$ s and $t = \frac{8}{3}$ s) to get $t = 0$ and $t = \frac{2}{3}$ s. Note that $\frac{8}{3} - 2 = \frac{8}{3} - \frac{6}{3} = \frac{2}{3}$.
- You can verify that $x = 0.5$ m when you plug either $t = 0$ or $t = \frac{2}{3}$ s into the original equation. This shows that these are correct answers.
- Make your answer more general by adding $2n$ seconds (since $T = 2.0$ s) to each of these answers.
- The banana will be at $x = 0.5$ m when $t = 2n$ s or $t = \frac{2}{3}$ s $+ 2n$ s, where n is a nonnegative integer (0, 1, 2, 3...). This means that the answers are $t = 0$, $t = \frac{2}{3}$ s, $t = 2.0$ s, $t = \frac{8}{3}$ s, $t = 4.0$ s, $t = \frac{14}{3}$ s, $t = 6.0$ s, $t = \frac{20}{3}$ s, and so on. This is because the sine wave repeats itself every $T = 2.0$ s.

---

3. First read the values of x_{max} (the vertical coordinate of a crest), x_{min} (the vertical coordinate of a trough), t_{1m} (the horizontal coordinate of the first crest), t_{2m} (the horizontal coordinate of the second crest), and x_0 (the initial vertical coordinate) directly from the graph.

- Check that you read the graph correctly: $x_{max} = 7.0$ m, $x_{min} = -5.0$ m, $t_{1m} \approx 1.6$ s, $t_{2m} \approx 6.6$ s, and $x_0 = -2.0$ m.
- The general equation for a sine wave is $x = A\sin(\omega_0 t + \varphi) + x_e$.
- We can write the equation for the sine wave once we find numerical values for A, ω_0, φ, and x_e. Find these one at a time.
- First find x_e. Use the equation $x_e = \frac{x_{max}+x_{min}}{2}$.
- The equilibrium position is $x_e = 1.0$ m.
- Next find A. Use the equation $A = x_{max} - x_e$.
- The amplitude is $A = 6.0$ m.
- Next find ω_0. You will need to find T first. Use the equation $T = t_{2m} - t_{1m}$.
- The period is $T = 5.0$ s.
- Now use the equation $\omega_0 = \frac{2\pi}{T}$.
- The angular frequency is $\omega_0 = \frac{2\pi}{5}$ rad/s.

- Finally, find φ. First determine the correct Quadrant using the method described on pages 10-11.
- Label the angles 0°, 90°, 180°, 270°, and 360° on the sine wave, as discussed on pages 10-11. See the diagram below.

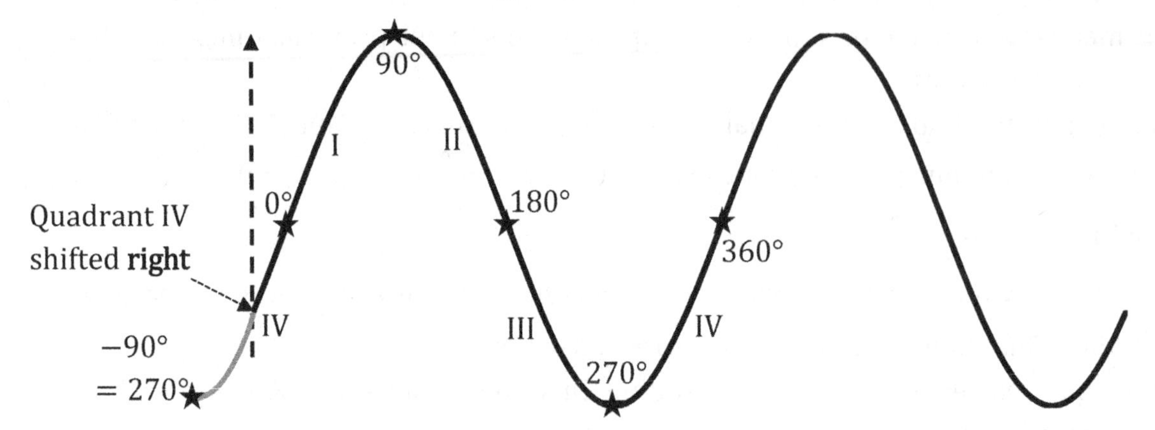

- This sine wave begins in Quadrant IV, between 270° and 360° (or, equivalently, between −90° and 0°).
- Use the equation $\varphi = \sin^{-1}\left(\frac{x_0 - x_e}{A}\right)$. Plug in numbers.
- You should get $\varphi = \sin^{-1}\left(-\frac{1}{2}\right)$. Note that $-2 - 1 = -3$ and $-\frac{3}{6} = -\frac{1}{2}$.
- The reference angle is $\varphi = 30°$, but this is **not** the phase angle because the phase angle lies in Quadrant IV.
- In Quadrant IV, subtract the reference angle from 360°. You should get 330°.
- Convert the phase angle to **radians**. Multiply by $\frac{\pi \text{ rad}}{180°}$.
- The phase angle is $\varphi = \frac{11\pi}{6}$ rad.
- Now we have $A = 6.0$ m, $\omega_0 = \frac{2\pi}{5}$ rad/s, $\varphi = \frac{11\pi}{6}$ rad, and $x_e = 1.0$ m.
- Plug these values into the general equation for a sine wave, $x = A\sin(\omega_0 t + \varphi) + x_e$.
- The equation for this sine wave is $x = 6\sin\left(\frac{2\pi t}{5} + \frac{11\pi}{6}\right) + 1$, where SI units have been suppressed to avoid clutter. It would also be correct to write the equation as $x = 6\sin\left(\frac{2\pi t}{5} - \frac{\pi}{6}\right) + 1$, since $\varphi = \frac{11\pi}{6}$ rad (or 330°) is equivalent to $\varphi = -\frac{\pi}{6}$ rad (or 30°), since you can add or subtract 2π rad (or 360°) to any angle and obtain an equivalent angle.
- We can think of this sine wave as being shifted $\frac{\pi}{6}$ rad to the **right**, since a **negative** phase angle ($\varphi = -\frac{\pi}{6}$ rad) shifts the graph to the right, or we can think of this sine wave as being shifted $\frac{11\pi}{6}$ rad to the **left**, since a **positive** phase angle ($\varphi = \frac{11\pi}{6}$ rad) shifts the graph to the left.

Chapter 2: Simple Harmonic Motion

4. First read the values of x_{max} (the vertical coordinate of a crest), x_{min} (the vertical coordinate of a trough), t_{1m} (the horizontal coordinate of the first crest), t_{2m} (the horizontal coordinate of the second crest), and x_0 (the initial vertical coordinate) directly from the graph.

- Check your intermediate answers: $x_{max} = 5.0$ m, $x_{min} = -9.0$ m, $t_{1m} \approx 0.7$ s, $t_{2m} \approx 4.7$ s, and $x_0 = 1.5$ m. (Note that x_0 is clearly less than 2.)
- Calculate the equilibrium position: $x_e = \frac{x_{max}+x_{min}}{2}$. The equilibrium position is $x_e = -2.0$ m. Note the **minus** sign.
- Calculate the amplitude: $A = x_{max} - x_e$. The amplitude is $A = 7.0$ m. Note that the two minus signs make a plus sign.
- Calculate the period: $T = t_{2m} - t_{1m}$. The period is $T = 4.0$ s.
- Calculate the angular frequency: $\omega_0 = \frac{2\pi}{T}$. The angular frequency is $\omega_0 = \frac{\pi}{2}$ rad/s.
- Determine the phase angle. Label the angles 0°, 90°, 180°, 270°, and 360° on the sine wave, as discussed on pages 10-11. See the diagram below.

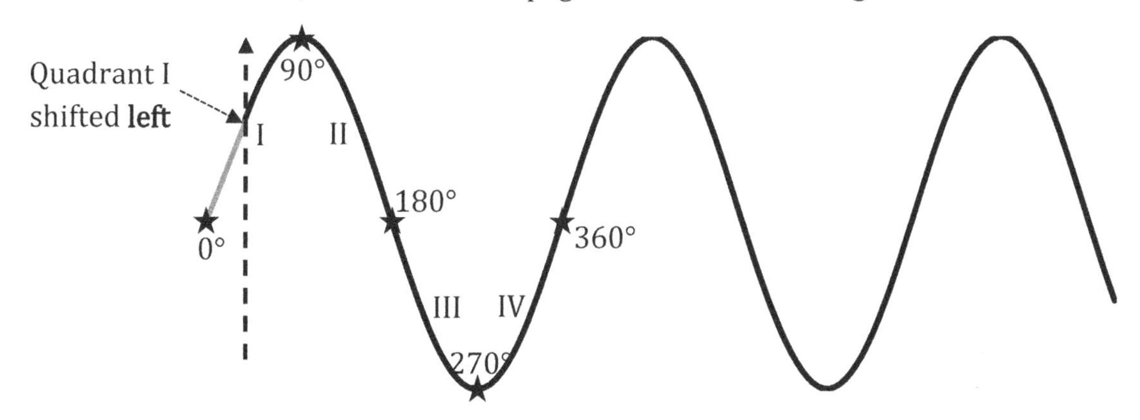

- This sine wave begins in Quadrant I, between 0° and 90°.
- Use the equation $\varphi = \sin^{-1}\left(\frac{x_0 - x_e}{A}\right)$. Plug in numbers. Recall that $x_0 = 1.5$ m.
- You should get $\varphi = \sin^{-1}\left(\frac{1}{2}\right)$. Note that $1.5 - (-2) = 1.5 + 2 = +3.5$ and $\frac{3.5}{7} = \frac{1}{2}$.
- The phase angle is $\varphi = 30°$ in Quadrant I.
- Convert the phase angle to **radians**. Multiply by $\frac{\pi \text{ rad}}{180°}$.
- The phase angle is $\varphi = \frac{\pi}{6}$ rad.

(A) Plug numbers into the equation $v_x = A\omega_0 \cos(\omega_0 t + \varphi)$. Set your calculator in **radians**.

- Recall that $A = 7.0$ m, $\omega_0 = \frac{\pi}{2}$ rad/s, and $\varphi = \frac{\pi}{6}$ rad.
- The velocity at $t = 6.0$ s is $v_x = -\frac{7\pi\sqrt{3}}{4}$ m/s. Using a calculator, $v_x = -9.5$ m/s.

(B) Use the equation $v_m = A\omega_0$.

- The maximum speed is $v_m = \frac{7\pi}{2}$ m/s. Using a calculator, $v_m = 11$ m/s.

(C) Use the equation $a_x = -A\omega_0^2 \sin(\omega_0 t + \varphi)$. Set your calculator in **radians**.

- The acceleration at $t = 6.0$ s is $a_x = \frac{7\pi^2}{8}$ m/s². Using a calculator, $a_x = 8.6$ m/s².

(D) Use the equation $a_m = A\omega_0^2$.

- The maximum acceleration is $a_m = \frac{7\pi^2}{4}$ m/s². Using a calculator, $a_m = 17$ m/s².

5. First read the values of v_m (the vertical coordinate of a crest of a **velocity** graph), t_{1m} (the horizontal coordinate of the first crest), t_{2m} (the horizontal coordinate of the second crest), and v_{x0} (the initial velocity) directly from the graph.

- Check your answers: $v_m = 6.0$ m/s, $t_{1m} \approx 0.12$ s, $t_{2m} \approx 0.62$ s, and $v_{x0} = 0$.
- Calculate the period: $T = t_{2m} - t_{1m}$. The period is $T = 0.50$ s.
- Calculate the angular frequency: $\omega_0 = \frac{2\pi}{T}$. The angular frequency is $\omega_0 = 4\pi$ rad/s.
- Determine the phase angle. Label the angles 0°, 90°, 180°, 270°, and 360° on the **cosine** wave, as discussed on pages 35-36. See the diagram below.
- **Note**: Although this graph looks just like a **sine** wave, it's really a shifted **cosine** wave. We know that because the equation for **velocity** involves **cosine**, **not** sine: $v_x = A\omega_0 \cos(\omega_0 t + \varphi)$. You must label the angles for **cosine**, **not** sine.

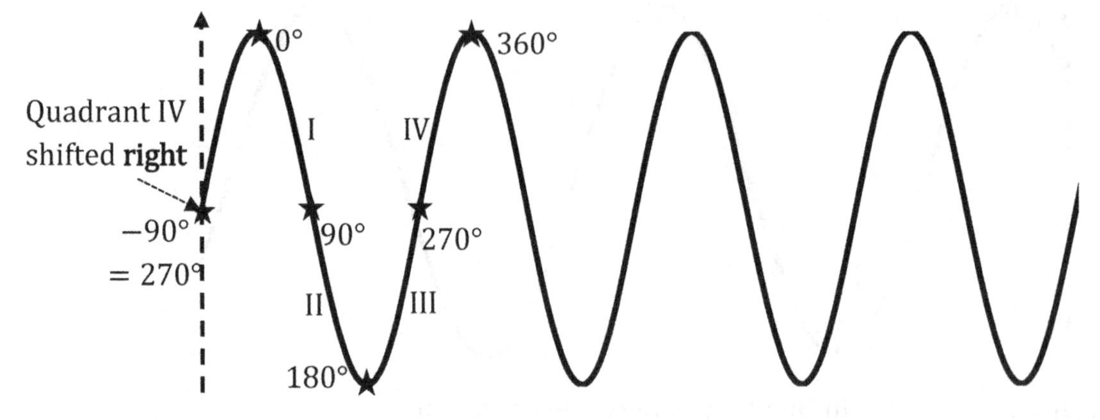

- This **cosine** wave begins at 270° (equivalent to −90°).
- **Note**: If this had been a position graph, it would have been a sine wave, and the phase angle would have been 0°. However, this is a **velocity** graph, which is a **cosine** wave. As a cosine wave, it is shifted right 90°, which makes the phase angle −90° (which is equivalent to 270°). Compare with the standard **cosine** graph below.

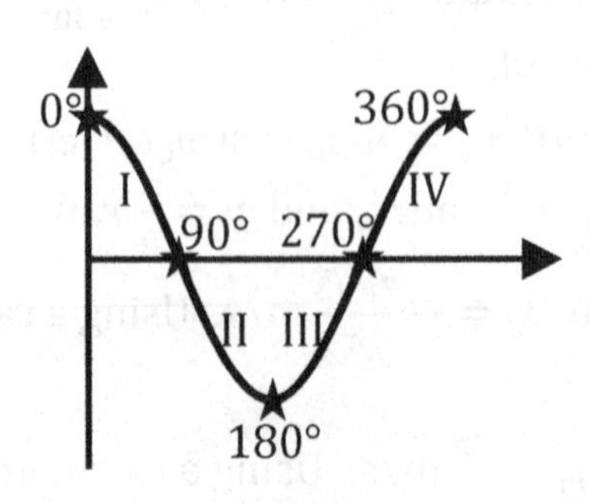

- Ordinarily, we would need to use the equation $\varphi = \cos^{-1}\left(\frac{v_{x0}}{A\omega_0}\right)$ for a velocity graph. However, if the answer happens to be 0°, 90°, 180°, 270°, and 360°, you can get the answer directly from the graph without having to use the above equation. In this case, the equation would yield $\varphi = \cos^{-1}(0)$, for which the two possible answers are 90° and 270°, since $\cos 90° = 0$ and $\cos 270° = 0$. The correct answer is 270° based on the angles that we drew on the previous page.
- The phase angle is $\varphi = 270°$.
- Convert the phase angle to **radians**. Multiply by $\frac{\pi \text{ rad}}{180°}$.
- The phase angle is $\varphi = \frac{3\pi}{2}$ rad.

(A) Simply read off the vertical value (v_x) when the horizontal value (t) equals 1.0 s.

- The velocity at $t = 1.0$ s is $v_x = 0$.

(B) Simply read off the vertical value (v_x) at a crest (where there is a maximum).

- The maximum speed is $v_m = 6.0$ m/s.

(C) Use the equation $a_x = -A\omega_0^2 \sin(\omega_0 t + \varphi)$. Set your calculator in **radians**.

- You will need to determine A before using the above equation.
- Use the equation $v_m = A\omega_0$ to find A. Divide both sides of the equation by ω_0. You should get $A = \frac{v_m}{\omega_0}$. Plug in numbers. You should get $A = \frac{3}{2\pi}$ m or $A = 0.48$ m.
- Plug numbers into the equation $a_x = -A\omega_0^2 \sin(\omega_0 t + \varphi)$. Recall that $A = \frac{3}{2\pi}$ m, $\omega_0 = 4\pi$ rad/s, and $\varphi = \frac{3\pi}{2}$ rad.
- The acceleration at $t = 1.0$ s is $a_x = 24\pi$ m/s^2. Using a calculator, $a_x = 75$ m/s^2.

(D) Use the equation $a_m = A\omega_0^2$.

- The maximum acceleration is $a_m = 24\pi$ m/s^2. Using a calculator, $a_m = 75$ m/s^2.

---

6. First read the values of a_m (the vertical coordinate of a crest of an **acceleration** graph), t_{1m} (the horizontal coordinate of the first crest), t_{2m} (the horizontal coordinate of the second crest), and a_{x0} (the initial acceleration) directly from the graph.

- Check your answers: $a_m = 4.0$ m/s^2, $t_{1m} = 0$, $t_{2m} = 2.0$ s, and $a_{x0} = 4.0$ m/s^2.
- Calculate the period: $T = t_{2m} - t_{1m}$. The period is $T = 2.0$ s.
- Calculate the angular frequency: $\omega_0 = \frac{2\pi}{T}$. The angular frequency is $\omega_0 = \pi$ rad/s.
- Determine the phase angle. Label the angles 0°, 90°, 180°, 270°, and 360° on the **negative sine** wave. See the diagram that follows.
- This is a little tricky: Acceleration is a **negative** sine wave, not a regular sine wave. That's because the equation for acceleration, $a_x = -A\omega_0^2 \sin(\omega_0 t + \varphi)$, includes a **minus sign**. We need to draw a negative sine wave, which means to flip a sine wave upside down. Then we need to figure out how much this negative sine wave is shifted in order to determine the phase angle.

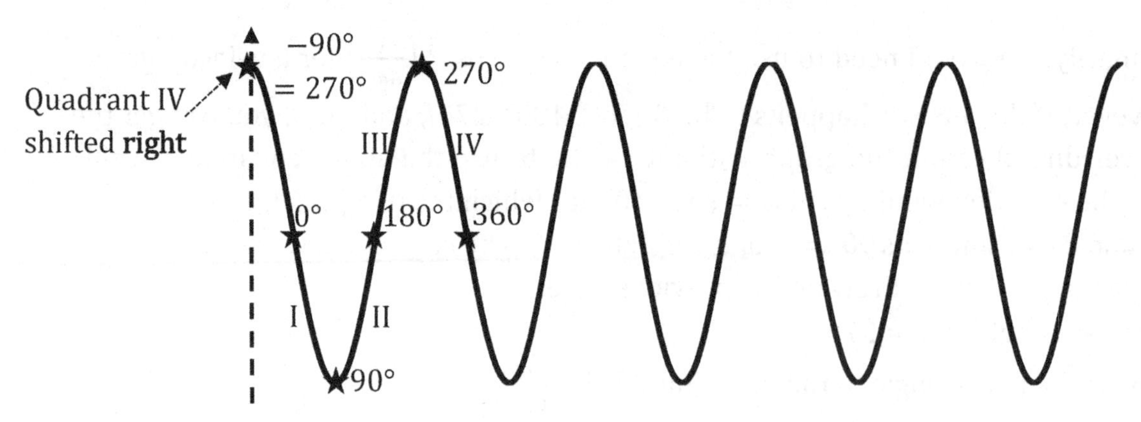

- This **negative** sine wave begins at 270° (equivalent to −90°).
- **Note**: If this had been a position graph, it would have been a sine wave, and the phase angle would have been 90°. However, this is an **acceleration** graph, which is a **negative** sine wave. As a negative sine wave, it is shifted right 90°, which makes the phase angle −90° (which is equivalent to 270°). For reference, below we drew a standard **negative** sine wave: Get this by flipping a sine wave upside down.

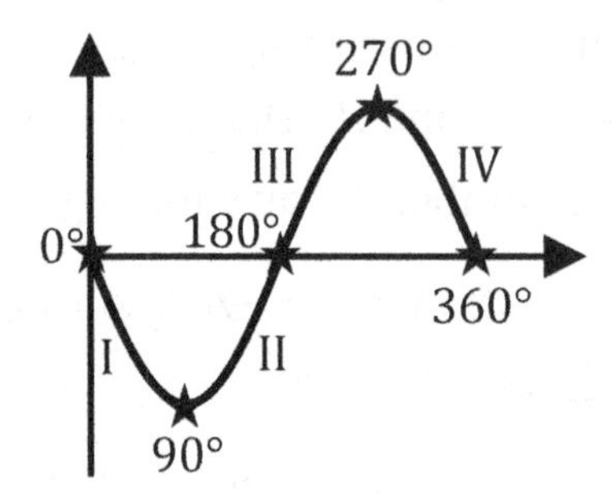

- Ordinarily, we would need to use the equation $\varphi = \sin^{-1}\left(-\frac{a_{x0}}{A\omega_0^2}\right)$ for an acceleration graph. However, if the answer happens to be 0°, 90°, 180°, 270°, and 360°, you can get the answer directly from the graph without having to use the above equation. In this case, the equation would yield $\varphi = \sin^{-1}(-1)$, for which the answer is 270°, since $\sin 270° = -1$.
- The phase angle is $\varphi = 270°$.
- Convert the phase angle to **radians**. Multiply by $\frac{\pi \text{ rad}}{180°}$.
- The phase angle is $\varphi = \frac{3\pi}{2}$ rad.

(A) Plug numbers into the equation $v_x = A\omega_0 \cos(\omega_0 t + \varphi)$. Set your calculator in **radians**.

- Use the equation $a_m = A\omega_0^2$ to find A. Divide both sides of the equation by ω_0^2. You should get $A = \frac{a_m}{\omega_0^2}$. Recall that $a_m = 4.0$ m/s^2 and $\omega_0 = \pi$ rad/s. You should get $A = \frac{4}{\pi^2}$ m or $A = 0.41$ m.
- Plug numbers into the equation $v_x = A\omega_0 \cos(\omega_0 t + \varphi)$. Recall that $\varphi = \frac{3\pi}{2}$ rad.
- The velocity at $t = 3.0$ s is $v_x = 0$.

(B) Use the equation $v_m = A\omega_0$. Recall that $A = \frac{4}{\pi^2}$ m and $\omega_0 = \pi$ rad/s.

- The maximum speed is $v_m = \frac{4}{\pi}$ m/s. Using a calculator, $v_m = 1.3$ m/s.

(C) Simply read off the vertical value (a_x) when the horizontal value (t) equals 3.0 s.

- The acceleration at $t = 3.0$ s is $a_x = -4.0$ m/s^2. Note the **minus sign**.

(D) Simply read off the vertical value (a_x) at a crest (where there is a maximum).

- The maximum acceleration is $a_m = 4.0$ m/s^2.

---

7. First compare the given equation with numbers, $x = 3\sin\left(2\pi t - \frac{\pi}{4}\right) - 4$, to the general equation for the **position** of a sine wave in symbols, $x = A\sin(\omega_0 t + \varphi) + x_e$, in order to identify the symbols A, ω_0, φ, and x_e.

- You should get $A = 3.0$ m, $\omega_0 = 2\pi$ rad/s, $\varphi = -\frac{\pi}{4}$ rad, and $x_e = -4.0$ m.

(A) Use the equation $v_x = A\omega_0\cos(\omega_0 t + \varphi)$. Put your calculator in **radians** mode.

- Subtract fractions with a **common denominator**.
- You should get $v_x = 6\pi\cos\left(\frac{3\pi}{4}\right)$.
- Note that $\cos\left(\frac{3\pi}{4}\right) = -\frac{\sqrt{2}}{2}$ if your calculator is in **radians** mode. (Alternatively, you could convert $\frac{3\pi}{4}$ rad to 135° to see that $\cos 135° = -\frac{\sqrt{2}}{2}$ in degrees mode.)
- The banana's velocity is $v_x = -3\pi\sqrt{2}$ m/s when $t = 0.50$ s. If you use a calculator, this works out to $v_x = -13$ m/s.

(B) Use the equation $v_m = A\omega_0$.

- The maximum speed is $v_m = 6\pi$ m/s. Using a calculator, $v_m = 19$ m/s.

(C) Use the equation $a_x = -A\omega_0^2\sin(\omega_0 t + \varphi)$. Put your calculator in **radians** mode.

- Subtract fractions with a **common denominator**.
- You should get $a_x = -12\pi^2\sin\left(\frac{3\pi}{4}\right)$.
- Note that $\sin\left(\frac{3\pi}{4}\right) = \frac{\sqrt{2}}{2}$ if your calculator is in **radians** mode. (Alternatively, you could convert $\frac{3\pi}{4}$ rad to 135° to see that $\sin 135° = \frac{\sqrt{2}}{2}$ in degrees mode.)
- The banana's acceleration is $a_x = -6\pi^2\sqrt{2}$ m/s^2 when $t = 0.50$ s. If you use a calculator, this works out to $a_x = -84$ m/s^2.

(D) Use the equation $a_m = A\omega_0^2$.

- The maximum acceleration is $a_m = 12\pi^2$ m/s^2. Using a calculator, $a_m = 118$ m/s^2. (That's about 12 gravities, which is quite much for a banana, but we selected numbers that make it possible to work out the math without a calculator, rather than choosing numbers that were more realistic.)

(E) Plug $v_x = \pm 3\pi$ m/s in the equation $v_x = A\omega_0 \cos(\omega_0 t + \varphi)$. Solve for t.

- The reason for the $\pm$ signs is that the banana could be moving either along $+x$ or $-x$ and still have a speed of 3π m/s. To find all of the answers, we must allow for all possibilities.
- First apply algebra to isolate the cosine function.
- You should get $\pm\frac{1}{2} = \cos\left(2\pi t - \frac{\pi}{4}\right)$.
- Take the inverse cosine of both sides of the equation.
- You should get $\cos^{-1}\left(\pm\frac{1}{2}\right) = 2\pi t - \frac{\pi}{4}$. Make sure your calculator is in **radians** mode.
- The four possible answers for the inverse cosine are $\frac{\pi}{3}$ rad, $\frac{2\pi}{3}$ rad, $\frac{4\pi}{3}$ rad, and $\frac{5\pi}{3}$ rad (corresponding to 60°, 120°, 240°, and 300°) because $\cos\left(\frac{\pi}{3}\right) = \frac{1}{2}$, $\cos\left(\frac{2\pi}{3}\right) = -\frac{1}{2}$, $\cos\left(\frac{4\pi}{3}\right) = -\frac{1}{2}$, and $\cos\left(\frac{5\pi}{3}\right) = \frac{1}{2}$.
- Plug all four possible angles ($\frac{\pi}{3}$ rad, $\frac{2\pi}{3}$ rad, $\frac{4\pi}{3}$ rad, and $\frac{5\pi}{3}$ rad) into the previous equation for time, which was $\cos^{-1}\left(\pm\frac{1}{2}\right) = 2\pi t - \frac{\pi}{4}$.
- You should get four equations:
$$\frac{\pi}{3} = 2\pi t - \frac{\pi}{4} \quad , \quad \frac{2\pi}{3} = 2\pi t - \frac{\pi}{4} \quad , \quad \frac{4\pi}{3} = 2\pi t - \frac{\pi}{4} \quad , \quad \frac{5\pi}{3} = 2\pi t - \frac{\pi}{4}$$
- Solve for time in each equation separately using algebra. Note that π cancels.
- You should get $t = \frac{7}{24}$ s, $t = \frac{11}{24}$ s, $t = \frac{19}{24}$ s, and $t = \frac{23}{24}$ s. We're **not** finished yet.
- There are more solutions than the four times ($t = \frac{7}{24}$ s, $t = \frac{11}{24}$ s, $t = \frac{19}{24}$ s, and $t = \frac{23}{24}$ s) that we have found thus far. The reason is that the cosine wave repeats itself. Every period, the cosine wave will complete a new cycle.
- Recall that $\omega_0 = 2\pi$ rad/s. Use the equation $\omega_0 = \frac{2\pi}{T}$ to solve for the period. Multiply both sides of the equation by T and divide by 2π. You should get $T = \frac{2\pi}{\omega_0}$.
- The period is $T = 1.0$ s.
- This means we can add or subtract multiples of 1.0 s to our previous times and also obtain valid answers.
- Add n seconds (since $T = 1.0$ s) to each of the answers.
- The banana will be have a speed of $|v_x| = 3\pi$ m/s when $t = \frac{7}{24}$ s $+ n$ s, $t = \frac{11}{24}$ s $+ n$ s, $t = \frac{19}{24}$ s $+ n$ s, or $t = \frac{23}{24}$ s $+ n$ s, where n is a nonnegative integer (0, 1, 2, 3...). This means that the answers are $t = \frac{7}{24}$ s, $t = \frac{11}{24}$ s, $t = \frac{19}{24}$ s, $t = \frac{23}{24}$ s, $t = \frac{31}{24}$ s, $t = \frac{35}{24}$ s, $t = \frac{43}{24}$ s, $t = \frac{47}{24}$ s, and so on. This is because the cosine wave repeats itself every $T = 1.0$ s.

8. First compare the given equation with numbers, $v_x = 2\cos\left(\frac{2\pi t}{3}\right)$, to the general equation for the **velocity** of an oscillating object in symbols, $v_x = A\omega_0 \cos(\omega_0 t + \varphi)$, in order to identify the symbols $v_m = A\omega_0$, ω_0, and φ.

- You should get $v_m = A\omega_0 = 2.0$ m/s, $\omega_0 = \frac{2\pi}{3}$ rad/s, and $\varphi = 0$.
- Use the equation $v_m = A\omega_0$ to solve for A. Divide both sides of the equation by ω_0. You should get $A = \frac{v_m}{\omega_0}$. Plug in numbers. You should get $A = \frac{3}{\pi}$ m. To divide by a fraction, multiply by its **reciprocal**. Note that the reciprocal of $\frac{2\pi}{3}$ is $\frac{3}{2\pi}$. Also note that $\frac{2}{2\pi/3} = 2 \div \frac{2\pi}{3} = 2 \times \frac{3}{2\pi} = \frac{3}{\pi}$. If you use a calculator, $A = 0.95$ m.

(A) Plug $t = 6.0$ s into the given equation. Put your calculator in **radians** mode.

- You should get $v_x = 2\cos(4\pi)$.
- Note that $\cos(4\pi) = 1$ if your calculator is in **radians** mode.
- The banana's velocity is $v_x = 2.0$ m/s when $t = 6.0$ s.

(B) The maximum speed is the coefficient of the cosine function in the given equation.

- The maximum speed is $v_m = 2.0$ m/s. (We found this earlier.)

(C) Use the equation $a_x = -A\omega_0^2 \sin(\omega_0 t + \varphi)$. Put your calculator in **radians** mode.

- You should get $a_x = -\frac{4\pi}{3}\sin(4\pi)$.
- Note that $\sin(4\pi) = 0$ if your calculator is in **radians** mode.
- The banana's acceleration is $a_x = 0$ when $t = 6.0$ s.

(D) Use the equation $a_m = A\omega_0^2$. Recall that $A = \frac{3}{\pi}$ m.

- The maximum acceleration is $a_m = \frac{4\pi}{3}$ m/s^2. Using a calculator, $a_m = 4.2$ m/s^2. Note that $\left(\frac{3}{\pi}\right)\left(\frac{2\pi}{3}\right)^2 = \left(\frac{3}{\pi}\right)\left(\frac{4\pi^2}{9}\right) = \frac{4\pi}{3}$.

(E) Plug $v_x = \pm\sqrt{3}$ m/s in the given equation. Solve for t.

- The reason for the $\pm$ signs is that the banana could be moving either along $+x$ or $-x$ and still have a speed of $\sqrt{3}$ m/s. To find all of the answers, we must allow for all possibilities.
- First apply algebra to isolate the cosine function.
- You should get $\pm\frac{\sqrt{3}}{2} = \cos\left(\frac{2\pi t}{3}\right)$.
- Take the inverse cosine of both sides of the equation.
- You should get $\cos^{-1}\left(\pm\frac{\sqrt{3}}{2}\right) = \frac{2\pi t}{3}$. Make sure your calculator is in **radians** mode.
- The four possible answers for the inverse cosine are $\frac{\pi}{6}$ rad, $\frac{5\pi}{6}$ rad, $\frac{7\pi}{6}$ rad, and $\frac{11\pi}{6}$ rad (corresponding to 30°, 150°, 210°, and 330°) because $\cos\left(\frac{\pi}{6}\right) = \frac{\sqrt{3}}{2}$, $\cos\left(\frac{5\pi}{6}\right) = -\frac{\sqrt{3}}{2}$, $\cos\left(\frac{7\pi}{6}\right) = -\frac{\sqrt{3}}{2}$, and $\cos\left(\frac{11\pi}{6}\right) = \frac{\sqrt{3}}{2}$.

- Plug all four possible angles ($\frac{\pi}{6}$ rad, $\frac{5\pi}{6}$ rad, $\frac{7\pi}{6}$ rad, and $\frac{11\pi}{6}$ rad) into the previous equation for time, which was $\cos^{-1}\left(\pm\frac{\sqrt{3}}{2}\right) = \frac{2\pi t}{3}$.
- You should get four equations:

$$\frac{\pi}{6} = \frac{2\pi t}{3} \quad , \quad \frac{5\pi}{6} = \frac{2\pi t}{3} \quad , \quad \frac{7\pi}{6} = \frac{2\pi t}{3} \quad , \quad \frac{11\pi}{6} = \frac{2\pi t}{3}$$

- Solve for time in each equation separately using algebra. Note that π cancels.
- You should get $t = \frac{1}{4}$ s, $t = \frac{5}{4}$ s, $t = \frac{7}{4}$ s, and $t = \frac{11}{4}$ s. We're **not** finished yet.
- There are more solutions than the four times ($t = \frac{1}{4}$ s, $t = \frac{5}{4}$ s, $t = \frac{7}{4}$ s, and $t = \frac{11}{4}$ s) that we have found thus far. The reason is that the cosine wave repeats itself. Every period, the cosine wave will complete a new cycle.
- Recall that $\omega_0 = \frac{2\pi}{3}$ rad/s. Use the equation $\omega_0 = \frac{2\pi}{T}$ to solve for the period. Multiply both sides of the equation by T and divide by 2π. You should get $T = \frac{2\pi}{\omega_0}$.
- To divide by a fraction, multiply by its **reciprocal**. Note that the reciprocal of $\frac{2\pi}{3}$ is $\frac{3}{2\pi}$. Also note that $\frac{2\pi}{2\pi/3} = 2\pi \div \frac{2\pi}{3} = 2\pi \times \frac{3}{2\pi} = 3$. The period is $T = 3.0$ s.
- This means we can add or subtract multiples of 3.0 s to our previous times and also obtain valid answers.
- Add $3n$ seconds (since $T = 3.0$ s) to each of the answers.
- The banana will be have a speed of $|v_x| = \sqrt{3}$ m/s when $t = \frac{1}{4}$ s $+ 3n$ s, $t = \frac{5}{4}$ s $+ 3n$ s, $t = \frac{7}{4}$ s $+ 3n$ s, or $t = \frac{11}{4}$ s $+ 3n$ s, where n is a nonnegative integer (0, 1, 2, 3...).

---

9. First compare the given equation with numbers, $a_x = -8\sin\left(\frac{\pi t}{2} + \frac{\pi}{2}\right)$, to the general equation for the **acceleration** of an oscillating object in symbols, $a_x = -A\omega_0^2 \sin(\omega_0 t + \varphi)$, in order to identify the symbols $a_m = A\omega_0^2$, ω_0, and φ.

- You should get $a_m = A\omega_0^2 = 8.0$ m/s^2, $\omega_0 = \frac{\pi}{2}$ rad/s, and $\varphi = \frac{\pi}{2}$ rad.
- Use the equation $a_m = A\omega_0^2$ to solve for A. Divide both sides of the equation by ω_0^2. You should get $A = \frac{a_m}{\omega_0^2}$. Plug in numbers. You should get $A = \frac{32}{\pi^2}$ m. To divide by a fraction, multiply by its **reciprocal**. Note that the reciprocal of $\frac{\pi}{2}$ is $\frac{2}{\pi}$. Also note that $\frac{8}{(\pi/2)^2} = 8 \div \frac{\pi^2}{2^2} = 8 \times \frac{2^2}{\pi^2} = \frac{32}{\pi^2}$. If you use a calculator, $A = 3.2$ m.

(A) Use the equation $v_x = A\omega_0 \cos(\omega_0 t + \varphi)$. Put your calculator in **radians** mode.

- You should get $v_x = \frac{16}{\pi}\cos\left(\frac{5\pi}{2}\right)$.
- Note that $\cos\left(\frac{5\pi}{2}\right) = 0$ if your calculator is in **radians** mode.
- The banana's velocity is $v_x = 0$ when $t = 4.0$ s.

(B) Use the equation $v_m = A\omega_0$.

- The maximum speed is $v_m = \frac{16}{\pi}$ m/s. Using a calculator, $v_m = 5.1$ m/s.

(C) Plug $t = 4.0$ s into the given equation. Put your calculator in **radians** mode.

- You should get $a_x = -8\sin\left(\frac{5\pi}{2}\right)$.
- Note that $\sin\left(\frac{5\pi}{2}\right) = 1$ if your calculator is in **radians** mode.
- The banana's acceleration is $a_x = -8.0$ m/s^2 when $t = 4.0$ s.

(D) The maximum acceleration is the (absolute value of the) coefficient of the sine function in the given equation.

- The maximum acceleration is $a_m = 8.0$ m/s^2. (We found this earlier.)

(E) Plug $v_x = \pm\frac{8}{\pi}$ m/s m/s in the equation $v_x = A\omega_0 \cos(\omega_0 t + \varphi)$. Solve for t.

- The reason for the $\pm$ signs is that the banana could be moving either along $+x$ or $-x$ and still have a speed of $\frac{8}{\pi}$ m/s. To find all of the answers, we must allow for all possibilities.
- First apply algebra to isolate the cosine function.
- You should get $\pm\frac{1}{2} = \cos\left(\frac{\pi t}{2} + \frac{\pi}{2}\right)$. Note that $\frac{8}{\pi} \div \frac{16}{\pi} = \frac{8}{\pi} \times \frac{\pi}{16} = \frac{1}{2}$.
- Take the inverse cosine of both sides of the equation.
- You should get $\cos^{-1}\left(\pm\frac{1}{2}\right) = \frac{\pi t}{2} + \frac{\pi}{2}$. Make sure your calculator is in **radians** mode.
- The four possible answers for the inverse cosine are $\frac{\pi}{3}$ rad, $\frac{2\pi}{3}$ rad, $\frac{4\pi}{3}$ rad, and $\frac{5\pi}{3}$ rad (corresponding to 60°, 120°, 240°, and 300°) because $\cos\left(\frac{\pi}{3}\right) = \frac{1}{2}$, $\cos\left(\frac{2\pi}{3}\right) = -\frac{1}{2}$, $\cos\left(\frac{4\pi}{3}\right) = -\frac{1}{2}$, and $\cos\left(\frac{5\pi}{3}\right) = \frac{1}{2}$.
- Plug all four possible angles ($\frac{\pi}{3}$ rad, $\frac{2\pi}{3}$ rad, $\frac{4\pi}{3}$ rad, and $\frac{5\pi}{3}$ rad) into the previous equation for time, which was $\cos^{-1}\left(\pm\frac{1}{2}\right) = \frac{\pi t}{2} + \frac{\pi}{2}$.
- You should get four equations:

$$\frac{\pi}{3} = \frac{\pi t}{2} + \frac{\pi}{2} \quad , \quad \frac{2\pi}{3} = \frac{\pi t}{2} + \frac{\pi}{2} \quad , \quad \frac{4\pi}{3} = \frac{\pi t}{2} + \frac{\pi}{2} \quad , \quad \frac{5\pi}{3} = \frac{\pi t}{2} + \frac{\pi}{2}$$

- Solve for time in each equation separately using algebra. Note that π cancels.
- You should get $t = -\frac{1}{3}$ s, $t = \frac{1}{3}$ s, $t = \frac{5}{3}$ s, and $t = \frac{7}{3}$ s. We're **<u>not</u>** finished yet.
- There are more solutions than the four times ($t = -\frac{1}{3}$ s, $t = \frac{1}{3}$ s, $t = \frac{5}{3}$ s, and $t = \frac{7}{3}$ s) that we have found thus far. The reason is that the cosine wave repeats itself. Every period, the cosine wave will complete a new cycle.
- Recall that $\omega_0 = \frac{\pi}{2}$ rad/s. Use the equation $\omega_0 = \frac{2\pi}{T}$ to solve for the period. Multiply both sides of the equation by T and divide by 2π. You should get $T = \frac{2\pi}{\omega_0}$.

- To divide by a fraction, multiply by its **reciprocal**. Note that the reciprocal of $\frac{\pi}{2}$ is $\frac{2}{\pi}$. Also note that $\frac{2\pi}{\pi/2} = 2\pi \div \frac{\pi}{2} = 2\pi \times \frac{2}{\pi} = 4$. The period is $T = 4.0$ s.
- This means we can add or subtract multiples of 4.0 s to our previous times and also obtain valid answers.
- Add $4n$ seconds (since $T = 4.0$ s) to each of the answers.
- In addition, add one period to the first time, $t = -\frac{1}{3}$ s, to find a positive time equivalent to this negative time. That is, $t = -\frac{1}{3}$ s is equivalent to $t = \frac{11}{3}$ s, since $-\frac{1}{3} + 4 = -\frac{1}{3} + \frac{12}{3} = \frac{11}{3}$.
- The banana will be have a speed of $|v_x| = \frac{8}{\pi}$ m/s when $t = \frac{1}{3}$ s $+ 4n$ s, $t = \frac{5}{3}$ s $+ 4n$ s, $t = \frac{7}{3}$ s $+ 4n$ s, or $t = \frac{11}{3}$ s $+ 4n$ s, where n is a nonnegative integer (0, 1, 2, 3...).

---

10. Apply the strategy that we used in the example on page 40. Study that example.
    - Solve for force in the given equation. You should get:
    $$F = -2\rho g A \Delta h$$
    - Apply Newton's second law: $\sum F_x = ma_x$.
    - Substitute the force into Newton's second law. You should get:
    $$-2\rho g A \Delta h = ma_x$$
    - Solve for acceleration in the previous equation. You should get:
    $$a_x = -\frac{2\rho g A \Delta h}{m}$$
    - Since the equation for acceleration has the same structure as $a_x = -\omega_0^2 \Delta h$ (where h is the position coordinate suitable for this problem), the system undergoes simple harmonic motion. That completes the first part of the solution.
    - Compare the two equations for acceleration above to determine ω_0^2. You should get:
    $$\omega_0^2 = \frac{2\rho g A}{m}$$
    - Density (ρ) equals mass (m) divided by volume (V): $\rho = \frac{m}{V}$. Substitute this expression in the previous equation. Mass cancels.
    $$\omega_0^2 = \frac{2gA}{V}$$
    - The volume of a cylinder equals height (h) times cross-sectional area (A): $V = Ah$. Substitute this into the previous equation. Area cancels.
    $$\omega_0^2 = \frac{2g}{h}$$
    - Squareroot both sides of the equation. The equation for angular frequency is $\omega_0 = \sqrt{\frac{2g}{h}}$.

11. Apply the strategy that we used in the example on page 40. Study that example.
   - The "trick" to this problem is to apply Gauss's law. Review Volume 2 of this series, Chapter 8, pages 125-128. **Note**: It would be **incorrect** to simply apply Newton's law of gravity, $F = -\frac{Gm_e m_b}{r^2}$, to the banana. That equation is true **outside** of the earth, but it is **not** true in the tunnel. You must apply Gauss's law to find an equation for gravitational force **inside** of the tunnel.
   - On pages 125-128 of Volume 2, we applied Gauss's law to a uniformly charged sphere and found the following equation for the electric field in Region I inside of the sphere. Find this equation on page 128 of Volume 2: It's one of the alternate forms for Region I ($r < a$).

$$\vec{\mathbf{E}} = \frac{kQr}{a^3}\hat{\mathbf{r}}$$

   - This problem involves a sphere of mass (the earth, approximately), which creates a gravitational field. Therefore, for this problem we will replace the electric field ($\vec{\mathbf{E}}$) with the gravitational field ($\vec{\mathbf{g}}$) and we will replace the charge (Q) of the sphere with the mass (m_e) of the earth. We will also use the symbol R_e for the radius of the earth in place of the symbol a for the radius of the charged sphere. One more change is that the proportionality constant will be the gravitational constant (G) from Newton's law of gravity (Volume 1, Chapter 18) instead of Coulomb's constant (k) from Coulomb's law (Volume 2, Chapter 1).

$$\vec{\mathbf{g}} = -\frac{Gm_e r}{R_e^3}\hat{\mathbf{r}}$$

   - Note the **minus sign**. The reason for the minus sign is that two masses (like the earth and the moon), which are both positive (all masses are positive), **attract**, whereas two positive charges **repel**.
   - Recall from Volume 1 that the gravitational force (which we call **weight**) exerted on the banana equals its mass times gravity: $\vec{\mathbf{F}} = m_b\vec{\mathbf{g}}$. (But note that g does **not** equal 9.81 m/s^2 as the banana falls through the tunnel: That value is only correct near earth's surface. The previous equation tells us the value of g at a distance r from earth's center, where r varies in the range $0 \le r \le R_e$.) Multiply both sides of the equation $\vec{\mathbf{g}} = -\frac{Gm_e r}{R_e^3}\hat{\mathbf{r}}$ by the mass of the banana (m_b) in order to find the force that the earth exerts on the banana.

$$\vec{\mathbf{F}} = -\frac{Gm_e m_b r}{R_e^3}\hat{\mathbf{r}}$$

   - Apply Newton's second law to the banana: $\sum \vec{\mathbf{F}} = m_b\vec{\mathbf{a}}$.
   - Substitute the force into Newton's second law. You should get:

$$-\frac{Gm_e m_b r}{R_e^3}\hat{\mathbf{r}} = m_b\vec{\mathbf{a}}$$

   - Note that the mass of the banana (m_b) cancels out.

$$-\frac{Gm_e r}{R_e^3}\hat{\mathbf{r}} = \vec{\mathbf{a}}$$

- Since the equation for acceleration has the same structure as $a = -\omega_0^2 r$ (where r is the position coordinate suitable for this problem), the system undergoes simple harmonic motion. That completes the first part of the solution.
- Compare the two equations for acceleration above to determine ω_0^2. You should get:

$$\omega_0^2 = \frac{Gm_e}{R_e^3}$$

- Squareroot both sides of the equation. Apply the equation $\omega_0 = \frac{2\pi}{T}$ in order to determine the period. You should get:

$$\omega_0 = \sqrt{\frac{Gm_e}{R_e^3}} = \frac{2\pi}{T}$$

- Multiply both sides of the equation by T and divide by $\sqrt{\frac{Gm_e}{R_e^3}}$. Note that $\frac{1}{\sqrt{\frac{Gm_e}{R_e^3}}} = \sqrt{\frac{R_e^3}{Gm_e}}$.
- The equation for the period is:

$$T = 2\pi\sqrt{\frac{R_e^3}{Gm_e}}$$

- If you plug in earth's average radius ($R_e = 6.384 \times 10^6$ meters), earth's mass ($m_e = 5.972 \times 10^{24}$ kilograms), and the gravitational constant ($G = 6.674 \times 10^{-11} \frac{\text{Nm}^2}{\text{kg}^2}$), the period comes out to $T = 5155$ seconds, which equates to $T = 86$ minutes. (Depending upon where the supergenius chimpanzees dig the tunnel, the value of R_e may need to be modified slightly. If also you want to account for the variation in earth's density or the actual non-spherical shape of the earth, the problem becomes much more complicated. And then if you want to allow the earth to rotate, you get the "fun" of dealing with the Coriolis force. Hopefully, also, the tunnel is so well insulated that the banana doesn't melt when it reaches the core zone. And so on.)

Chapter 3: Oscillating Spring

12. First identify the known symbols in SI units.
    - The suspended mass is $m = 0.250$ kg. Recall that 1 kg = 1000 g.
    - The spring constant is $k = 9.00$ N/m.
    - Use the equation $T = 2\pi\sqrt{\frac{m}{k}}$.
    - If you're not using a calculator, it's convenient to write $\frac{.25}{9} = \frac{1/4}{9} = \frac{1}{4} \div 9 = \frac{1}{4} \times \frac{1}{9} = \frac{1}{36}$.
    - The period is $T = \frac{\pi}{3}$ s. If you use a calculator, the period works out to $T = 1.05$ s.

13. First identify the known symbols in SI units.
    - The suspended mass is $m = 0.500$ kg. Recall that 1 kg = 1000 g.
    - The spring constant is $k = 8.0$ N/m.
    - The total time is $t = 20\pi$ s. This is **not** the period.
    - Use the equation $T = 2\pi\sqrt{\frac{m}{k}}$.
    - If you're not using a calculator, it's convenient to write $\frac{.5}{8} = \frac{1/2}{8} = \frac{1}{2} \div 8 = \frac{1}{2} \times \frac{1}{8} = \frac{1}{16}$.
    - The period is $T = \frac{\pi}{2}$ s.
    - Use the following equation.

$$\text{number of oscillations} = \frac{\text{total time}}{T}$$

    - Note that $\frac{20}{1/2} = 20 \div \frac{1}{2} = 20 \times 2 = 40$.
    - The number of oscillations completed is 40.

---

14. First identify the known symbols in SI units.
    - The spring constant is $k = 2.0$ N/m.
    - The period is $T = \frac{\pi}{2}$ s.
    - Use the equation $T = 2\pi\sqrt{\frac{m}{k}}$. Solve for the suspended mass.
    - Divide both sides of the equation by 2π. Square both sides of the equation. Multiply both sides of the equation by the spring constant. You should get $m = \frac{T^2 k}{4\pi^2}$.
    - The mass is $m = 0.125$ kg, which is the same as $m = \frac{1}{8}$ kg.

---

15. Study the example of a comparison problem on page 56.
    - The period for the first mass is $T_1 = 2.0$ s.
    - The ratio of the suspended masses is $\frac{m_2}{m_1} = 2$.
    - You are looking for the period with the new mass. Call this T_2.
    - Write down one equation for the period of each banana with appropriate subscripts. The masses (m_1 and m_2) and periods (T_1 and T_2) are different, but the spring constant (k) is the same in each case.

$$T_1 = 2\pi\sqrt{\frac{m_1}{k}} \quad , \quad T_2 = 2\pi\sqrt{\frac{m_2}{k}}$$

    - Divide the two equations. To divide a fraction, multiply by its **reciprocal**. The reciprocal of $\sqrt{\frac{m_1}{k}}$ is $\sqrt{\frac{k}{m_1}}$. The 2π's and the k's will cancel out. You should get:

$$\frac{T_2}{T_1} = \sqrt{\frac{m_2}{m_1}}$$

- Plug in numbers and simplify. You should get $T_2 = T_1\sqrt{2}$.
- The period of the new system is $T_2 = 2\sqrt{2}$ s. If you use a calculator, $T_2 = 2.8$ s.

---

16. Examine the equation for the period of a spring closely.
    - Does the period of a spring depend on gravitational acceleration (g)?
    - No, it doesn't. The equation for the period of a spring, $T = 2\pi\sqrt{\frac{m}{k}}$, only involves mass ($m$) and spring constant ($k$).
    - Neither the suspended mass nor the spring constant change when the system is transported to the moon. (The block of wood would have less **weight** on the moon compared to the earth, but it would have the **same mass**. Recall that mass is a measure of **inertia**, or resistance to acceleration, whereas weight is a measure of the gravitational pull. It may help to review Volume 1, Chapter 13, Problem 89, part A.)
    - The period of the new system is exactly the same as the period of the old system. The period is $T = \frac{5}{6}$ s. If you use a calculator, this works out to $T = 0.83$ s.

---

17. Study the example of a comparison problem on page 56.
    - The period for the first spring is $T_1 = 1.5$ s.
    - The ratio of the spring constants is $\frac{k_2}{k_1} = 9$.
    - You are looking for the period with the new spring. Call this T_2.
    - Write down one equation for the period of each spring with appropriate subscripts. The spring constants (k_1 and k_2) and periods (T_1 and T_2) are different, but the suspended mass (m) is the same in each case.

$$T_1 = 2\pi\sqrt{\frac{m}{k_1}} \quad , \quad T_2 = 2\pi\sqrt{\frac{m}{k_2}}$$

    - Divide the two equations. To divide a fraction, multiply by its **reciprocal**. The reciprocal of $\sqrt{\frac{m}{k_1}}$ is $\sqrt{\frac{k_1}{m}}$. The 2π's and the m's will cancel out. You should get:

$$\frac{T_2}{T_1} = \sqrt{\frac{k_1}{k_2}}$$

    - Plug in numbers and simplify. You should get $T_2 = \frac{T_1}{3}$. Note that $\frac{k_1}{k_2} = \frac{1}{9}$ since $\frac{k_2}{k_1} = 9$.
    - The period of the new system is $T_2 = 0.50$ s, which is the same as $T = \frac{1}{2}$ s.

Chapter 4: Oscillating Pendulum

18. First identify the known symbols in SI units.
    - The length of the pendulum is $L = 40$ m. (That's very long. No wonder the monkey is on the roof of a tall building.)
    - You should also know that gravitational acceleration is $g = 9.81 \text{ m/s}^2 \approx 10 \text{ m/s}^2$.
    - The suspended mass, $m = 0.500$ kg, is **not needed** to solve the problem. The period of a simple pendulum, $T \approx 2\pi\sqrt{\frac{L}{g}}$, does **not** depend on mass.
    - Use the equation $T \approx 2\pi\sqrt{\frac{L}{g}}$ for a **simple** pendulum.
    - The period is $T \approx 4\pi$ s. If you use a calculator, $T \approx 13$ s to two significant figures.

---

19. First identify the known symbols in SI units.
    - The suspended mass for the spring is $m_1 = 3.0$ kg.
    - The mass of the pendulum bob, $m_2 = 1.5$ kg, is **not needed** to solve the problem. The period of a simple pendulum, $T \approx 2\pi\sqrt{\frac{L}{g}}$, does **not** depend on mass.
    - The spring constant is $k = 6.0$ N/m.
    - You should also know that gravitational acceleration is $g = 9.81 \text{ m/s}^2 \approx 10 \text{ m/s}^2$.
    - Use the equations $T = 2\pi\sqrt{\frac{m_1}{k}}$ and $T \approx 2\pi\sqrt{\frac{L}{g}}$.
    - The problem states that the period is the same for the spring and the pendulum.
    - Set the two periods equal to one another. Divide both sides by 2π. The 2π's cancel out. You should get:

$$\sqrt{\frac{m_1}{k}} \approx \sqrt{\frac{L}{g}}$$

    - Square both sides of the equation. You should get:

$$\frac{m_1}{k} \approx \frac{L}{g}$$

    - Multiply both sides of the equation by g. You should get:

$$L \approx \frac{m_1 g}{k}$$

    - The length of the pendulum is $L \approx 5.0$ m. Be sure to use the suspended mass for the **spring**, which is $m_1 = 3.0$ kg. (**Don't** use $m_2 = 1.5$ kg, which is for the pendulum.)

20. First identify the known symbols in SI units.
    - The mass of the rod is $m = 4.0$ kg. Since this is a **physical** pendulum (**not** a simple pendulum), the mass will appear in the formula. However, you don't really need to know the mass of the rod: It would cancel out in the equation after you plug in the formula for the moment of inertia.
    - The length of the rod is $L = 0.60$ m.
    - You should also know that gravitational acceleration is $g = 9.81 \text{ m/s}^2 \approx 10 \text{ m/s}^2$.
    - Use the equation $T \approx 2\pi\sqrt{\frac{I}{mgd}}$ for a **physical** pendulum.
    - The distance d that we need in the formula equals $d = \frac{L}{2} = \frac{0.6}{2} = 0.30$ m. That's because d is measured from the hinge to the **center of mass** of the rod, which is one-half the length of the rod from the hinge.
    - The moment of inertia of a uniform rod about an axis that is perpendicular to the rod and passing through its end (that's the case here) is $I = \frac{mL^2}{3}$. (You can find this formula in Volume 1, Chapter 32, or any standard physics textbook.)
    - If you plug the formula for moment of inertia into the equation for period, you get:

$$T \approx 2\pi\sqrt{\frac{L^2}{3gd}}$$

    - Notice how the mass of the rod cancels out this way. (You could first calculate the moment of inertia as a number and then plug it into the original period equation. You will get the same answer, provided that your math is correct.)
    - Note that $\sqrt{0.04} = 0.2$ and that $0.4 = \frac{2}{5}$.
    - The period is $T \approx \frac{2\pi}{5}$ s. If you use a calculator, this works out to $T \approx 1.3$ s.

---

21. First identify the known symbols in SI units.
    - The spring constant is $k = 20$ N/m.
    - The length of the pendulum is $L = 5.0$ m.
    - The period for the spring is $T_1 = 3.0$ s.
    - The period for the pendulum is $T_2 = 2.0$ s.
    - You **don't** know gravitational acceleration because it isn't the earth!
    - Use the equations $T_1 = 2\pi\sqrt{\frac{m}{k}}$ and $T_2 \approx 2\pi\sqrt{\frac{L}{g}}$.
    - Divide T_2 by T_1. The 2π's cancel out. Simplify. To divide by a fraction, multiply by its reciprocal. The reciprocal of $\sqrt{\frac{m}{k}}$ is $\sqrt{\frac{k}{m}}$. You should get:

$$\frac{T_2}{T_1} \approx \sqrt{\frac{Lk}{gm}}$$

- Square both sides of the equation. You should get:

$$\left(\frac{T_2}{T_1}\right)^2 \approx \frac{Lk}{gm}$$

- Multiply both sides of the equation by mg and divide both sides by $\left(\frac{T_2}{T_1}\right)^2$. Note that dividing by $\left(\frac{T_2}{T_1}\right)^2$ is the same as multiplying by $\left(\frac{T_1}{T_2}\right)^2$. You should get:

$$mg \approx Lk\left(\frac{T_1}{T_2}\right)^2$$

- The weight of the rock on planet Monk is $mg \approx 225$ N.
- **Note**: **Weight** (W) equals **mass** (m) times gravitational acceleration (g): $W = mg$. If you solved for m, you're not finished yet. You still need to multiply by g. Note that g is **not** equal to 9.81 m/s^2 because it's not the earth. You need to solve for g.
- **Alternate solution**: Another way to solve this problem is to first solve for mass in the equation $T_1 = 2\pi\sqrt{\frac{m}{k}}$, then solve for gravitational acceleration in the equation $T_2 \approx 2\pi\sqrt{\frac{L}{g}}$, and then to multiply m and g together. If you do it this way, you should get $m = \frac{45}{\pi^2}$ kg ≈ 4.6 kg, $g \approx 5\pi^2$m/s$^2 \approx 49$m/s^2, and $mg \approx 225$ N.

---

22. Study the example of a comparison problem on page 67.

- The period for the first cord is $T_1 = \sqrt{2}$ s.
- The ratio of the lengths is $\frac{L_2}{L_1} = 2$.
- You are looking for the period with the new cord. Call this T_2.
- Write down one equation for the period of each pendulum with appropriate subscripts. The lengths (L_1 and L_2) and periods (T_1 and T_2) are different, but the gravitational acceleration (g) is the same in each case.

$$T_1 \approx 2\pi\sqrt{\frac{L_1}{g}} \quad , \quad T_2 \approx 2\pi\sqrt{\frac{L_2}{g}}$$

- Divide the two equations. To divide a fraction, multiply by its **reciprocal**. The reciprocal of $\sqrt{\frac{L_1}{g}}$ is $\sqrt{\frac{g}{L_1}}$. The 2π's and the g's will cancel out. You should get:

$$\frac{T_2}{T_1} \approx \sqrt{\frac{L_2}{L_1}}$$

- Plug in numbers and simplify. You should get $T_2 \approx T_1\sqrt{2}$.
- The period of the new system is $T_2 \approx 2.0$ s. Note that $\sqrt{2}\sqrt{2} = 2$.

23. Examine the equation for the period of a simple pendulum closely.
   - Does the period of a simple pendulum depend on mass (m)?
   - No, it doesn't. The equation for the period of a simple pendulum, $T \approx 2\pi\sqrt{\frac{L}{g}}$, only involves length ($L$) and gravitational acceleration (g).
   - Neither the length of the cord nor the gravitational acceleration change when the suspended mass is changed. (This assumes that the cord doesn't stretch. Assume that the cord doesn't stretch significantly unless a problem states otherwise.)
   - The period of the new pendulum is exactly the same as the period of the old pendulum. The period is $T = 1.6$ s.

---

24. Study the example of a comparison problem on page 67.
   - The period on the earth is $T_1 = \sqrt{6}$ s.
   - The ratio of the gravitational accelerations is $\frac{g_2}{g_1} = \frac{1}{6}$ (since the problem states that gravity is reduced by a factor of 6 on the moon – so it has 1/6 earth's gravity).
   - You are looking for the period on the moon. Call this T_2.
   - Write down one equation for the period of each pendulum with appropriate subscripts. The gravitational accelerations (g_1 and g_2) and periods (T_1 and T_2) are different, but the length of the pendulum (L) is the same in each case.

$$T_1 \approx 2\pi\sqrt{\frac{L}{g_1}} \quad , \quad T_2 \approx 2\pi\sqrt{\frac{L}{g_2}}$$

   - Divide the two equations. To divide a fraction, multiply by its **reciprocal**. The reciprocal of $\sqrt{\frac{L}{g_1}}$ is $\sqrt{\frac{g_1}{L}}$. The 2π's and the L's will cancel out. You should get:

$$\frac{T_2}{T_1} \approx \sqrt{\frac{g_1}{g_2}}$$

   - Plug in numbers and simplify. You should get $T_2 \approx T_1\sqrt{6}$. Note that $\sqrt{\frac{1}{1/6}} = \sqrt{6}$.
   - The period of the new system is $T_2 \approx 6.0$ s. Note that $\sqrt{6}\sqrt{6} = 6$.
   - **Note**: Compare the ratio $\sqrt{\frac{g_1}{g_2}}$ for this problem to the ratio $\sqrt{\frac{L_2}{L_1}}$ from Problem 22. Notice how the 1's and 2's are swapped. The reason for this is that L is in the numerator of $T \approx 2\pi\sqrt{\frac{L}{g}}$, whereas g is in the denominator. Increasing L increases the period, whereas increasing g decreases the period. Since the moon's surface gravity is weaker than earth's surface gravity, the period is longer on the moon (the smaller g in the denominator makes T larger).

Chapter 5: Wave Motion

25. First identify the known symbols in SI units.
    - There are 30 oscillations in a total time of 60 seconds. (The equation for this is written out in words, as you will see below.)
    - The wavelength is $\lambda = 6.0$ m.
    - The vertical crest-to-trough distance is $y_{max} - y_{min} = 0.50$ m.

(A) The wavelength is given in the problem: $\lambda = 6.0$ m.

(B) Use the following equation.

$$T = \frac{\text{total time}}{\text{number of oscillations}}$$

- The period is $T = 2.0$ s.

(C) Use the equation $f = \frac{1}{T}$.

- The frequency is $f = \frac{1}{2}$ Hz $= 0.50$ Hz.

(D) The amplitude is **one-half** of the vertical distance from crest to trough. (It is the distance from crest to **equilibrium**, but the equilibrium position wasn't specified in this problem.)

- Use the equation $A = \frac{y_{max} - y_{min}}{2}$. (You can find this equation in Chapter 1, for example, but note that what we called x in Chapter 1 corresponds to what we are calling y in Chapter 5.)
- The amplitude is $A = 0.25$ m.

(E) Use the equation $k_0 = \frac{2\pi}{\lambda}$.

- The angular wave number is $k_0 = \frac{\pi}{3}$ rad/m. If you use a calculator, $k_0 = 1.0$ rad/m.

(F) Use the equation $v = \frac{\lambda}{T}$.

- The wave speed is $v = 3.0$ m/s.

---

26. First convert the linear mass density ($\mu = 25$ g/cm) to SI units.
    - Do the conversion in two steps. First convert the grams to kilograms.
    - You should get $\mu = 0.025$ kg/cm. (Don't use this value.)
    - Now convert the $\frac{1}{\text{cm}}$ to $\frac{1}{\text{m}}$. Don't divide by 100. **Multiply** by 100. Why? Because centimeters are in the **denominator**. If we were converting from cm to m, we would divide by 100, but to convert from $\frac{1}{\text{cm}}$ to $\frac{1}{\text{m}}$, we multiply by 100. See below.

$$\frac{1}{\text{cm}} = \frac{1}{\text{cm}} \times \frac{100 \text{ cm}}{1 \text{ m}} = 100 \frac{1}{\text{m}}$$

- The linear mass density is $\mu = 2.5$ kg/m in SI units.
- Use the equation $v = \sqrt{\frac{F}{\mu}}$.
- The wave speed is $v = 6.0$ m/s. Note that $\frac{90}{2.5} = 36$.

27. First convert the temperature from Celsius to Kelvin.
    - Use the equation $T_K = T_C + 273.15$.
    - We will round 273.15 to 273 in order to solve the problem without a calculator. We will also approximate the universal gas constant as $R \approx \frac{25}{3} \frac{\text{J}}{\text{mol}\cdot\text{K}}$.
    - The absolute temperature is $T = T_K = 300$ K.
    - Use the equation $v = \sqrt{\frac{\gamma RT}{M}}$.
    - The adiabatic constant is $\gamma = \frac{7}{5}$ because oxygen is diatomic (O_2).
    - The molar mass of O_2 is $M = 0.032$ kg/mol.
    - Note that the molar mass of oxygen is 16 g/mol. Divide by 1000 to convert this to kg/mol to get 0.016 kg/mol. Then double this to get 0.032 kg/mol for diatomic oxygen gas (which consists of O_2 molecules).
    - Note that $\frac{1}{0.032} = \frac{125}{4}$ such that $\frac{\left(\frac{7}{5}\right)\left(\frac{25}{3}\right)(300)}{0.032} = \left(\frac{7}{5}\right)\left(\frac{25}{3}\right)(300)\left(\frac{125}{4}\right)$.
    - This simplifies to $(7)(25)(625) = (7)(25)(625)$.
    - Note that $\sqrt{25} = 5$, $\sqrt{625} = 25$, and $(5)(25) = 125$.
    - The wave speed is $v \approx 125\sqrt{7}$ m/s. If you use a calculator, $v = 331$ m/s.

---

28. Follow the steps from the examples on page 80.
    - First subtract the two values in decibels: 100 dB − 40 dB = 60 dB.
    - Divide this by 10 decibels: $\frac{60 \text{ dB}}{10 \text{ dB}} = 6$.
    - Make this number a power of ten: $10^6 = 1{,}000{,}000$.
    - A 100-dB sound is 1,000,000 (one **million**) times louder than a 40-dB sound.

Chapter 6: Doppler Effect and Shock Waves

29. Identify the source and the observer, and list the known quantities.
    - The police car is the **source**. The speed of the source is $v_s = 90$ m/s.
    - The chimpanzee is the **observer**. The speed of the observer is $v_o = 65$ m/s.
    - The speed of sound in air is $v = 340$ m/s.
    - The unshifted frequency is $f_0 = 800$ Hz.
    - Draw a diagram with the police officer in pursuit of the chimpanzee.

      police car chimpanzee
      90 m/s ⟶ ⟶ 65 m/s

    - Since the chimpanzee (the observer) is heading **away from** the police car (the source), we use the **lower** sign in the numerator, which is **negative** (−).
    - Since the police car (the source) is heading **towards** the chimpanzee (the observer), we use the **upper** sign in the denominator, which is **negative** (−).

- In this problem, the equation for the Doppler effect becomes:

$$f = f_0 \frac{v - v_o}{v - v_s}$$

- The chimpanzee hears the police siren at a frequency of $f = 880$ Hz.

---

30. Identify the source and the observer, and list the known quantities.
   - The train is the **source**. The speed of the source is $v_s = 40$ m/s.
   - The gorilla is the **observer**. The speed of the observer is $v_o = 20$ m/s.
   - The speed of sound in air is $v = 340$ m/s.
   - The unshifted frequency is $f_0 = 500$ Hz.
   - Draw a diagram with the gorilla and train heading toward one another.

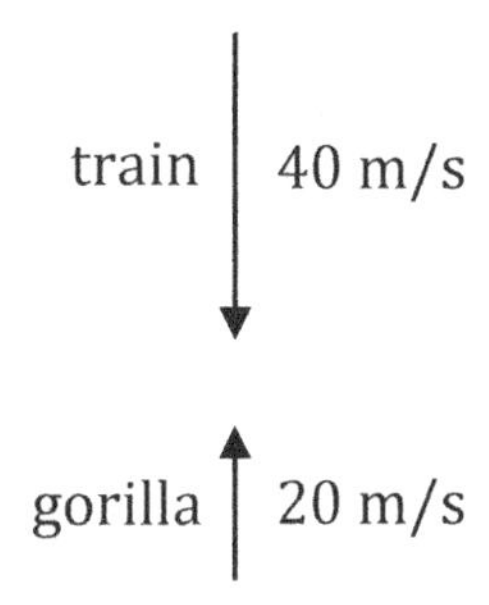

   - Since the gorilla (the observer) is heading **towards** the train (the source), we use the **upper** sign in the numerator, which is **positive** (+).
   - Since the train (the source) is heading **towards** the gorilla (the observer), we use the **upper** sign in the denominator, which is **negative** (−).
   - In this problem, the equation for the Doppler effect becomes:

$$f = f_0 \frac{v + v_o}{v - v_s}$$

   - The gorilla hears the train whistle at a frequency of $f = 600$ Hz.

---

31. Identify the source and the observer, and list the known quantities.
   - The ambulance is the **source**. The speed of the source is $v_s = 60$ m/s.
   - The orangutan is the **observer**. The speed of the observer is $v_o = 40$ m/s.
   - The speed of sound in air is $v = 340$ m/s.
   - The unshifted frequency is $f_0 = 1200$ Hz.
   - Draw a diagram with the orangutan and ambulance heading away from each other.

ambulance orangutan

60 m/s ←——— ———→ 40 m/s

   - Since the orangutan (the observer) is heading **away from** the ambulance (the source), we use the **lower** sign in the numerator, which is **negative** (−).
   - Since the ambulance (the source) is heading **away from** the orangutan (the observer), we use the **lower** sign in the denominator, which is **positive** (+).

- In this problem, the equation for the Doppler effect becomes:

$$f = f_0 \frac{v - v_o}{v + v_s}$$

- The orangutan hears the siren at a frequency of $f = 900$ Hz.

---

32. Identify the source and the observer, and list the known quantities.
    - The train is the **source.** We are solving for the speed of the source (v_s).
    - The chimpanzee is the **observer.** The speed of the observer is $v_o = 60$ m/s.
    - The speed of sound in air is $v = 340$ m/s.
    - The unshifted frequency of the train whistle (the source) is $f_0 = 850$ Hz.
    - The shifted frequency heard by the chimpanzee (the observer) is $f = 800$ Hz.
    - Draw a diagram with the chimpanzee and train heading in the same direction, with the chimpanzee heading towards the train.

chimpanzee train

60 m/s ⟶ ⟶ v_s

- Since the chimpanzee (the observer) is heading **towards** the train (the source), we use the **upper** sign in the numerator, which is **positive** (+).
- Since the train (the source) is heading **away from** the chimpanzee (the observer), we use the **lower** sign in the denominator, which is **positive** (+).
- In this problem, the equation for the Doppler effect becomes:

$$f = f_0 \frac{v + v_o}{v + v_s}$$

- Plug numbers into the equation. You should get:

$$800 = 850 \frac{400}{340 + v_s}$$

- Multiply both sides of the equation by $(340 + v_s)$ and divide by 800. You should get:

$$340 + v_s = \frac{850}{800}(400)$$

- Solve for v_s. Note that $\frac{850}{800}(400) = 425$.
- The speed of the train (the source) is $v_s = 85$ m/s.
- Conceptual note: Even though the chimpanzee is heading towards the train, the chimpanzee hears the whistle at a lower frequency (800 Hz compared to 850 Hz) because the train is getting further away from the chimpanzee (because the train is traveling faster than the chimpanzee, as 85 m/s is greater than 60 m/s).

33. The "trick" to this problem is to setup two different equations for the two different situations described in the problem.

- In both cases, the train is the source. We are solving for the speed of the source (v_s).
- In both cases, the monkey is the observer. The monkey is at rest: $v_o = 0$.
- The speed of sound in air is $v = 340$ m/s.
- We **don't know** the unshifted frequency (f_0) for either case, but it won't matter. As we will see, f_0 will cancel out in the algebra later.
- First write down an equation for when the train is approaching the monkey.

train monkey

v_s → (at rest)

- When the train approaches the monkey, the shifted frequency is $f_1 = 750$ Hz and we use the **upper** sign in the denominator, which is **negative** ($-$). For this situation, the equation for the Doppler effect is (remember that $v_o = 0$):

$$f_1 = f_0 \frac{v}{v - v_s}$$

- We used a subscript in f_1 to distinguish it from the shifted frequency for the second situation (when the train is getting further away, after passing the monkey).
- Plug numbers into the above equation. You should get:

$$750 = f_0 \frac{340}{340 - v_s}$$

- Now write down an equation for when the train is heading away from the monkey.

monkey train

(at rest) v_s →

- When the train is heading away from the monkey, the shifted frequency is $f_2 = 500$ Hz and we use the **lower** sign in the denominator, which is **positive** ($+$). For this situation, the equation for the Doppler effect is (remember that $v_o = 0$):

$$f_2 = f_0 \frac{v}{v + v_s}$$

- Plug numbers into the above equation. You should get:

$$500 = f_0 \frac{340}{340 + v_s}$$

- There are now two equations. Solve for v_s.
- It's convenient to divide the two equations. This will make f_0 cancel out.

$$\frac{750}{500} = \frac{f_0 \frac{340}{340 - v_s}}{f_0 \frac{340}{340 + v_s}}$$

- Note that $\frac{750}{500} = 1.5$. To divide by a fraction, multiply by its **reciprocal**. The reciprocal of $\frac{340}{340+v_s}$ is $\frac{340+v_s}{340}$. Note that f_0 and two of the 340's cancel.

- You should get:

$$1.5 = \frac{340 + v_s}{340 - v_s}$$

- Multiply both sides of the equation by $(340 - v_s)$. That is, **cross multiply**. Then distribute the 1.5 to both terms. You should get:

$$510 - 1.5v_s = 340 + v_s$$

- **Combine like terms**. Add $1.5v_s$ to both sides of the equation and subtract 340.
- You should get $170 = 2.5v_s$.
- The speed of the train (the source) is $v_s = 68$ m/s. (One way to do this without a calculator is to write 2.5 as $\frac{5}{2}$.)

---

34. Identify the known quantities.
- The time is $t = 5\sqrt{2}$ s.
- The half-angle of the conical shock wave is $\theta = 45°$.
- The speed of sound in air is $v = 340$ m/s.
- Draw a diagram. The same situation is drawn on pages 86-87.

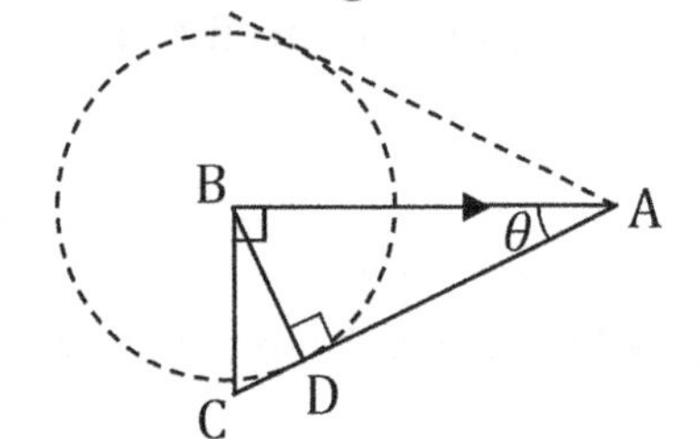

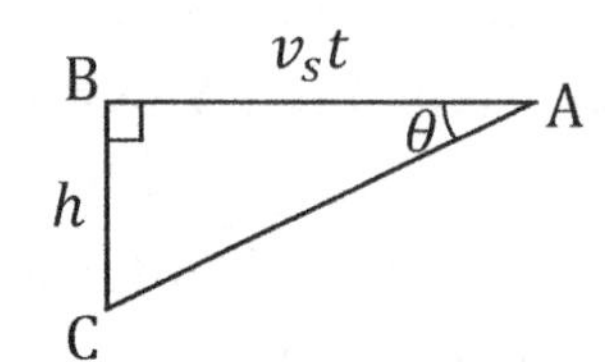

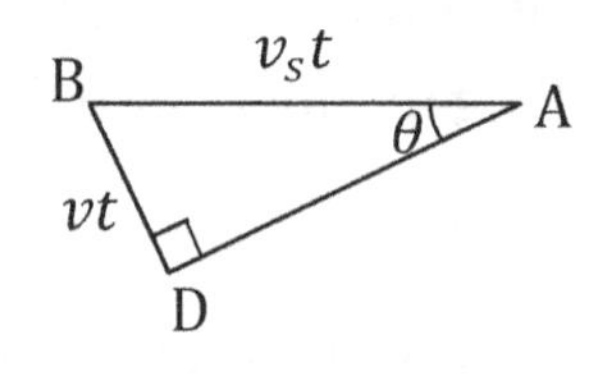

(A) Use the equation $\sin\theta = \frac{v}{v_s}$.
- Multiply both sides of the equation by v_s and divide by $\sin\theta$.
- You should get $v_s = \frac{v}{\sin\theta}$.
- Note that $\frac{340}{\sqrt{2}/2} = \frac{680}{\sqrt{2}} = \frac{680\sqrt{2}}{\sqrt{2}\sqrt{2}} = 340\sqrt{2}$. To divide by a fraction, multiply by its **reciprocal**. We multiplied by $\frac{\sqrt{2}}{\sqrt{2}}$ to **rationalize** the denominator. Note that $\sqrt{2}\sqrt{2} = 2$.
- The speed of the jet is $v_s = 340\sqrt{2}$ m/s. Note that the answer is the same as $v_s = \frac{680}{\sqrt{2}}$ m/s since $\frac{1}{\sqrt{2}} = \frac{1}{\sqrt{2}}\frac{\sqrt{2}}{\sqrt{2}} = \frac{\sqrt{2}}{2}$. If you use a calculator, $v_s = 481$ m/s.

(B) Use the equation $\tan\theta = \frac{h}{v_s t}$.
- Multiply both sides of the equation by $v_s t$. You should get $h = v_s t \tan\theta$.
- Recall from trig that $\tan 45° = 1$. Note that $\sqrt{2}\sqrt{2} = 2$.
- The altitude of the jet is $h = 3.4$ km, which is the same as $h = 3400$ m.

GET A DIFFERENT ANSWER?

If you get a different answer and can't find your mistake even after consulting the hints and explanations, what should you do?

Please contact the author, Dr. McMullen.

How? Visit one of the author's blogs (see below). Either use the Contact Me option, or click on one of the author's articles and post a comment on the article.

www.monkeyphysicsblog.wordpress.com
www.improveyourmathfluency.com
www.chrismcmullen.wordpress.com

Why?

- If there happens to be a mistake (although much effort was put into perfecting the answer key), the correction will benefit other students like yourself in the future.
- If it turns out not to be a mistake, **you may learn something** from Dr. McMullen's reply to your message.

99.99% of students who walk into Dr. McMullen's office believing that they found a mistake with an answer discover one of two things:

- They made a mistake that they didn't realize they were making and learned from it.
- They discovered that their answer was actually the same. This is actually fairly common. For example, the answer key might say $t = \frac{\sqrt{3}}{3}$ s. A student solves the problem and gets $t = \frac{1}{\sqrt{3}}$ s. These are actually the same: Try it on your calculator and you will see that both equal about 0.57735. Here's why: $\frac{1}{\sqrt{3}} = \frac{1}{\sqrt{3}}\frac{\sqrt{3}}{\sqrt{3}} = \frac{\sqrt{3}}{3}$.

Two experienced physics teachers solved every problem in this book to check the answers, and dozens of students used this book and provided feedback before it was published. Every effort was made to ensure that the final answer given to every problem is correct.

But all humans, even those who are experts in their fields and who routinely aced exams back when they were students, make an occasional mistake. So if you believe you found a mistake, you should report it just in case. Dr. McMullen will appreciate your time.

Chapter 7: Standing Waves

35. Study the example on pages 99-100.

(A) First identify the **boundary conditions**.

- Since both ends are clamped, there will be a **node** at each end.
- Visualize the long train of sine waves below.

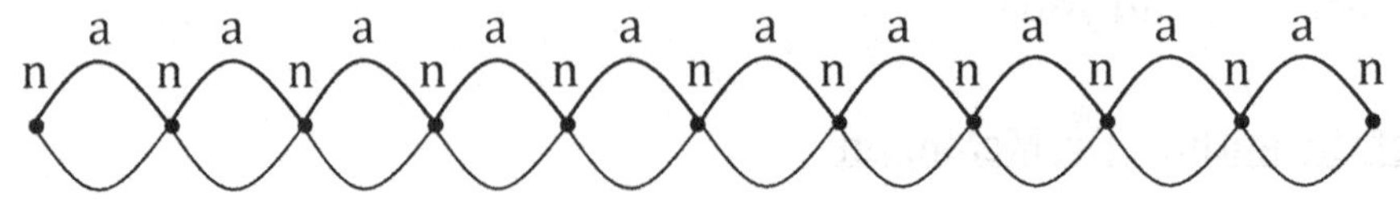

- The shortest section of this train that has a node (n) at both ends is drawn below. This is the **fundamental** (or **first harmonic**).

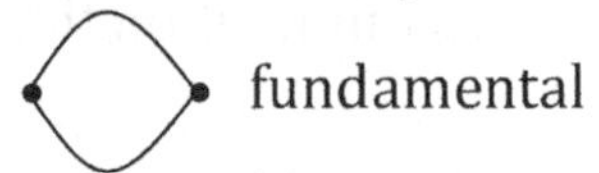

- The second shortest section that has a node (n) at both ends is drawn below. This is the **first overtone** (or **second harmonic**).

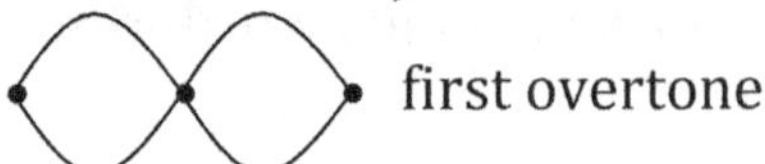

- The third shortest section that has a node (n) at both ends is drawn below. This is the **second overtone** (or **third harmonic**).

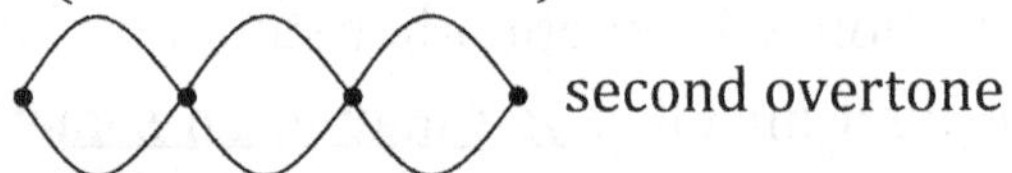

(B) Apply the following equation to each standing wave that you drew.

$$\begin{pmatrix}\text{number of} \\ \text{cycles}\end{pmatrix} \lambda = L$$

- Compare each standing wave above to a single cycle drawn below.

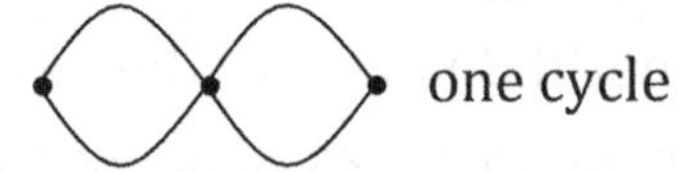

- Count the number of "footballs" and divide by 2 to get the number of cycles.
- You should get $\frac{1}{2}$ cycle, 1 cycle, and $\frac{3}{2}$ cycles. Plug these into the above equation and solve for wavelength. The active length of the string is $L = 1.5$ m.
- The wavelengths of the fundamental and first two overtones are $\lambda_0 = 3.0$ m, $\lambda_1 = 1.5$ m, and $\lambda_2 = 1.0$ m. Note, for example, that $\lambda_0 = 2L = 2(1.5) = 3.0$ m.

(C) Apply the equation $v = \sqrt{\frac{F}{\mu}}$. First convert the linear mass density to kg/m.

- Divide by 1000 to convert from g to kg, and multiply by 100 to convert $\frac{1}{\text{cm}}$ to $\frac{1}{\text{m}}$. The linear mass density is $\mu = 0.030$ kg/m.
- The speed of the wave is $v = 30$ m/s. Note that $\sqrt{\frac{27}{0.03}} = \sqrt{900} = 30$.
- Use the equation $f = \frac{v}{\lambda}$.
- The first three resonance frequencies are $f_0 = 10$ Hz, $f_1 = 20$ Hz, and $f_2 = 30$ Hz.

36. Study the example on pages 101-102.

(A) First identify the **boundary conditions**.

- There will be an **anti-node** at the left end because the left end is **open**.
- There will be a **node** at the right end because the right end is **closed**. (The air at the very right edge doesn't have the freedom to vibrate horizontally.)
- Visualize the long train of sine waves below.

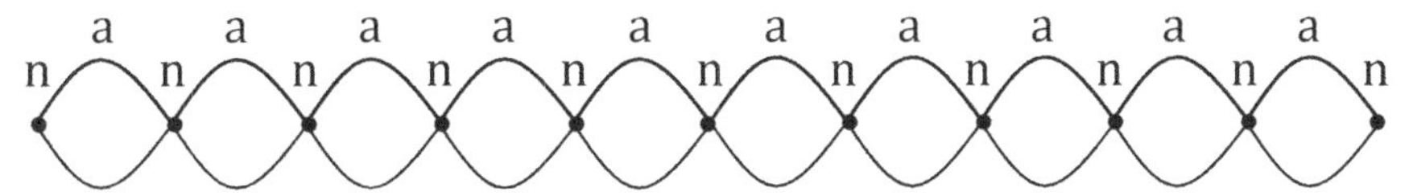

- The shortest section of this train that has an anti-node (a) at the left end and a node (n) at the right end is shown below. This is the **fundamental** (or **first harmonic**).

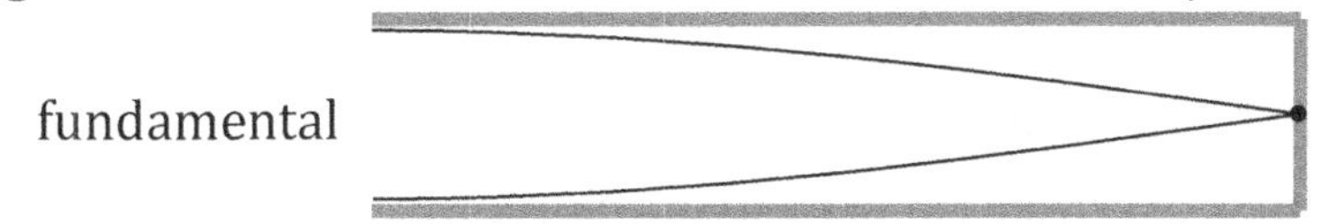

- The second shortest section that has an anti-node (a) at the left end and a node (n) at the right end is shown below. This is the **first overtone** (or **second harmonic**).

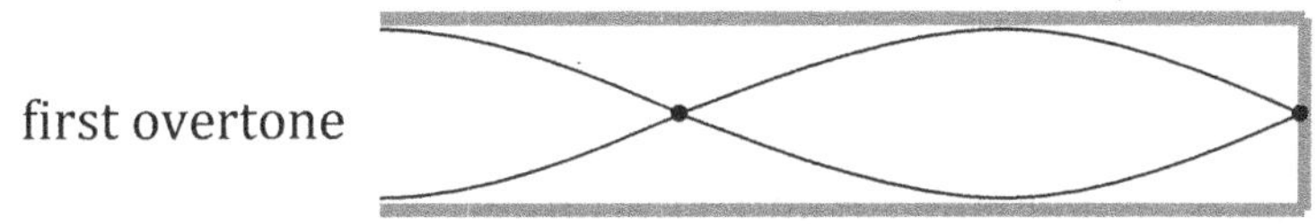

- The third shortest section that has an anti-node (a) at the left end and a node (n) at the right end is shown below. This is the **second overtone** (or **third harmonic**).

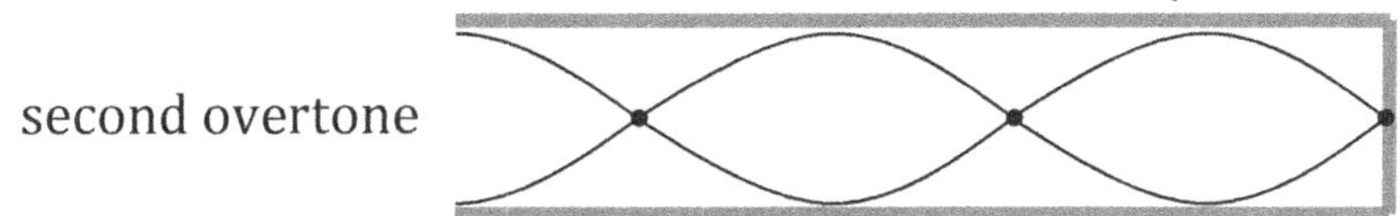

(B) Apply the following equation to each standing wave that you drew.

$$\begin{pmatrix}\text{number of} \\ \text{cycles}\end{pmatrix} \lambda = L$$

- Compare each standing wave above to a single cycle drawn below.

- Count the number of "footballs" and divide by 2 to get the number of cycles.
- You should get $\frac{1}{4}$ cycle, $\frac{3}{4}$ cycle, and $\frac{5}{4}$ cycles. Plug these into the above equation and solve for wavelength. The length of the pipe is $L = \frac{34}{100}$ m $= 0.34$ m.
- The wavelengths of the fundamental and first two overtones are $\lambda_0 = \frac{34}{25}$ m, $\lambda_1 = \frac{34}{75}$ m, and $\lambda_2 = \frac{34}{125}$ m. Note, for example, that $\lambda_0 = 4L = 4\left(\frac{34}{100}\right) = \frac{34}{25}$ m.

(C) Use the equation $f = \frac{v}{\lambda}$. The speed of sound in air is $v = 340$ m/s.

- The first three resonance frequencies are $f_0 = 250$ Hz, $f_1 = 750$ Hz, and $f_2 = 1250$ Hz. Note, for example, that $\frac{340}{34/25} = (340)\left(\frac{25}{34}\right) = (10)(25) = 250$.

37. Study the example on pages 101-102.

(A) First identify the **boundary conditions.**

- Since both ends are **open**, there will be an **anti-node** at each end.
- Visualize the long train of sine waves below.

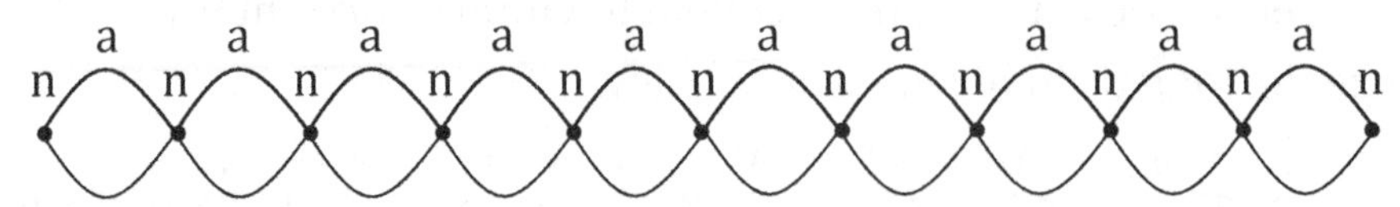

- The shortest section of this train that has an anti-node (a) at both ends is drawn below. This is the **fundamental** (or **first harmonic**).

fundamental 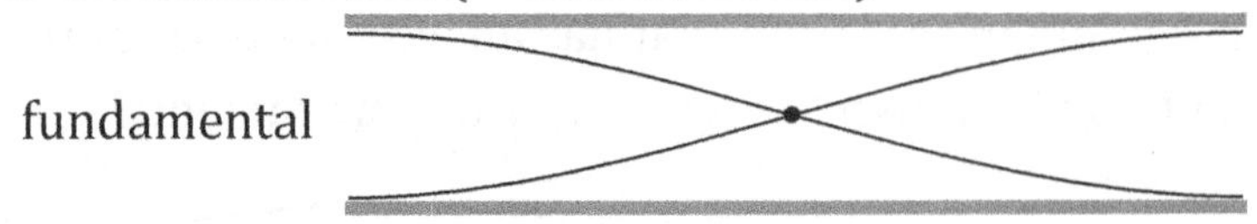

- The second shortest section that has an anti-node (a) at both ends is drawn below. This is the **first overtone** (or **second harmonic**).

first overtone 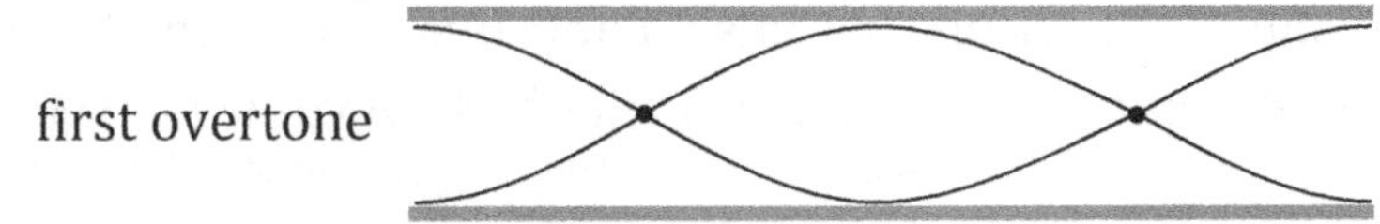

- The third shortest section that has an anti-node (a) at both ends is drawn below. This is the **second overtone** (or **third harmonic**).

second overtone 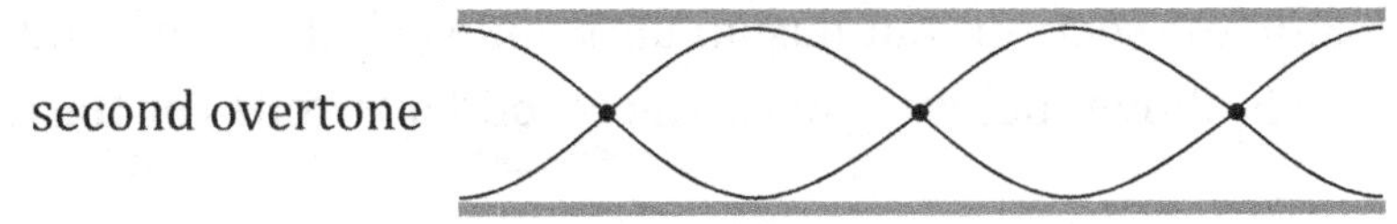

(B) Apply the following equation to each standing wave that you drew.

$$\begin{pmatrix}\text{number of} \\ \text{cycles}\end{pmatrix} \lambda = L$$

- Compare each standing wave above to a single cycle drawn below.

 one cycle

- Count the number of "footballs" and divide by 2 to get the number of cycles.
- You should get $\frac{1}{2}$ cycle, 1 cycle, and $\frac{3}{2}$ cycles. Plug these into the above equation and solve for wavelength. The length of the pipe is $L = 4.5$ m.
- The wavelengths of the fundamental and first two overtones are $\lambda_0 = 9.0$ m, $\lambda_1 = 4.5$ m, and $\lambda_2 = 3.0$ m. Note, for example, that $\lambda_0 = 2L = 2(4.5) = 9.0$ m.

(C) Use the equation $f = \frac{v}{\lambda}$. The wave speed given in the problem is $v = 300$ m/s.

- The first three resonance frequencies are $f_0 = \frac{100}{3}$ Hz, $f_1 = \frac{200}{3}$ Hz, and $f_2 = 100$ Hz. Note, for example, that $\frac{300}{9} = \frac{100}{3}$. If you use a calculator, $f_0 = 33$ Hz, $f_1 = 67$ Hz, and $f_2 = 100$ Hz.

38. Study the example on pages 103-104.
    - First identify the **boundary conditions**.
    - Since both ends are **free**, there will be an **anti-node** at each end.
    - There will also be a **node** at the location of the clamp, which is located one-fourth the length of the rod from its left end. See the diagram below.

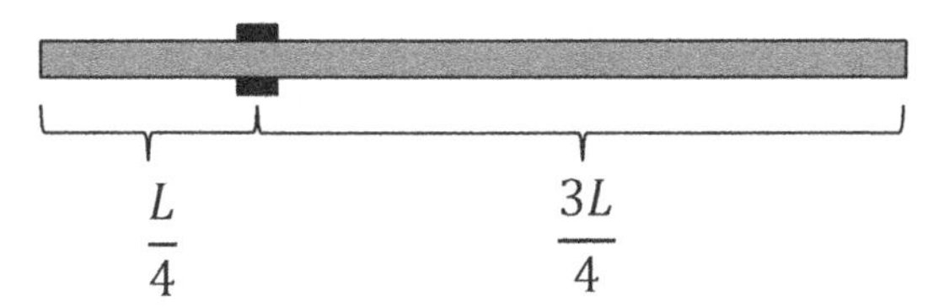

    - Visualize the long train of sine waves below.

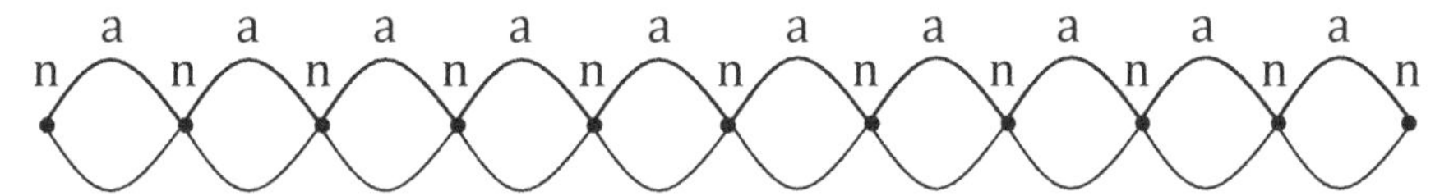

    - The shortest section of this train that has an anti-node (a) at the left end, another anti-node (a) at the right end, and a node (n) exactly one-quarter of the length from the left end is drawn below. (Anything shorter would either not have the required node at the exact location of the clamp or wouldn't have full anti-nodes at both ends.) This is the **fundamental**.

fundamental 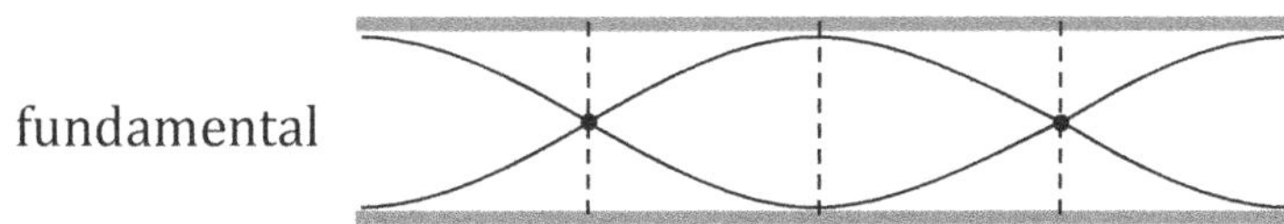

    - The second shortest section that has an anti-node (a) at the left end, another anti-node (a) at the right end, and a node (n) exactly one-quarter of the length from the left end is drawn below. (Anything shorter – other than the fundamental shown above – would either not have the required node at the exact location of the clamp or wouldn't have full anti-nodes at both ends.) This is the **first overtone**.

first overtone 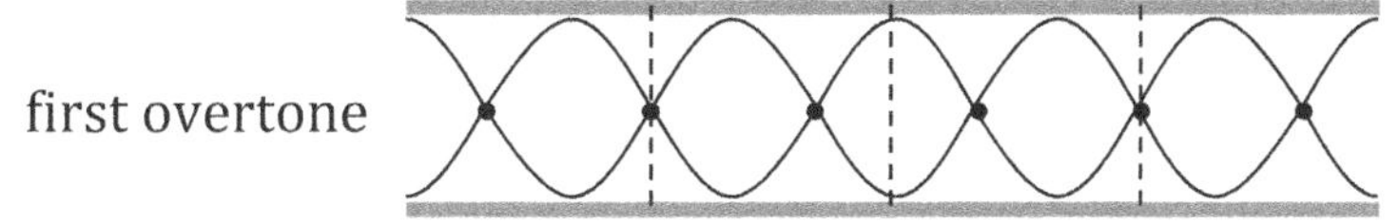

    - Apply the following equation to each standing wave that you drew.

$$\begin{pmatrix}\text{number of}\\ \text{cycles}\end{pmatrix}\lambda = L$$

    - Compare each standing wave above to a single cycle drawn below.

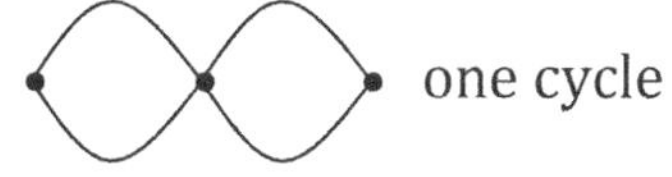 one cycle

    - Count the number of "footballs" and divide by 2 to get the number of cycles.
    - You should get 1 cycle and 3 cycles. Plug these into the above equation and solve for wavelength. You should get $\lambda_0 = L$ and $\lambda_1 = \frac{L}{3}$. The length of the pipe is $L = 6.0$ m.
    - The wavelengths of the fundamental and first overtone are $\lambda_0 = 6.0$ m and $\lambda_1 = 2.0$ m.

- Use the equation $f = \frac{v}{\lambda}$. The wave speed is given in the problem as $v = 6000$ m/s.
- The first two resonance frequencies are $f_0 = 1.0$ kHz and $f_1 = 3.0$ kHz, which is the same as $f_0 = 1000$ Hz and $f_1 = 3000$ Hz.

Chapter 8: Density

39. First use the formula for the volume of a rectangular box: $V = LWH$.
    - The volume of the banana cake is $V = 0.040\text{m}^3$.
    - Use the equation for density: $\rho = \frac{m}{V}$.
    - The density of the banana cake is $\rho = 800\ \frac{\text{kg}}{\text{m}^3}$.

---

40. Study the example on page 113.
    - Write down the density equation for each banana.
    - Use subscripts for mass and density, which are different for the two bananas. **Don't** use subscripts for volume, which is the same for both bananas in this problem. We will use Pb for lead and Al for aluminum. (These are the symbols for these elements on chemistry's periodic table.)
    - The equations are $\rho_{Pb} = \frac{m_{Pb}}{V}$ and $\rho_{Al} = \frac{m_{Al}}{V}$.
    - Since both bananas have the same volume in this problem, solve for volume. Multiply both sides of each equation by V and divide by ρ.
    - You should get $V = \frac{m_{Pb}}{\rho_{Pb}}$ and $V = \frac{m_{Al}}{\rho_{Al}}$. Set the two expressions for volume equal to one another.
    - You should get $\frac{m_{Pb}}{\rho_{Pb}} = \frac{m_{Al}}{\rho_{Al}}$.
    - The problem states that the lead (Pb) banana is four times as dense as the aluminum (Al) banana. Express this in an equation.
    - You should get $\rho_{Pb} = 4\rho_{Al}$. Substitute this into the previous equation. Simplify.
    - You should get $m_{Pb} = 4m_{Al}$, or, equivalently, $m_{Al} = \frac{m_{Pb}}{4}$.
    - Either way, the lead banana has 4 times the mass (and therefore also 4 times the weight) as the aluminum banana. Since the lead banana is **more dense**, and since the two bananas have the same volume, the lead banana has **more mass** (and weight).

---

41. Study the example on page 113.
    - Write down the density equation for each banana.
    - Use subscripts for density and volume, which are different for the two bananas. **Don't** use subscripts for mass, which is the same for both bananas in this problem. We will use Pb for lead and Al for aluminum. (These are the symbols for these elements on chemistry's periodic table.)

- The equations are $\rho_{Pb} = \frac{m}{V_{Pb}}$ and $\rho_{Al} = \frac{m}{V_{Al}}$.
- Since both bananas have the same mass in this problem, solve for mass. Multiply both sides of each equation by the respective volume.
- You should get $m = \rho_{Pb} V_{Pb}$ and $m = \rho_{Al} V_{Al}$. Set the two expressions for mass equal to one another.
- You should get $\rho_{Pb} V_{Pb} = \rho_{Al} V_{Al}$.
- The problem states that the lead (Pb) banana is four times as dense as the aluminum (Al) banana. Express this in an equation.
- You should get $\rho_{Pb} = 4\rho_{Al}$. Substitute this into the previous equation. Simplify.
- You should get $4V_{Pb} = V_{Al}$, or, equivalently, $V_{Pb} = \frac{V_{Al}}{4}$.
- Either way, the aluminum banana has 4 times the volume as the lead banana. Since the aluminum banana is **less dense**, and since the two bananas have the same mass, the aluminum banana occupies **more space**.

---

42. Read about density on page 109.
    - Density is a characteristic property of a substance, which depends on the size and mass of its atoms, along with the way that the atoms are arranged.
    - The entire rod is made out of the same substance: aluminum. Therefore, the density of the rod is the same throughout.
    - Whether you have half a rod or the whole rod, it still has the density of aluminum. Therefore, both the original rod and the half of the rod that the monkey keeps (and even the half of the rod that the monkey threw in the trash can) have the same density.
    - The formula for density is $\rho = \frac{m}{V}$. When the monkey cuts the rod in half, the half-rod has half the mass and half the volume of the original rod. However, the two one-half's cancel out in the formula: $\rho = \frac{m/2}{V/2} = \frac{m}{V}$. The density remains the **same**.

Chapter 9: Pressure

43. First determine the weight of the box and the area of the specified side.
    - Use the formula $F = mg$ to find the weight and $A = LW$ to find the area.
    - The smallest side has $L = 0.20$ m and $W = 0.10$ m.
    - You should get $F = 200$ N (if you round gravity to 10 m/s^2) and $A = 0.020$ m^2.
    - Use the equation $P = \frac{F}{A}$.
    - The pressure is $P = 10$ kPa $= 10{,}000$ Pa $= 1.0 \times 10^4$ Pa.

44. Consider the formula $P = P_0 + \rho g h$.
    - Which (if any) of the following symbols are different for the pressure at the bottom of the liquid in each container: P_0, ρ, g, h?
    - P_0 is the air pressure above the liquid. It's the same for each.
    - ρ is the density of the liquid. Each container has the same liquid, and the same ρ.
    - $g = 9.81$ m/s^2 for all three containers.
    - h is the depth from the liquid level to the bottom (since the question asks about the pressure at the bottom). The problem states that h is the same for each.
    - Since P_0, ρ, g, and h are the same for each, the pressure at the bottom ($P = P_0 + \rho g h$) must also be the **same** for each.
    - The main concept is that the pressure in a fluid depends on **depth**, and the depth is the same for all three cases.

---

45. Use the equation $P = P_0 + \rho g h$, with $P_0 \approx 1.0 \times 10^5$ Pa and $g = 9.81$ m/s$^2 \approx 10$ m/s^2.
(A) The pressure is $P = 600$ kPa $= 600{,}000$ Pa $= 6.0 \times 10^5$ Pa.
(B) First determine the radius of the circular window.
    - **Convert** the **diameter** to meters. Use the formula $R = \frac{D}{2}$ to find the **radius.**
    - You should get $R = 0.050$ m.
    - Find the area of the circular window.
    - Use the formula $A = \pi R^2$.
    - You should get $A = 0.0025\pi$ m^2. If you use a calculator, $A = 0.0079$ m^2.
    - Use the formula $P = \frac{F}{A}$. Solve for force.
    - You should get $F = PA$. Use the pressure from part (A).
    - The force is $F = 1500\pi$ N $= 1.5\pi$ kN. If you use a calculator, $F = 4.7$ kN $= 4700$ N.

Chapter 10: Archimedes' Principle

46. Study the top example on page 125.
    - First identify the given information in SI units.
    - The object has a mass of $m_o = 5.0$ kg.
    - The density of the water is $\rho_f = 1000 \ \frac{\text{kg}}{\text{m}^3}$.
    - The acceleration is $a_y = -6.0$ m/s^2 because it is **downward** (taking $+y$ to be up).
    - You should also know gravitational acceleration:[1] $g = 9.81$ m/s$^2 \approx 10$ m/s^2.

---

[1] Gravitational acceleration (g) is **always** positive, whereas the object's acceleration (a_y) can be negative. For example, in the equation for weight ($W = mg$), we use $g = 9.81$ m/s^2, but for an object in free fall, we use $a_y = -g = -9.81$ m/s^2 in the equations of one-dimensional uniform acceleration. In this problem, $a_y = -6.0$ m/s^2 (and <u>not</u> $a_y = -9.81$ m/s^2) because the object is <u>not</u> freely falling (there is an upward buoyant force affecting the object's acceleration).

(A) Apply Newton's second law, as we did on page 125. Our signs are based on choosing $+y$ to be upward (as labeled below).

- The buoyant force ($\vec{\mathbf{F}}_B$) pushes up while weight of the object ($m_o\vec{\mathbf{g}}$) pulls down.

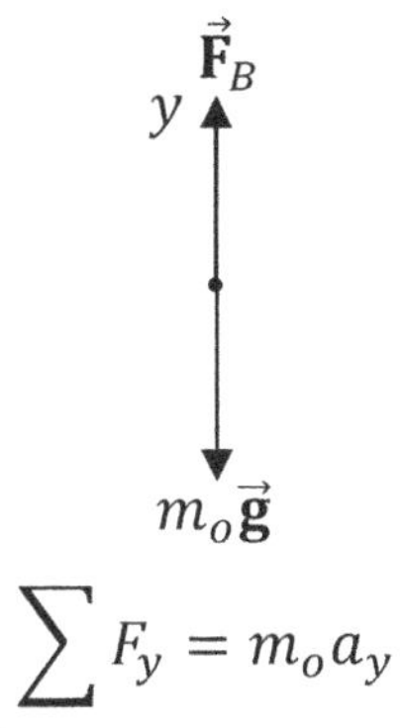

$$\sum F_y = m_o a_y$$

$$F_B - m_o g = m_o a_y$$

- Add the weight of the object to both sides of the equation.
- You should get $F_B = m_o g + m_o a_y$. Note that a_y is **negative**.
- The buoyant force is $F_B = 20$ N.

(B) Use the equation for a **fully submerged** object: $\rho_f g - \rho_o g = \rho_o a_y$.

- Add $\rho_o g$ to both sides of the equation and **factor** out the ρ_o.
- You should get $\rho_f g = \rho_o(a_y + g)$. Divide both sides of the equation by $(a_y + g)$.
- You should get:

$$\rho_o = \frac{\rho_f g}{a_y + g}$$

- The density of the object is $\rho_o = 2500 \frac{\text{kg}}{\text{m}^3}$. Recall that a_y is **negative**.

---

47. Study the second example on page 125.

- First identify the given information in SI units.
- The fraction of the ice cube sticking out is $\frac{1}{4}$. Note that $\frac{1}{4} = 0.25 = 25\%$.
- The density of the ice cube is $\rho_o = 900 \frac{\text{kg}}{\text{m}^3}$.
- Use the following equation (from page 124).

$$\begin{matrix}\text{the fraction of an ice cube} \\ \text{sticking out of the liquid}\end{matrix} = 1 - \frac{\rho_o}{\rho_f}$$

- You should get $\frac{1}{4} = 1 - \frac{900}{\rho_f}$. Solve for ρ_f.
- Add $\frac{900}{\rho_f}$ to both sides and subtract $\frac{1}{4}$. You should get $\frac{900}{\rho_f} = \frac{3}{4}$.
- **Cross multiply**. Isolate ρ_f.
- The density of the banana juice is $\rho_f = 1200 \frac{\text{kg}}{\text{m}^3}$.

48. Study the example on pages 126-127.
    - First identify the given information in SI units.
    - The **actual** weight of the object is $W_o = 20$ N.
    - The **apparent** weight of the object is $W_a = 12$ N.
    - The density of the water is $\rho_f = 1000 \frac{\text{kg}}{\text{m}^3}$.
    - You should also know gravitational acceleration: $g = 9.81 \text{ m/s}^2 \approx 10 \text{ m/s}^2$.
    - Use the equation $W_o = m_o g$. Solve for the mass of the object. Divide both sides of the equation by g. You should get $m_o = \frac{W_o}{g}$.
    - The object's mass is $m_o = 2.0$ kg.
    - A scale would measure **apparent** weight the same way that we found the final tension in part (B) of the example on pages 126-127. Draw a FBD and apply Newton's second law just as in part (B) of that example.

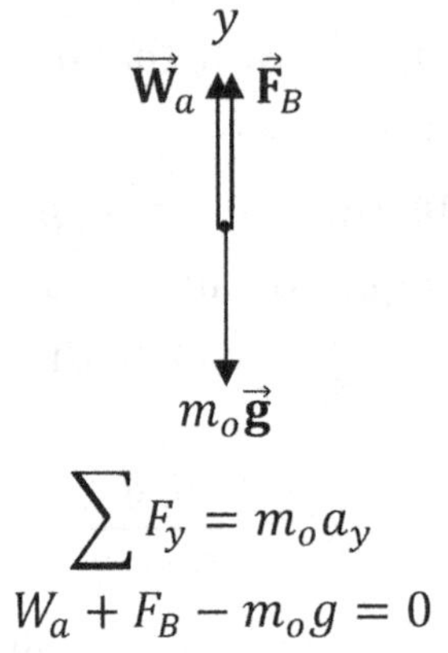

$$\sum F_y = m_o a_y$$
$$W_a + F_B - m_o g = 0$$

- Solve for the buoyant force. Add $m_o g$ to both sides and subtract W_a.
- You should get $F_B = m_o g - W_a$.
- The buoyant force is $F_B = 8.0$ N.
- Apply Archimedes' principle: $F_B = m_f g = \rho_f V_f g$.
- Since the object is **fully submerged** in the water, the volume of the object equals the volume of the **displaced** fluid ($V_o = V_f$). Therefore, $F_B = \rho_f V_o g$.
- Use the equation $\rho_o = \frac{m_o}{V_o}$ to solve for the volume of the object. Multiply both sides of the equation by V_o and divide by ρ_o. You should get $V_o = \frac{m_o}{\rho_o}$.
- Substitute this expression for volume into the equation $F_B = \rho_f V_o g$.
- You should get $F_B = \frac{\rho_f m_o g}{\rho_o}$. Solve for ρ_o. Multiply both sides by ρ_o and divide by F_B.
- You should get $\rho_o = \frac{\rho_f m_o g}{F_B}$. Recall that $m_o = 2.0$ kg and $F_B = 8.0$ N.
- The density of the object is $\rho_o = 2500 \frac{\text{kg}}{\text{m}^3}$.

49. Study the second example on page 125.
- First identify the given information in SI units.
- The density of the wooden cube is $\rho_o = 640 \frac{\text{kg}}{\text{m}^3}$.
- The density of the water[2] is $\rho_f = 1024 \frac{\text{kg}}{\text{m}^3}$.
- The edge length of the wooden cube is $L = 0.25$ m.

(A) Use the following equation (from page 124).

$$\text{the fraction of a floating cube sticking out of the liquid} = 1 - \frac{\rho_o}{\rho_f}$$

- This comes out to 0.375 as a decimal, which equates to 37.5% as a percentage. (If you prefer to express it as a fraction, it is $\frac{375}{1000}$, which reduces to $\frac{3}{8}$.)

(B) See the note regarding a floating cube on page 124.
- If additional mass (m_a) is placed on the floating cube, use the following equation.

$$\rho_f V_f = m_o + m_a$$

- We're solving for the added mass (m_a). Of these symbols, we presently know only the density of water (ρ_f). We must find V_f and m_o before we can use this equation.
- Use the equation $V_o = L^3$ to find the volume of the wooden cube.
- The volume of the wooden cube is $V_o = \frac{1}{64}$ m^3. If you're not using a calculator, it will be convenient to leave this number as a fraction. If you are using a calculator, the volume of the wooden cube comes out to $V_o = 0.015625$ m^3.
- In part (B), the wooden cube is **fully submerged** (just barely, but that's enough). This means that the volume of the object equals the volume of the **displaced** fluid ($V_o = V_f$). Thus, the volume of the displaced fluid is $V_f = \frac{1}{64}$ m^3.
- Use the equation $\rho_o = \frac{m_o}{V_o}$ to find the mass of the wooden cube. Multiply both sides of the equation by V_o. You should get $m_o = \rho_o V_o$.
- The mass of the wooden cube is $m_o = 10$ kg.
- Now you can use the equation $\rho_f V_f = m_o + m_a$ with $\rho_f = 1024 \frac{\text{kg}}{\text{m}^3}$, $V_f = \frac{1}{64}$ m^3, and $m_o = 10$ kg.
- Subtract m_o from both sides of the equation. You should get $m_a = \rho_f V_f - m_o$.
- The mass of the monkey is $m_o = 6.0$ kg. (This monkey is relatively small.)

[2] Yes, this value is slightly different than what we have used in other problems. The density of water does depend on factors such as temperature and purity, so in reality, there isn't just "one" number for "the" density of water. We've used 1000 in most problems to make the arithmetic simple to carry out without a calculator, but in this problem it turns out that 1024 actually makes the arithmetic simpler (if you try it with 1000 instead of 1024, you'll get 5.625 kg instead of a nice round 6.0 kg).

50. Study the examples from this chapter.
- First identify the given information in SI units.
- The mass of the balloon plus its load is $m_L = 1.64$ kg.
- The volume of the object is $V_o = 2.0 \text{ m}^3$.
- The density of the helium is $\rho_{He} = 0.18 \frac{\text{kg}}{\text{m}^3}$.
- The density of the air is $\rho_f = 1.3 \frac{\text{kg}}{\text{m}^3}$. (Air is the **fluid** that the object is submerged in.)
- You should also know gravitational acceleration: $g = 9.81 \text{ m/s}^2 \approx 10 \text{ m/s}^2$.

(A) Apply Archimedes' principle: $F_B = m_f g = \rho_f V_f g$.
- Since the balloon is **fully submerged** in the air, the volume of the object equals the volume of the **displaced** fluid ($V_o = V_f$). Therefore, $F_B = \rho_f V_o g$.
- The buoyant force is $F_B = 26$ N.

(B) First find the mass of the helium (He) contained in the balloon.
- Use the equation $\rho_{He} = \frac{m_{He}}{V_o}$. Multiply both sides of the equation by V_o.
- You should get $m_{He} = \rho_{He} V_o$.
- The mass of the helium is $m_{He} = 0.36$ kg.
- The total mass of the balloon, its load, and the helium is: $m_o = m_L + m_{He}$.
- The mass of the object is $m_o = 2.0$ kg.
- Apply Newton's second law, as we did on page 125.
- The buoyant force ($\vec{\mathbf{F}}_B$) pushes up while weight of the object ($m_o \vec{\mathbf{g}}$) pulls down.

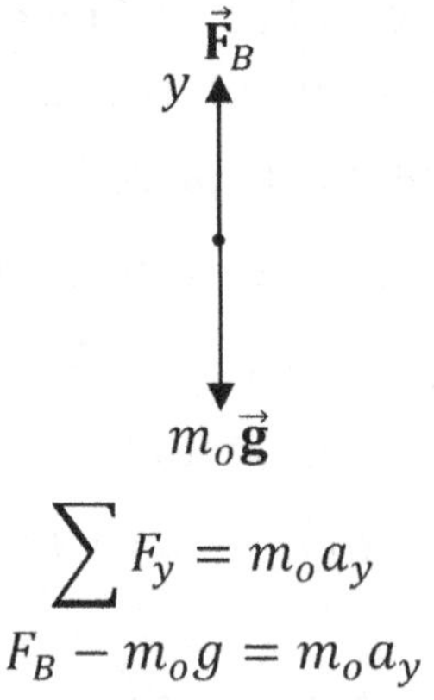

$$\sum F_y = m_o a_y$$
$$F_B - m_o g = m_o a_y$$

- Solve for the acceleration. Divide both sides of the equation by m_o.
- You should get:

$$a_y = \frac{F_B - m_o g}{m_o}$$

- Be sure to use $m_o = 2.0$ kg (the **total** mass of the object, including the balloon and the helium inside of it).
- The acceleration is $a_y = 3.0 \text{ m/s}^2$. It is directed upward.

Chapter 11: Fluid Dynamics

51. Identify the given information in SI units.
- The diameter is $D_1 = 0.050$ m at the first position.
- The speed of flow is $v_1 = \frac{16}{\pi}$ m/s at the first position.
- The diameter is $D_2 = 0.020$ m at the second position.
- Divide each diameter by 2 in order to find the corresponding radius.
- You should get $R_1 = 0.025$ m and $R_2 = 0.010$ m.
- Use the formulas $A_1 = \pi R_1^2$ and $A_2 = \pi R_2^2$.
- You should get $A_1 = 0.000625\pi$ m^2 and $A_2 = 0.0001\pi$ m^2.
- If you're not using a calculator, it's more convenient to work with the fraction, $A_1 = \frac{\pi}{1600}$ m^2. Note that $(0.025)^2 = \left(\frac{1}{40}\right)^2 = \frac{1}{1600}$ is the same as 0.000625.
- If you're using a calculator, $A_1 = 0.00196$ m^2 and $A_2 = 0.000314$ m^2.

(A) Use the equation for **flow rate**: $Q = A_1 v_1$.
- You should get $Q = 0.010 \frac{\text{m}^3}{\text{s}}$. Note that the π's cancel.
- Convert $\frac{\text{m}^3}{\text{s}}$ to $\frac{\text{m}^3}{\text{min.}}$. Normally, we would use SI units, but the question specified otherwise. Multiply by 60.
- If we were converting from second to minutes, we would divide by 60. However, we're instead converting from $\frac{1}{\text{s}}$ to $\frac{1}{\text{min.}}$. Note that $\frac{1}{\text{s}} = \frac{1}{\text{s}} \times \frac{60 \text{ s}}{1 \text{ min.}} = 60 \frac{1}{\text{min.}}$.
- The flow rate is $Q = 0.60 \frac{\text{m}^3}{\text{min.}}$, which is the same as $Q = \frac{3}{5} \frac{\text{m}^3}{\text{min.}}$.

(B) Apply the **continuity equation**: $A_1 v_1 = A_2 v_2$. Solve for v_2.
- You should get $v_2 = \frac{A_1}{A_2} v_1$.
- The speed of flow is $v_2 = \frac{100}{\pi}$ m/s in the section of the pipe with a diameter of 2.0 cm. If you use a calculator, this works out to $v_2 = 32$ m/s.

---

52. Identify the given information in SI units.
- The water level is $y_1 = 0.50$ m above the bottom of the container.
- The hole is $y_2 = 0.30$ m above the bottom of the container.
- The diameter of the hole is $D = 0.0010$ m. Note that this was given in mm, **not** cm. Recall that 1 mm = 0.001 m and that 1 cm = 0.01 m.

(A) First determine the **depth** of the hole below the water level.
- Subtract the two heights to find the depth: $h = y_1 - y_2$. Recall that h represents **depth** (below the water level), whereas y represents height (above the bottom).
- The depth is $h = 0.20$ m.
- Apply **Torricelli's law**: $v = \sqrt{2gh}$. Note that $(20)(0.2) = 4$.
- The water escapes through the hole with a speed of $v = 2.0$ m/s.

(B) Divide the diameter by 2 in order to find the radius.

- You should get $R = 0.00050$ m.
- Use the formula $A = \pi R^2$. You should get $A = 0.00000025\pi$ m^2. If you use a calculator, this works out to $A = 0.000000785$ m^2.
- Use the equation for **flow rate**: $Q = Av$. Recall from part (A) that $v = 2.0$ m/s.
- You should get $Q = 0.00000050\pi \frac{\text{m}^3}{\text{s}} = 5\pi \times 10^{-7} \frac{\text{m}^3}{\text{s}}$. If you use a calculator, this works out to $Q = 1.6 \times 10^{-6} \frac{\text{m}^3}{\text{s}}$.
- Convert $\frac{\text{m}^3}{\text{s}}$ to $\frac{\text{cm}^3}{\text{min.}}$. Normally, we would use SI units, but the question specified otherwise. Multiply by 60 and multiply by 10^6.
- If we were converting from second to minutes, we would divide by 60. However, we're instead converting from $\frac{1}{\text{s}}$ to $\frac{1}{\text{min.}}$. Note that $\frac{1}{\text{s}} = \frac{1}{\text{s}} \times \frac{60 \text{ s}}{1 \text{ min.}} = 60 \frac{1}{\text{min.}}$.
- If we were converting from m to cm, we would multiply by 100. However, we're instead converting from m^3 to cm^3. Note that $1 \text{ m}^3 = (1 \text{ cm})^3 = 100^3 \text{ cm}^3 = 10^6 \text{ cm}^3$.
- The flow rate is $Q = 30\pi \frac{\text{cm}^3}{\text{min.}}$. If you use a calculator, $Q = 94 \frac{\text{cm}^3}{\text{min.}}$.

---

53. Identify the given information in SI units.

- The diameter is $D_1 = 0.02\sqrt{2}$ m at the first position.
- The diameter is $D_2 = 0.0200$ m at the second position.
- The pressure is $P_1 = 82{,}500$ Pa at the first position.
- The speed of flow is $v_1 = 5.0$ m/s at the first position.
- The density of the water is $\rho = 1000 \frac{\text{kg}}{\text{m}^3}$.
- Divide each diameter by 2 in order to find the corresponding radius.
- You should get $R_1 = 0.01\sqrt{2}$ m and $R_2 = 0.0100$ m.
- Use the formulas $A_1 = \pi R_1^2$ and $A_2 = \pi R_2^2$.
- You should get $A_1 = 0.0002\pi$ m^2 and $A_2 = 0.0001\pi$ m^2. Note that $\left(\sqrt{2}\right)^2 = 2$.

(A) Apply the **continuity equation**: $A_1 v_1 = A_2 v_2$. Solve for v_2.

- You should get $v_2 = \frac{A_1}{A_2} v_1$.
- The speed of flow is $v_2 = 10$ m/s in the constricted portion of the tube.

(B) Apply **Bernoulli's equation**.

$$P_1 + \frac{1}{2}\rho v_1^2 + \rho g y_1 = P_2 + \frac{1}{2}\rho v_2^2 + \rho g y_2$$

- Set $y_1 = y_2 = 0$ since the pipe is horizontal (the height doesn't change). Solve for P_2. You should get:

$$P_2 = P_1 + \frac{1}{2}\rho v_1^2 - \frac{1}{2}\rho v_2^2$$

- Recall that we found $v_2 = 10$ m/s in part (B).
- The pressure is $P_2 = 45{,}000$ Pa $= 45$ kPa in the constricted portion of the pipe.

(C) Use the equation for **flow rate**: $Q = A_1 v_1$ or $Q = A_2 v_2$.

- You should get $Q = 0.001\pi \frac{\text{m}^3}{\text{s}} = \pi \times 10^{-3} \frac{\text{m}^3}{\text{s}}$. If you use a calculator, this works out to $Q = 0.0031 \frac{\text{m}^3}{\text{s}} = 3.1 \times 10^{-3} \frac{\text{m}^3}{\text{s}}$.

Chapter 12: Pascal's Law

54. Study the example on page 143.
- Identify the given information in SI units.
- Use the subscript 1 for the narrow piston and the subscript 2 for the wide piston.
- The diameter of the narrow piston is $D_1 = 0.10$ m.
- The diameter of the wide piston is $D_2 = 0.40$ m.
- A force as large as $F_1 = 200$ N can be applied to the narrow piston.
- Earth's surface gravity is $g = 9.81 \text{ m/s}^2 \approx 10 \text{ m/s}^2$.
- Divide each diameter by 2 in order to find the corresponding radius.
- You should get $R_1 = 0.05$ m and $R_2 = 0.20$ m.
- Use the formulas $A_1 = \pi R_1^2$ and $A_2 = \pi R_2^2$.
- You should get $A_1 = 0.0025\pi \text{ m}^2$ and $A_2 = 0.04\pi \text{ m}^2$.
- Apply **Pascal's law**.

$$\frac{F_1}{A_1} = \frac{F_2}{A_2}$$

- Solve for F_2. Multiply both sides of the equation by A_2.
- You should get:

$$F_2 = \frac{A_2}{A_1} F_1$$

- Note that the π's cancel. Also note that $\frac{0.04}{0.0025} = \frac{400}{25} = 16$. (In the first step, we multiplied the numerator and denominator both by 10,000 in order to remove the decimals.)
- The maximum lift force is $F_2 = 3200$ N.
- Apply the equation $F_2 = m_2 g$.
- Solve for m_2. Divide both sides of the equation by g.
- You should get $m_2 = \frac{F_2}{g}$.
- The maximum mass that can be lifted is $m_2 = 320$ kg. If you don't round 9.81 to 10, the maximum mass is $m_2 = 326$ kg.

Chapter 13: Temperature Conversions

55. Use the formula that has the desired temperature on the left-hand side and the given temperature on the right-hand side.

(A) Use the formula $T_C = T_K - 273.15$ to convert Kelvin to Celsius.
- The temperature in Celsius is $T_C = -73$°C. The answer is **negative.**
- Note that the value in Kelvin (200) is **greater** than the value in Celsius (−73).

(B) Use the formula $T_K = T_C + 273.15$ to convert Celsius to Kelvin.
- The temperature in Kelvin is $T_K = 473$ K.
- Note that the value in Kelvin (473) is **greater** than the value in Celsius (200).

(C) Use the formula $T_C = \frac{5}{9}(T_F - 32)$ to convert Fahrenheit to Celsius.
- Note that $14 - 32 = -18$ is **negative.**
- The temperature in Celsius is $T_C = -10$°C. The answer is **negative.**

(D) Use the formula $T_F = \frac{9}{5}T_C + 32$ to convert Celsius to Fahrenheit.
- Be sure to keep the **minus sign** in −15°C. Note that $\frac{9}{5}(-15) = -27$.
- The temperature in Fahrenheit is $T_F = 5.0$°F. This answer is **positive.**

(E) Convert Kelvin to Fahrenheit in two steps.
- First use the formula $T_C = T_K - 273.15$ to convert Kelvin to Celsius.
- You should get $T_C = -35$°C.
- Next use the formula $T_F = \frac{9}{5}T_C + 32$ to convert Celsius to Fahrenheit.
- The temperature in Fahrenheit is $T_F = -31$°F. The answer is **negative.**

(F) Convert Fahrenheit to Kelvin in two steps.
- First use the formula $T_C = \frac{5}{9}(T_F - 32)$ to convert Fahrenheit to Celsius.
- You should get $T_C = 35$°C. It's **not** the same as part (E) – this is **positive.**
- Next use the formula $T_K = T_C + 273.15$ to convert Celsius to Kelvin.
- The temperature in Kelvin is $T_K = 308$ K.
- Note that the value in Kelvin (308) is **greater** than the value in Celsius (35).

---

56. Study the last example on page 148.
- Since this problem involves Kelvin and Fahrenheit, you will need two equations.
- We will use the equations $T_K = T_C + 273.15$ and $T_F = \frac{9}{5}T_C + 32$.
- Set $T_K = T_F$ in both of these equations.
- You should get $T_F = T_C + 273.15$ and $T_F = \frac{9}{5}T_C + 32$.
- Since $T_C + 273.15$ equals T_F and since $\frac{9}{5}T_C + 32$ also equals T_F, we may set $T_C + 273.15$ equal to $\frac{9}{5}T_C + 32$.

- That is, $T_C + 273.15 = \frac{9}{5}T_C + 32$.
- **Combine like terms.** Note that $\frac{9}{5} - 1 = \frac{9}{5} - \frac{5}{5} = \frac{9-5}{5} = \frac{4}{5}$.
- You should get $241.15 = \frac{4}{5}T_C$.
- Solve for T_C. Multiply both sides of the equation by 5/4.
- You should get $T_C = \frac{1205}{4}$°C if you're not using a calculator. Otherwise, $T_C = 301.4$°C.
- Note that T_C is **not** the final answer. We need to find T_K or T_F (note that T_K and T_F have the same numerical value in this problem).
- Plug T_C into the equation $T_K = T_C + 273.15$.
- The temperature in Kelvin is $T_K \approx \frac{2297}{4}$ K ≈ 574 K if you don't use a calculator. If you use a calculator and don't round 273.15 to 273, you should get $T_K = 575$ K.
- Note that the value in Kelvin (575) is **greater** than the value in Celsius (301).
- **Check your answer**: Plug T_C into the equation $T_F = \frac{9}{5}T_C + 32$.
- You should get $T_F = 575$°F, which has the same numerical value as $T_K = 575$ K.

Chapter 14: Thermal Expansion

57. First identify the given information. It's okay to use Celsius for thermal expansion.
    - The coefficient of thermal expansion is $\alpha = 25.000 \times 10^{-6}$ /°C.
    - The length of the pole is $L_0 = 30.000$ m when the temperature is $T_0 = -20.000$°C.
    - We want to find the length L when the temperature is $T = 40.000$°C.
    - Apply the formula for **linear** expansion: $\Delta L = \alpha L_0 \Delta T$.
    - Note that $\Delta L = L - L_0$ and $\Delta T = T - T_0$, such that $L - L_0 = \alpha L_0(T - T_0)$.
    - Add L_0 to both sides of the equation: $L = L_0 + \alpha L_0(T - T_0)$.
    - Note that $T - T_0 = 40 - (-20) = 40 + 20 = 60$°C.
    - The length of the pole is $L = 30.045$ m at 40.000°C.

---

58. Compare the wording for Problems 57 and 58 closely. What's different?
    - The numbers are identical, but the wording is different. The difference matters.
    - In Problem 57, the pole expanded and the tape measure didn't.
    - In Problem 58, the tape measure expands and the pole doesn't.
    - The tape measure initially reads $L_0 = 30.000$ m at $T_0 = -20.000$°C.
    - In Problem 57, we found that $\Delta L = 0.045$ m (that's why L is 0.045 m longer than L_0).
    - This same value, $\Delta L = 0.045$ m, applies to this problem, except that we must interpret it differently because **the tape measure expands rather than the pole**.
    - Since the tape measure expands in this problem, at 40.000°C the 30.000 m mark will be **taller than the pole** by 0.045 m. So we **subtract** 0.045 m from 30.000 m.
    - The tape measure will read $L = 30.000 - 0.045 = 29.955$ m at the top of the pole.

59. First identify the given information. It's okay to use Celsius for thermal expansion.

- The coefficient of thermal expansion is $\alpha = 40.000 \times 10^{-6}$ /°C.
- The radius of the ball is $R_0 = 3.0000$ m when the temperature is $T_0 = 10.000$°C.
- We want to find the volume V and radius R when the temperature is $T = 35.000$°C.

(A) First calculate the initial volume: $V_0 = \frac{4}{3}\pi R_0^3$.

- You should get $V_0 = 36\pi$ m^3. If you use a calculator, $V_0 = 113.10$ m^3.
- Apply the formula for **volume** expansion: $\Delta V = 3\alpha V_0 \Delta T$.
- Note that $\Delta V = V - V_0$ and $\Delta T = T - T_0$, such that $V - V_0 = 3\alpha V_0(T - T_0)$.
- Add V_0 to both sides of the equation: $V = V_0 + 3\alpha V_0(T - T_0)$.
- The volume of the ball is $V = 36.108\pi$ m^3 at 35.000°C. If you use a calculator, this works out to $V = 113.44$ m^3.

(B) Apply the formula for **linear** expansion: $\Delta R = \alpha R_0 \Delta T$.

- We used R for radius instead of L for length, but the formula is the same.
- Note that $\Delta R = R - R_0$ and $\Delta T = T - T_0$, such that $R - R_0 = \alpha R_0(T - T_0)$.
- Add R_0 to both sides of the equation: $R = R_0 + \alpha R_0(T - T_0)$.
- The radius of the ball is $R = 3.0030$ m at 35.000°C.

(C) Plug your answer for part (B), $R = 3.0030$ m, into the formula $V = \frac{4}{3}\pi R^3$.

- You should get the same answer as part (A): $V = 36.108\pi$ m^3 = 113.44 m^3.

---

60. First identify the given information. It's okay to use Celsius for thermal expansion.

- The coefficient of thermal expansion is $\alpha = 30.000 \times 10^{-6}$ /°C.
- The edge length is $L_0 = 2.0000$ m when the temperature is $T_0 = 5.000$°C.
- In part (B), we want to find information when the temperature is $T = 55.000$°C.

(A) To find the surface area of a cube, first find the area of one square face and then multiply by the number of faces.

- Since there are 6 faces, the initial surface area is $A_0 = 6L_0^2$.
- The initial volume is $V_0 = L_0^3$.
- The surface area is $A_0 = 24.000$ m^2 and the volume is $V_0 = 8.0000$ m^3 at 5.000°C.

(B) Apply the formulas for **surface** and **volume** expansion: $\Delta A = 2\alpha A_0 \Delta T$ and $\Delta V = 3\alpha V_0 \Delta T$.

- Rewrite these as $A - A_0 = 2\alpha A_0(T - T_0)$ and $V - V_0 = 3\alpha V_0(T - T_0)$.
- Solve for A and V: $A = A_0 + 2\alpha A_0(T - T_0)$ and $V = V_0 + 3\alpha V_0(T - T_0)$.
- The surface area is $A = 24.072$ m^2 and the volume is $V = 8.0360$ m^3 at 55.000°C.

Note: You could alternatively use the formula for **linear** expansion, $\Delta L = \alpha L_0 \Delta T$, to determine that $L = 2.0030$ m at 55.000°C, and then apply the formulas $A = 6L^2$ and $V = L^3$. You would get the same answers this way.

Chapter 15: Heat Transfer

61. Identify the given information in suitable units.
    - The specific heat of water is $C = 4200 \frac{\text{J}}{\text{kg}\cdot°\text{C}}$.
    - The latent heat of fusion for water is $L_f = 3.3 \times 10^5 \frac{\text{J}}{\text{kg}}$.
    - The latent heat of vaporization for water is $L_v = 2.3 \times 10^6 \frac{\text{J}}{\text{kg}}$.
    - The mass of the water is $m = 0.500$ kg.
    - The initial temperature of the water is $T_0 = 25°\text{C}$. **Note**: It's okay to use degrees Celsius (°C) instead of Kelvin (K) in this problem because the equation involves ΔT.

(A) First the water cools from $T_0 = 25°\text{C}$ to $T = 0°\text{C}$. Use the equation $Q_1 = mC\Delta T$.
- The final temperature $T = 0°\text{C}$ is the **freezing point** of water (see Chapter 13).
- You should get $Q_1 = -52{,}500\text{J} = -52.5$ kJ. This is **not** the answer to the question.
- Note that Q_1 is **negative** because **cooling** is exothermic: $T - T_0 = -25°\text{C}$.
- Next the water freezes into ice. Use $Q_2 = -mL_f$ for this phase transition.
- Note that Q_2 is **negative** because **freezing** is exothermic. You must put this minus sign in the equation yourself.
- You should get $Q_2 = -165{,}000\text{J} = -165$ kJ. This is **not** the answer to the question.
- The net heat equals $Q_{net} = Q_1 + Q_2$.
- The final answer is $Q_{net} = -217{,}500$ J $= -218$ kJ $= -0.22$ MJ $= -2.2 \times 10^5$ J.
- The water releases 2.2×10^5 J (to two significant figures) of heat energy. (We interpret the **minus** sign to mean that heat is **released**.)

(B) First the water warms from $T_0 = 25°\text{C}$ to $T = 100°\text{C}$. Use the equation $Q_1 = mC\Delta T$.
- The final temperature $T = 100°\text{C}$ is the **boiling point** of water (see Chapter 13).
- You should get $Q_1 = 157{,}500\text{J} = 157.5$ kJ. This is **not** the answer to the question.
- Note that Q_1 is **positive** because **warming** is endothermic: $T - T_0 = +75°\text{C}$.
- Next the water freezes into ice. Use $Q_2 = mL_v$ for this phase transition.
- Note that Q_2 is **positive** because **boiling** is endothermic.
- You should get $Q_2 = 1{,}150{,}000\text{J} = 1{,}150$ kJ. This is **not** the answer to the question.
- The net heat equals $Q_{net} = Q_1 + Q_2$.
- The final answer[1] is $Q_{net} = 1{,}307{,}500$ J $= 1308$ kJ $= 1.3$ MJ $= 1.3 \times 10^6$ J.
- The water absorbs 1.3×10^6 J (to two significant figures) of heat energy. (We interpret the **plus** sign to mean that heat is **absorbed**.)

[1] Only two of these digits are significant. The given values – $4200 \frac{\text{J}}{\text{kg}\cdot°\text{C}}$, $3.3 \times 10^5 \frac{\text{J}}{\text{kg}}$, and $2.3 \times 10^6 \frac{\text{J}}{\text{kg}}$ – were rounded to two significant figures, so our final answer can't be more precise than that.

62. This is a **calorimetry** problem. First identify the given information in suitable units.
- The mass of the liquid is $m_\ell = 400$ g. (Grams are okay for calorimetry problems.)
- The mass of the solid is $m_s = 100$ g. Note: "ℓ" is for liquid, "s" is for solid.
- The initial temperature of the liquid is $T_\ell = 30$°C.
- The initial temperature of the solid is $T_s = 110$°C.
- The specific heat of the liquid is $C_\ell = 2400 \frac{\text{J}}{\text{kg}\cdot°\text{C}}$.
- The specific heat of the solid is $C_s = 640 \frac{\text{J}}{\text{kg}\cdot°\text{C}}$.
- Set the sum of the heat changes equal to zero: $Q_\ell + Q_s = 0$.
- Use the equations $Q_\ell = m_\ell C_\ell(T_e - T_\ell)$ and $Q_s = m_s C_s(T_e - T_s)$.
- You should get: $m_\ell C_\ell(T_e - T_\ell) + m_s C_s(T_e - T_s) = 0$.
- Plug in numbers and simplify. Note that $\frac{960{,}000}{64{,}000} = 15$.
- You should get: $15(T_e - 30) = -(T_e - 110)$.
- **Distribute** the 15 and distribute the minus sign. Note that $(-1)(-110) = +110$.
- You should get: $15T_e - 450 = -T_e + 110$.
- **Combine like terms.** Add T_e and 450 to both sides of the equation.
- You should get: $16T_e = 560$.
- Divide both sides of the equation by 16. Note that $\frac{560}{16} = 35$.
- The equilibrium temperature is $T_e = 35$°C.

---

63. This is a **calorimetry** problem. First identify the given information in suitable units.
- The mass of the liquid is $m_\ell = 600$ g. (Grams are okay for calorimetry problems.)
- The mass of the ice cube is $m_s = 50$ g. Note: "ℓ" is for liquid, "s" is for solid.
- The initial temperature of the liquid is $T_\ell = 50$°C.
- The initial temperature of the ice cube is $T_s = 0$°C.
- The specific heat of the liquid is $C_\ell = 2150 \frac{\text{J}}{\text{kg}\cdot°\text{C}}$.
- The specific heat of water is $C_W = 4200 \frac{\text{J}}{\text{kg}\cdot°\text{C}}$.
- The latent heat of fusion for water is $L_W = 3.3 \times 10^5 \frac{\text{J}}{\text{kg}}$.
- Set the sum of the heat changes equal to zero: $Q_\ell + Q_s + Q_{melt} = 0$.
- Use the equations $Q_\ell = m_\ell C_\ell(T_e - T_\ell)$, $Q_s = m_s C_W(T_e - T_s)$, and $Q_{melt} = m_s L_W$.
- Note: $Q_{melt} = m_s L_W$ is for ice melting, while $Q_s = m_s C_W(T_e - T_s)$ is for the warming of the ice water after the ice melts, and $Q_\ell = m_\ell C_\ell(T_e - T_\ell)$ is for the liquid cooling.
- You should get: $m_\ell C_\ell(T_e - T_\ell) + m_s C_W(T_e - T_s) + m_s L_W = 0$.
- Plug in numbers and simplify. Note that four zeroes cancel in each term.
- You should get $129(T_e - 50) + 21T_e + 1650 = 0$.
- **Distribute** the 129.

- You should get $129T_e - 6450 + 21T_e + 1650 = 0$.
- **Combine like terms.** Add 6450 to both sides, subtract 1650, and simplify.
- You should get $150T_e = 4800$.
- Divide both sides of the equation by 150. Note that $\frac{4800}{150} = 32$.
- The equilibrium temperature is $T_e = 32$°C.

---

64. This is a **calorimetry** problem. First identify the given information in suitable units.
- The mass of the water is $m_\ell = 0.200$ kg. Note: Normally, grams are okay for calorimetry problems, but since not all of the ice melts, this solution is a little different. It will be convenient to use kilograms to calculate Q in SI units.
- The mass of the ice cube is $m_s = 0.100$ kg. Note: "ℓ" is for liquid, "s" is for solid.
- The initial temperature of the water is $T_\ell = 33$°C.
- The initial temperature of the ice cube is $T_s = 0$°C.
- The specific heat of water is $C_w = 4200 \frac{\text{J}}{\text{kg}\cdot°\text{C}}$.
- The latent heat of fusion for water is $L = 3.3 \times 10^5 \frac{\text{J}}{\text{kg}}$.

(A) The question implies that all of the ice doesn't melt.
- Let's explore why all of the ice might not melt.
- Begin with the heat change for the phase transition: $Q_{melt} = m_s L$.
- You should get $Q_{melt} = 33{,}000$ J. (We used to kilograms for this calculation.)
- Suppose the water cools all the way to the temperature of the ice (0°C). If this happens, the heat change for the water will be $Q_\ell = m_\ell C_\ell (0 - T_\ell)$. Plug in numbers to calculate Q_ℓ for this case.
- You should get $Q_\ell = -27{,}720$ J.
- This is the problem: The ice requires more heat to melt, 33,000 J, than the water can possibly supply by cooling all the way down to 0°C (the water can only supply up to 27,720 J of heat to the ice). That's why only a **fraction** of the ice melts.
- In the previous problem, we used the equation[2] $Q_\ell + Q_s + Q_{melt} = 0$. However, in this problem, $Q_s = 0$. Since only a fraction of the ice melts, the ice will still be at 0°C (whereas in the previous problem, all of the ice melted, and then it warmed).
- In this problem, the water cools all the way down to 0°C, and a **fraction** of the ice melts. This is represented by the equation $Q_\ell + Q_{melt} = 0$, where $Q_{melt} = m_{melt} L$. The mass m_{melt} is just for the ice that **melts**, which is less than 100 g.
- Recall that we already found that $Q_\ell = -27{,}720$ J. Therefore, $m_{melt} L = 27{,}720$ J.
- The ice that melts has a mass of $m_{melt} = 0.084$ kg $= 84$ g.
- Compare 84 g to the original 100 g to see that 84% of the ice melts.

---

[2] If you try to use the equation $m_\ell C_\ell (T_e - T_\ell) + m_s C_w (T_e - T_s) + m_s L_w = 0$ for this problem, you will get a negative value for T_e, which is **impossible**: T_e **<u>can't</u>** be less than the initial temperature (0°C) of the ice.

(B) We already reasoned in part (A) that the water cools all the way down to 0°C because only a fraction of the ice melts (so the ice doesn't warm at all) in this problem.

- The equilibrium temperature is $T_e = 0$°C.

---

65. Begin with the integral for thermal conduction with uniform thermal conductivity.

$$\Delta T = \frac{P}{k} \int_{i}^{f} \frac{ds}{A}$$

(A) The **heat flow coordinate**, s, runs along the direction of the heat flow from the higher temperature surface (at T_h) to the lower temperature surface (at T_c). In this problem, the heat flows horizontally to the right. In cylindrical coordinates, z runs along the axis of the cylinder, so in this problem, $s = z$.

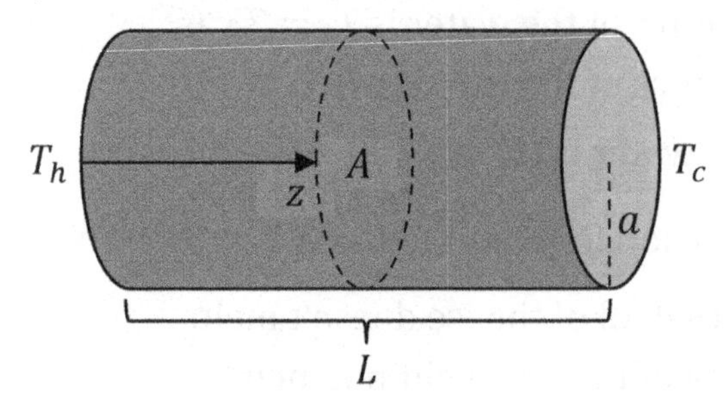

- The **cross-sectional area** (A) is perpendicular to the heat flow coordinate, as shown above. The cross-sectional area is the area of a circle with radius a: $A = \pi a^2$.
- Plug $ds = dz$ and $A = \pi a^2$ into the thermal conduction integral. The limits are from $z = 0$ (at the left face with T_h) to $z = L$ (at the right face with T_c).

$$\Delta T = \frac{P}{k} \int_{z=0}^{L} \frac{dz}{\pi a^2}$$

- Since a is a constant, you may pull it out of the integral.

$$\Delta T = \frac{P}{\pi k a^2} \int_{z=0}^{L} dz$$

- You should get:

$$\Delta T = \frac{PL}{\pi k a^2}$$

- Solve for the rate of heat transfer (P). Multiply both sides by $\pi k a^2$ and divide by L.

$$P = \frac{\pi k a^2 \Delta T}{L}$$

(B) The R value is related to the thermal conductivity by $R = \frac{L}{k}$.

- Replace $\frac{k}{L}$ with $\frac{1}{R}$ in the answer for part (A).

$$P = \frac{\pi a^2 \Delta T}{R}$$

66. Begin with the integral for thermal conduction with uniform thermal conductivity.

$$\Delta T = \frac{P}{k}\int_{i}^{f} \frac{ds}{A}$$

- The **heat flow coordinate**, s, runs along the direction of the heat flow from the higher temperature surface (at T_h) to the lower temperature surface (at T_c). In this problem, the heat flows radially outward along r_c of cylindrical coordinates (review Volume 2, Chapter 6, pages 83-90): $s = r_c$.
- The **cross-sectional area** (A) is perpendicular to the heat flow coordinate. In this problem, the cross-sectional area is the surface area of the body of a **cylinder** with radius r_c and length L. Use the equation $A = 2\pi r_c L$.
- Plug $ds = dr_c$ and $A = 2\pi r_c L$ into the thermal conduction integral. The limits are from $r_c = a$ (at the inner surface with T_h) to $r_c = b$ (at the outer surface with T_c).

$$\Delta T = \frac{P}{k}\int_{r_c=a}^{b} \frac{dr_c}{2\pi r_c L}$$

- Since L is a constant, you may pull it out of the integral.

$$\Delta T = \frac{P}{2\pi k L}\int_{r_c=a}^{b} \frac{dr_c}{r_c}$$

- Recall from calculus that $\int \frac{dx}{x} = \ln(x)$ and that $\ln(b) - \ln(a) = \ln\left(\frac{b}{a}\right)$. (For a quick review of **logarithms**, see Volume 2, Chapter 17.) You should get:

$$\Delta T = \frac{P}{2\pi k L}\ln\left(\frac{b}{a}\right)$$

- Multiply both sides of the equation by $2\pi k L$ and divide by $\ln\left(\frac{b}{a}\right)$.

$$P = \frac{2\pi k L \Delta T}{\ln\left(\frac{b}{a}\right)}$$

---

67. Review the example on page 172, which involved thermal conducting slabs in **series**.

- This question is different, but it will still help to use equations from that example.
- Use the equations $\Delta T_1 = \frac{P_1 L_1}{k_1 A_1}$ and $\Delta T_2 = \frac{P_2 L_2}{k_2 A_2}$. As in the similar example, $P_1 = P_2$ and $A_1 = A_2$. This problem also states that $L_1 = L_2$.
- Rewrite the equations as $\Delta T_1 = \frac{PL}{k_1 A}$ and $\Delta T_2 = \frac{PL}{k_2 A}$.
- Solve for $\frac{PL}{A}$ in each equation. You should get $\frac{PL}{A} = k_1 \Delta T_1$ and $\frac{PL}{A} = k_2 \Delta T_2$.
- Therefore, $k_1 \Delta T_1 = k_2 \Delta T_2$.
- Note that $\Delta T_1 = 65°\text{C} - T_s$ and $\Delta T_2 = T_s - 25°\text{C}$, where T_s is the temperature of the shared wall. Plug these expressions into the previous equation.
- You should get: $k_1(65 - T_s) = k_2(T_s - 25)$.

- Plug in the numerical values for the thermal conductivities.
- You should get: $400(65 - T_s) = 240(T_s - 25)$.
- Divide both sides of the equation by 80 (so you can work with smaller numbers).
- You should get: $5(65 - T_s) = 3(T_s - 25)$.
- **Distribute** the 5 and 3. You should get: $325 - 5T_s = 3T_s - 75$.
- **Combine like terms.** Add $5T_s$ and 75 to both sides of the equation.
- You should get: $400 = 8T_s$. Divide both sides of the equation by 8.
- The temperature of the shared surface is $T_s = 50$°C.
- (There is more than one way to solve this problem. This is just one method.)

Chapter 16: Ideal Gases

68. Study the examples on page 185.
    - Apply the ratio form of the ideal gas law: $\frac{P_1V_1}{T_1} = \frac{P_2V_2}{T_2}$.
    - Since temperature is held **constant**, $T_1 = T_2$ and temperature cancels out.
    - You should get $P_1V_1 = P_2V_2$. Solve for the final pressure.
    - You should get $P_2 = \left(\frac{V_1}{V_2}\right) P_1$.
    - Since the volume quadruples, $V_2 = 4V_1$. Solve for the ratio $\frac{V_1}{V_2}$.
    - You should get $\frac{V_1}{V_2} = \frac{1}{4}$. (It's the same as $\frac{V_2}{V_1} = 4$, but we need $\frac{V_1}{V_2}$, **not** $\frac{V_2}{V_1}$.)
    - The final pressure is $P_2 = 0.20$ atm, which is one-fourth the initial pressure.

---

69. Study the examples on page 185.
    - Apply the ratio form of the ideal gas law: $\frac{P_1V_1}{T_1} = \frac{P_2V_2}{T_2}$.
    - Since pressure is held **constant**, $P_1 = P_2$ and pressure cancels out.
    - You should get $\frac{V_1}{T_1} = \frac{V_2}{T_2}$. Solve for the final volume.
    - You should get $V_2 = \left(\frac{T_2}{T_1}\right) V_1$.
    - Since the temperature doubles, $T_2 = 2T_1$. Solve for the ratio $\frac{T_2}{T_1}$.
    - You should get $\frac{T_2}{T_1} = 2$.
    - The final volume is $V_2 = 120$ cc, which is twice the initial volume.

---

70. Study the examples on page 185.
    - Apply the ratio form of the ideal gas law: $\frac{P_1V_1}{T_1} = \frac{P_2V_2}{T_2}$.
    - Since volume is held **constant**, $V_1 = V_2$ and volume cancels out.
    - You should get $\frac{P_1}{T_1} = \frac{P_2}{T_2}$. Solve for the final temperature.

- You should get $T_2 = \left(\frac{P_2}{P_1}\right) T_1$.
- Since the pressure halves, $P_2 = \frac{P_1}{2}$. Solve for the ratio $\frac{P_2}{P_1}$.
- You should get $\frac{P_2}{P_1} = \frac{1}{2}$.
- The final temperature is $T_2 = 200$ K, which is half the initial temperature.

---

71. Study the examples on page 185.
    - Apply the ratio form of the ideal gas law: $\frac{P_1 V_1}{T_1} = \frac{P_2 V_2}{T_2}$.
    - All three variables change in this problem. **Nothing** cancels out. That's okay: We still solve the problem the same way. (Notice that we were given more information in this problem than in the previous problems. That's why nothing needs to cancel.)
    - Solve for the final temperature.
    - You should get $T_2 = \left(\frac{P_2}{P_1}\right)\left(\frac{V_2}{V_1}\right) T_1$.
    - Since the pressure halves, $P_2 = \frac{P_1}{2}$. Solve for the ratio $\frac{P_2}{P_1}$. You should get $\frac{P_2}{P_1} = \frac{1}{2}$.
    - Since the volume triples, $V_2 = 3V_1$. Solve for the ratio $\frac{V_2}{V_1}$. You should get $\frac{V_2}{V_1} = 3$.
    - **Important note**: Convert the temperature to **Kelvin** before you plug in numbers. The equation does **not** work if you use Celsius.
    - Use the equation $T_K = T_C + 273.15 \approx T_C + 273$.
    - The initial temperature, $T_1 = 27°\text{C}$, equals $T_1 = 300$ K in proper units. Be sure to plug 300 K into the formula, and **not** 27°C.
    - The final temperature is $T_2 = 450$ K, which is three-halves the initial temperature in Kelvin. Use the equation $T_C = T_K - 273.15 \approx T_K - 273$ to convert the final answer to degrees Celsius. The final temperature in Celsius is $T_2 = 177°\text{C}$.

Chapter 17: Van der Waals Fluids

72. First solve for pressure in the equation for a van der Waals fluid.
    - You should get:

$$P = \frac{RT}{\left(\frac{V}{n} - b\right)} - \frac{an^2}{V^2}$$

    - Take a partial derivative of P with respect to V, treating every other symbol (including n and T) as a constant. You should get:

$$\frac{\partial P}{\partial V} = -\frac{RT}{n\left(\frac{V}{n} - b\right)^2} + \frac{2an^2}{V^3}$$

    - We applied the rules $\frac{d}{dx}\frac{1}{(cx-b)} = -\frac{c}{(cx-b)^2}$ with $c = \frac{1}{n}$ and $\frac{d}{dx}\frac{1}{x^2} = -\frac{2}{x^3}$.

- Take a second partial derivative with respect to volume. You should get:

$$\frac{\partial^2 P}{\partial V^2} = \frac{2RT}{n^2\left(\frac{V}{n} - b\right)^3} - \frac{6an^2}{V^4}$$

- For the critical point, $\frac{\partial P}{\partial V} = 0$ and $\frac{\partial^2 P}{\partial V^2} = 0$. Set the partial derivatives equal to zero, and add a subscript "c" (for "critical point") to the symbols P, V, and T.
- You should get:

$$-\frac{RT_c}{n\left(\frac{V_c}{n} - b\right)^2} + \frac{2an^2}{V_c^3} = 0 \quad \text{and} \quad \frac{2RT_c}{n^2\left(\frac{V_c}{n} - b\right)^3} - \frac{6an^2}{V_c^4} = 0$$

- Bring one term to the right-hand side of each equation and **cross multiply**. You should get:

$$RT_cV_c^3 = 2an^3\left(\frac{V_c}{n} - b\right)^2 \quad \text{and} \quad 2RT_cV_c^4 = 6an^4\left(\frac{V_c}{n} - b\right)^3$$

- It is convenient to divide the right equation by the left equation. You should get:

$$2V_c = 3n\left(\frac{V_c}{n} - b\right)$$

- Note that RT_c **cancels**. Also note that $\frac{V_c^4}{V_c^3} = V_c$, $\frac{n^4}{n^3} = n$, and $\frac{\left(\frac{V_c}{n} - b\right)^3}{\left(\frac{V_c}{n} - b\right)^2} = \left(\frac{V_c}{n} - b\right)$.
- Divide both sides of the equation by $3n$. You should get:

$$\frac{2V_c}{3n} = \frac{V_c}{n} - b$$

- Note that the question specified **molar** volume. The molar volume is $\frac{V}{n}$.
- Solve for $\frac{V_c}{n}$ in the previous equation. Add b to both sides and subtract $\frac{2V_c}{3n}$.
- Note that $\frac{V_c}{n} - \frac{2V_c}{3n} = \frac{3V_c}{3n} - \frac{2V_c}{3n} = \frac{V_c}{3n}$. (That is, $1 - \frac{2}{3} = \frac{1}{3}$.)
- The critical molar volume is $\frac{V_c}{n} = 3b$.
- Plug $\frac{V_c}{n} = 3b$ into one of the previous equations that have both V_c and T_c, but **<u>not</u>** P.
- We chose the equation $RT_cV_c^3 = 2an^3\left(\frac{V_c}{n} - b\right)^2$. It's convenient to divide by n^3 to get $RT_c\left(\frac{V_c^3}{n^3}\right) = 2a\left(\frac{V_c}{n} - b\right)^2$, which is the same as $RT_c\left(\frac{V_c}{n}\right)^3 = 2a\left(\frac{V_c}{n} - b\right)^2$.
- When you plug $\frac{V_c}{n} = 3b$ into this equation, you should get:

$$27RT_cb^3 = 8ab^2$$

- Note that $(3b)^3 = 3^3b^3 = 27b^3$ and $(3b - b)^2 = (2b)^2 = 2^2b^2 = 4b^2$.
- Solve for T_c. Divide both sides of the equation by $27Rb^3$.
- The critical temperature is $T_c = \frac{8a}{27bR}$.
- Plug $\frac{V_c}{n} = 3b$ and $T_c = \frac{8a}{27bR}$ into the original van der Waals equation. You should get:

$$P_c = \frac{4a}{27b^2} - \frac{a}{9b^2}$$

- Note that $\frac{V_c}{n} - b = 3b - b = 2b$, $RT_c = \frac{8a}{27b}, \frac{8a}{27b} \div 2b = \frac{4a}{27b^2}$, and $\frac{n^2}{V_c^2} = \frac{1}{(3b)^2} = \frac{1}{9b^2}$.
- Make a **common denominator** to subtract the fractions. Note that $\frac{a}{9b^2} = \frac{3a}{27b^2}$.
- The critical pressure is $P_c = \frac{a}{27b^2}$. We have now shown that $P_c = \frac{a}{27b^2}$, $\frac{V_c}{n} = 3b$, and $T_c = \frac{8a}{27bR}$, which completes the solution.

Chapter 18: The Laws of Thermodynamics

73. Study the examples on pages 199-201.

(A) Perform the thermodynamic work integral: $W_1 = \int_{V=V_0}^{V} P\,dV$.

- Along the **isobar**, the **pressure is constant**. Pull the pressure out of the integral.
- You should get $W_1 = P_0 \int_{V=V_0}^{V} dV = P_0(V - V_0)$.
- Convert the initial pressure from kPa to Pa: $P_0 = 40{,}000$ Pa.
- The work done along the isobar is $W_1 = 200{,}000$ J $= 200$ kJ. It is **positive** (meaning that work is done <u>by</u> the gas) because the **volume increased**.
- Along the **isochor**, the **volume is constant**. <u>No</u> work is done for this: $W_2 = 0$.
- The total work is $W_{tot} = W_1 + W_2 = 200{,}000$ J $= 200$ kJ.

(B) Along the **isobar** (where **pressure** is constant), use the equation for the molar specific heat at constant **pressure**: $dQ_1 = nc_P dT$. Along the **isochor** (where **volume** is constant), use the equation for the molar specific heat at constant **volume**: $dQ_2 = nc_V dT$.

- Since this is a **monatomic** ideal gas, $c_P = \frac{5}{2}R$ and $c_V = \frac{3}{2}R$. (See page 197.)
- You should get $Q_1 = \frac{5}{2}nR\Delta T_1$ and $Q_2 = \frac{3}{2}nR\Delta T_2$.
- You don't know temperature, so use the "trick" shown on page 204.
- For an **ideal gas**, $PV = nRT$. Along the **isobar** (constant **pressure**), $P_0\Delta V = nR\Delta T_1$. Along the **isochor** (constant **volume**), $(\Delta P)V = nR\Delta T_2$.
- Use these equations to eliminate temperature from the previous equations for heat.
- You should get $Q_1 = \frac{5}{2}P_0\Delta V$ and $Q_2 = \frac{3}{2}(\Delta P)V$.
- The heat changes are $Q_1 = 500{,}000$ J $= 500$ kJ and $Q_2 = 480{,}000$ J $= 480$ kJ.
- The total heat change is $Q_{tot} = Q_1 + Q_2$.
- The gas **absorbs** (since Q_{tot} is **positive**) $Q_{tot} = 980{,}000$ J $= 980$ kJ of heat energy.

(C) Apply the **first law** of thermodynamics: $\Delta U_{tot} = Q_{tot} - W_{tot}$.

- The internal energy of the gas increases by $\Delta U_{tot} = 780{,}000$ J $= 780$ kJ.
- Alternatively, find $\Delta U_1 = \frac{3}{2}nR\Delta T_1 = \frac{3}{2}P_0\Delta V$ and $\Delta U_2 = \frac{3}{2}nR\Delta T_2 = \frac{3}{2}(\Delta P)V$, and then use $\Delta U_{tot} = \Delta U_1 + \Delta U_2$. If you use this method, you should get $U\Delta_1 = 300$ kJ, $\Delta U_2 = 480$ kJ, and $\Delta U_{tot} = 780$ kJ.

74. Study the examples on pages 199-201.

(A) Perform the thermodynamic work integral: $W = \int_{V=V_0}^{V} P\,dV$.

- Since the pressure is **not** constant, **don't** pull it out of the integral.
- Use the equation for an adiabatic expansion of an ideal gas to express the pressure in terms of the volume.
- For an **adiabatic** expansion of an **ideal gas**, $P_0 V_0^\gamma = PV^\gamma$. (See page 198.)
- For a monatomic ideal gas, $\gamma = \frac{c_P}{c_V} = \frac{5}{3}$. (See page 197.)
- Combine the last two equations to get $P_0 V_0^{5/3} = PV^{5/3}$. Solve for P.
- You should get $P = \frac{P_0 V_0^{5/3}}{V^{5/3}}$. Plug this into the work integral. You should get:

$$W = P_0 V_0^{5/3} \int_{V=V_0}^{V} \frac{dV}{V^{5/3}}$$

- Note that $\frac{1}{V^{5/3}} = V^{-5/3}$. Therefore, $W = P_0 V_0^{5/3} \int_{V=V_0}^{V} V^{-5/3}\,dV$.
- Recall from calculus that $\int x^c\,dx = \frac{x^{c+1}}{c+1}$ (except when $c = -1$). Let $c = -\frac{5}{3}$.
- You should get $\int x^{-5/3}\,dx = \frac{x^{-5/3+1}}{-5/3+1}$. Note that $-\frac{5}{3} + 1 = -\frac{5}{3} + \frac{3}{3} = -\frac{2}{3}$ and $\frac{1}{-2/3} = -\frac{3}{2}$.
- The anti-derivative is $\int x^{-5/3}\,dx = -\frac{3}{2}x^{-2/3}$. Note that $x^{-2/3} = \frac{1}{x^{2/3}}$.
- Therefore, $\int x^{-5/3}\,dx = -\frac{3}{2x^{2/3}}$. Using this for the work integral, you should get:

$$W = -\frac{3P_0 V_0^{5/3}}{2}\left[\frac{1}{V^{2/3}}\right]_{V=1}^{8}$$

- The limits of integration are from $V_0 = 1.0\text{ m}^3$ to $V = 8.0\text{ m}^3$. Evaluate the anti-derivative over the limits. Note that $8^{2/3} = \left(\sqrt[3]{8}\right)^2 = 2^2 = 4$. Alternatively, enter 8^(2/3) on your calculator to see that $8^{2/3} = 4$.
- You should get $\left[\frac{1}{V^{2/3}}\right]_{V=1}^{8} = \frac{1}{4} - 1 = -\frac{3}{4}$. For the work done, you should get:

$$W = \frac{9P_0 V_0^{5/3}}{8}$$

- Note that $\frac{3}{2}\left(\frac{3}{4}\right) = \frac{9}{8}$ and that the two minus signs make a plus sign. Plug in the initial pressure and volume. Convert the initial pressure from kPa to Pa: $P_0 = 16{,}000$ Pa.
- The work done along the adiabat is $W = 18{,}000\text{ J} = 18\text{ kJ}$.

(B) **No** heat is exchanged along an **adiabat**: $Q = 0$.

(C) Apply the **first law** of thermodynamics: $\Delta U = Q - W$.

- The internal energy of the gas changes by $\Delta U = -18{,}000\text{ J} = -18\text{ kJ}$.
- Alternatively, find $\Delta U = \frac{3}{2}nR\Delta T = \frac{3}{2}(PV - P_0V_0)$ and use $P_0 V_0^{5/3} = PV^{5/3}$ to find P. With this method, $P = 500$ Pa and $\Delta U_{tot} = -18{,}000\text{ J} = -18\text{ kJ}$.

75. Study the example on page 201.

Perform the thermodynamic work integral: $W = \int_{V=V_0}^{V} P\,dV$.

- Since the pressure is **not** constant, **don't** pull it out of the integral.
- Use the given equation in order to express the pressure in terms of the volume. It will be similar to what we did in the example on page 201.
- The given equation is $PV^2 =$ const. (where "const." is short for "constant"). Rewrite this equation as $P_0 V_0^2 = PV^2$. Solve for P.
- You should get $P = \frac{P_0 V_0^2}{V^2}$. Plug this into the work integral. You should get:

$$W = P_0 V_0^2 \int_{V=V_0}^{V} \frac{dV}{V^2}$$

- Note that $\frac{1}{V^2} = V^{-2}$. Therefore, $W = P_0 V_0^2 \int_{V=V_0}^{V} V^{-2}\,dV$.
- Recall from calculus that $\int x^c\,dx = \frac{x^{c+1}}{c+1}$ (except when $c = -1$). Let $c = -2$.
- You should get $\int x^{-2}\,dx = \frac{x^{-2+1}}{-2+1}$. Note that $-2 + 1 = -1$.
- The anti-derivative is $\int x^{-2}\,dx = -x^{-1}$. Note that $x^{-1} = \frac{1}{x}$.
- Therefore, $\int x^{-2}\,dx = -\frac{1}{x}$. Using this for the work integral, you should get:

$$W = -P_0 V_0^2 \left[\frac{1}{V}\right]_{V=3}^{6}$$

- The limits of integration are from $V_0 = 3.0 \text{ m}^3$ to $V = 6.0 \text{ m}^3$. Evaluate the anti-derivative over the limits.
- You should get $\left[\frac{1}{V}\right]_{V=3}^{6} = \frac{1}{6} - \frac{1}{3} = \frac{1}{6} - \frac{2}{6} = -\frac{1}{6}$. For the work done, you should get:

$$W = \frac{P_0 V_0^2}{6}$$

- Note that the two minus signs make a plus sign. Plug in the initial pressure and volume. Convert the initial pressure from kPa to Pa: $P_0 = 90{,}000$ Pa.
- The work done is $W = 135{,}000 \text{ J} = 135 \text{ kJ}$.

---

76. Study the example on the top of page 202.

- The work done equals the **area** under the P-V path.
- On the diagram on the next page, you can see that the area under the straight line path can be divided up into a triangle and rectangle.
- (It may help to review Volume 1, Chapter 6.)
- The work done equals the area of the triangle plus the area of the rectangle.
- Use the equation $W = -(A_{tri} + A_{rect})$. It's **negative** because the volume **decreases.**
- The area of the triangle is $A_{tri} = \frac{1}{2}bh$ and the area of the rectangle is $A_{rect} = LW$.

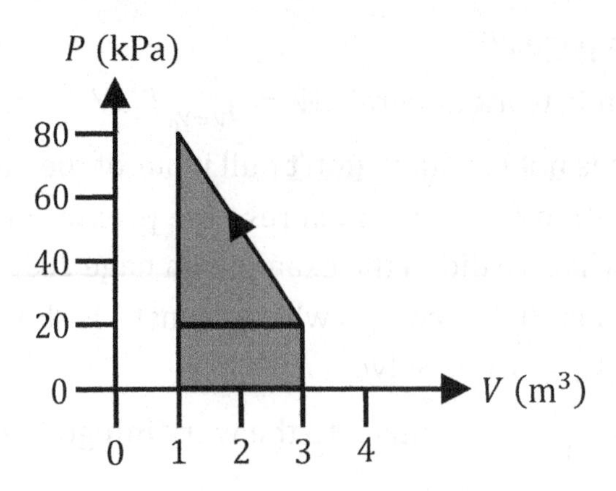

- Check the values that you read from the graph: $b = 3.0 - 1.0 = 2.0$ m^3, $h = 80 - 20 = 60$ kPa, $L = 3.0 - 1.0 = 2.0$ m^3, and $W = 20 - 0 = 20$ kPa.
- Convert 60 kPa and 20 kPa to Pascals: $h = 60{,}000$ Pa and $W = 20{,}000$ Pa.
- The areas are $A_{tri} = 60{,}000$ J and $A_{rect} = 40{,}000$ J.
- The work done is $W = -100{,}000$ J $= -100$ kJ. The work is **negative** because the **volume decreases** (the path goes to the **left**).

---

77. Study the example on the bottom of page 202.
    - The **net** work equals the **area** of the triangle.
    - The base is $b = 8.0 - 2.0 = 6.0$ m^3 and the height is $h = 40 - 10 = 30$ kPa.
    - Convert h to Pascals: $h = 30{,}000$ Pa.
    - Use the equation $W_{net} = A_{tri} = \frac{1}{2}bh$.
    - The **net** work done is $W_{net} = 90{,}000$ J $= 90$ kJ. The net work is **positive** because the path is **clockwise.**

---

78. Study the example on pages 203-206.

(A) Path AB is an **isochor**: The **volume** is constant ($V_B = V_A$).
- Since the volume doesn't change, **no** work is done along path AB: $W_{AB} = 0$.
- Along an **isochor** (where **volume** is constant), use the equation for the molar specific heat at constant **volume**: $dQ = nc_V dT$.
- Since this is a **monatomic** ideal gas, $c_V = \frac{3}{2}R$. (See page 197.)
- You should get $Q_{AB} = \frac{3}{2}nR(T_B - T_A)$.
- You don't know temperature, so use the "trick" shown on page 204.
- You should get $nR(T_B - T_A) = P_B V_B - P_A V_A$ and $Q_{AB} = \frac{3}{2}(P_B V_B - P_A V_A)$.
- Read the initial and final values directly from the graph: $V_A = 8.0$ m^3, $P_A = 20$ kPa, $V_B = 8.0$ m^3, and $P_B = 5$ kPa.
- The heat change is $Q_{AB} = -180{,}000$ J $= -180$ kJ. Since Q_{AB} is **negative,** the gas **releases** heat.

- Apply the **first law** of thermodynamics: $\Delta U_{AB} = Q_{AB} - W_{AB}$.
- The internal energy change is $\Delta U_{AB} = -180{,}000 \text{ J} = -180 \text{ kJ}$.

(B) Path BC is an **isobar**: The **pressure** is constant ($P_C = P_B$).

- Perform the thermodynamic work integral: $W_{BC} = \int_{V=V_B}^{V_C} P\, dV$.
- Since the **pressure is constant**, we may pull the pressure out of the integral.
- You should get $W_{BC} = P_B(V_C - V_B)$.
- Read the initial and final values directly from the graph: $V_B = 8.0 \text{ m}^3$, $P_B = 5$ kPa, $V_C = 4.0 \text{ m}^3$, and $P_C = 5$ kPa.
- The work done is $W_{BC} = -20{,}000 \text{ J} = -20 \text{ kJ}$. The work done is **negative** because the **volume decreased** (path BC goes to the **left**).
- Along an **isobar** (where **pressure** is constant), use the equation for the molar specific heat at constant **pressure**: $dQ = nc_P dT$.
- Since this is a **monatomic** ideal gas, $c_P = \frac{5}{2}R$. (See page 197.)
- You should get $Q_{BC} = \frac{5}{2}nR(T_C - T_B)$.
- You don't know temperature, so use the "trick" shown on page 204.
- You should get $nR(T_C - T_B) = P_C V_C - P_B V_B$ and $Q_{BC} = \frac{5}{2}(P_C V_C - P_B V_B)$.
- Recall that $V_B = 8.0 \text{ m}^3$, $P_B = 5$ kPa, $V_C = 4.0 \text{ m}^3$, and $P_C = 5$ kPa.
- The heat change is $Q_{BC} = -50{,}000 \text{ J} = -50 \text{ kJ}$. Since Q_{BC} is **negative**, the gas **releases** heat.
- Apply the **first law** of thermodynamics: $\Delta U_{BC} = Q_{BC} - W_{BC}$.
- The internal energy change is $\Delta U_{BC} = -30{,}000 \text{ J} = -30 \text{ kJ}$. Note that $-50{,}000 - (-20{,}000) = -50{,}000 + 20{,}000 = -30{,}000$.

(C) Path CD is an **isochor**: The **volume** is constant ($V_D = V_C$).

- Since the volume doesn't change, **<u>no</u>** work is done along path CD: $W_{CD} = 0$.
- Along an **isochor** (where **volume** is constant), use the equation for the molar specific heat at constant **volume**: $dQ = nc_V dT$.
- Since this is a **monatomic** ideal gas, $c_V = \frac{3}{2}R$. (See page 197.)
- You should get $Q_{CD} = \frac{3}{2}nR(T_D - T_C)$.
- You don't know temperature, so use the "trick" shown on page 204.
- You should get $nR(T_D - T_C) = P_D V_D - P_C V_C$ and $Q_{CD} = \frac{3}{2}(P_D V_D - P_C V_C)$.
- Read the initial and final values directly from the graph: $V_C = 4.0 \text{ m}^3$, $P_C = 5$ kPa, $V_D = 4.0 \text{ m}^3$, and $P_D = 20$ kPa.
- The heat change is $Q_{CD} = 90{,}000 \text{ J} = 90 \text{ kJ}$. Since Q_{CD} is **positive**, the gas **absorbs** heat.
- Apply the **first law** of thermodynamics: $\Delta U_{CD} = Q_{CD} - W_{CD}$.
- The internal energy change is $\Delta U_{CD} = 90{,}000 \text{ J} = 90 \text{ kJ}$.

(D) Path DA is an **isobar**: The **pressure** is constant ($P_D = P_A$).

- Perform the thermodynamic work integral: $W_{DA} = \int_{V=V_D}^{V_A} P\, dV$.
- Since the **pressure is constant**, we may pull the pressure out of the integral.
- You should get $W_{DA} = P_D(V_A - V_D)$.
- Read the initial and final values directly from the graph: $V_D = 4.0 \text{ m}^3$, $P_D = 20$ kPa, $V_A = 8.0 \text{ m}^3$, and $P_A = 20$ kPa.
- The work done is $W_{DA} = 80{,}000 \text{ J} = 80$ kJ. The work done is **positive** because the **volume increased** (path DA goes to the **right**).
- Along an **isobar** (where **pressure** is constant), use the equation for the molar specific heat at constant **pressure**: $dQ = nc_P dT$.
- Since this is a **monatomic** ideal gas, $c_P = \frac{5}{2}R$. (See page 197.)
- You should get $Q_{DA} = \frac{5}{2}nR(T_A - T_D)$.
- You don't know temperature, so use the "trick" shown on page 204.
- You should get $nR(T_A - T_D) = P_A V_A - P_D V_D$ and $Q_{DA} = \frac{5}{2}(P_A V_A - P_D V_D)$.
- Recall that $V_D = 4.0 \text{ m}^3$, $P_D = 20$ kPa, $V_A = 8.0 \text{ m}^3$, and $P_A = 20$ kPa.
- The heat change is $Q_{DA} = 200{,}000 \text{ J} = 200$ kJ. Since Q_{DA} is **positive**, the gas **absorbs** heat.
- Apply the **first law** of thermodynamics: $\Delta U_{DA} = Q_{DA} - W_{DA}$.
- The internal energy change is $\Delta U_{DA} = 120{,}000 \text{ J} = 120$ kJ.

(E) Use the equations $W_{net} = W_{AB} + W_{BC} + W_{CD} + W_{DA}$, $Q_{net} = Q_{AB} + Q_{BC} + Q_{CD} + Q_{DA}$, and $\Delta U_{net} = \Delta U_{AB} + \Delta U_{BC} + \Delta U_{CD} + \Delta U_{DA}$.

- The net work is $W_{net} = 60{,}000 \text{ J} = 60$ kJ. Alternatively, you could find the area of the rectangle ($A_{rect} = LW$, where $L = 4.0 \text{ m}^3$ and $W = 15$ kPa). You would get the same answer. The **net** work is **positive** because the path is **clockwise**.
- The net heat change is $Q_{net} = 60{,}000 \text{ J} = 60$ kJ.
- The net internal energy change is **zero**: $\Delta U_{net} = 0$.
- Note: $\Delta U_{net} = 0$ and $Q_{net} = W_{net}$ for **any complete** cycle (**closed** path).

---

79. Study the example on pages 203-206.

(A) Perform the thermodynamic work integral: $W_{AB} = \int_{V=V_0}^{V} P\, dV$.

- Since the pressure is **not** constant, **don't** pull it out of the integral.
- Use the equation for an adiabatic expansion of an ideal gas to express the pressure in terms of the volume.
- For an **adiabatic** expansion of an **ideal gas**, $P_A V_A^\gamma = PV^\gamma$. (See page 198.)
- For a monatomic ideal gas, $\gamma = \frac{c_P}{c_V} = \frac{5}{3}$. (See page 197.)
- Combine the last two equations to get $P_A V_A^{5/3} = PV^{5/3}$. Solve for P.

- You should get $P = \frac{P_A V_A^{5/3}}{V^{5/3}}$. Plug this into the work integral. You should get:

$$W_{AB} = P_A V_A^{5/3} \int_{V=V_0}^{V} \frac{dV}{V^{5/3}}$$

- Note that $\frac{1}{V^{5/3}} = V^{-5/3}$. Therefore, $W_{AB} = P_A V_A^{5/3} \int_{V=V_0}^{V} V^{-5/3}\, dV$.
- Recall from calculus that $\int x^c\, dx = \frac{x^{c+1}}{c+1}$ (except when $c = -1$). Let $c = -\frac{5}{3}$.
- You should get $\int x^{-5/3}\, dx = \frac{x^{-5/3+1}}{-5/3+1}$. Note that $-\frac{5}{3} + 1 = -\frac{5}{3} + \frac{3}{3} = -\frac{2}{3}$ and $\frac{1}{-2/3} = -\frac{3}{2}$.
- The anti-derivative is $\int x^{-5/3}\, dx = -\frac{3}{2} x^{-2/3}$. Note that $x^{-2/3} = \frac{1}{x^{2/3}}$.
- Therefore, $\int x^{-5/3}\, dx = -\frac{3}{2x^{2/3}}$. Using this for the work integral, you should get:

$$W_{AB} = -\frac{3 P_A V_A^{5/3}}{2} \left[\frac{1}{V^{2/3}}\right]_{V=1}^{8}$$

- The limits of integration are from $V_A = 1.0 \text{ m}^3$ to $V_B = 8.0 \text{ m}^3$. Evaluate the anti-derivative over the limits. Note that $8^{2/3} = \left(\sqrt[3]{8}\right)^2 = 2^2 = 4$. Alternatively, enter 8^(2/3) on your calculator to see that $8^{2/3} = 4$.
- You should get $\left[\frac{1}{V^{2/3}}\right]_{V=1}^{8} = \frac{1}{4} - 1 = -\frac{3}{4}$. For the work done, you should get:

$$W_{AB} = \frac{9 P_A V_A^{5/3}}{8}$$

- Note that $\frac{3}{2}\left(\frac{3}{4}\right) = \frac{9}{8}$ and that the two minus signs make a plus sign. Plug in the initial pressure and volume. Convert the initial pressure from kPa to Pa: $P_A = 32{,}000$ Pa.
- The work done along the adiabat is $W_{AB} = 36{,}000 \text{ J} = 36 \text{ kJ}$.
- Path AB is an **adiabat**: No **heat** is exchanged along an adiabat: $Q_{AB} = 0$.
- Apply the **first law** of thermodynamics: $\Delta U_{AB} = Q_{AB} - W_{AB}$.
- The internal energy change is $\Delta U_{AB} = -36{,}000 \text{ J} = -36 \text{ kJ}$.

(B) Path BC is an **isobar**: The **pressure** is constant ($P_C = P_B$).

- Perform the thermodynamic work integral: $W_{BC} = \int_{V=V_B}^{V_C} P\, dV$.
- Since the **pressure is constant**, we may pull the pressure out of the integral.
- You should get $W_{BC} = P_B(V_C - V_B)$.
- Read the initial and final values directly from the graph:[3] $V_B = 8.0 \text{ m}^3$, $P_B = 1$ kPa, $V_C = 1.0 \text{ m}^3$, and $P_C = 1$ kPa.
- The work done is $W_{BC} = -7{,}000 \text{ J} = -7.0 \text{ kJ}$. The work done is **negative** because the **volume decreased** (path BC goes to the **left**).

---

[3] One way to determine that $P_B = 1$ kPa is to apply the equation $P_A V_A^{5/3} = P_B V_B^{5/3}$ for the adiabat (for this monatomic ideal gas), using $P_A = 32$ kPa, $V_A = 1.0 \text{ m}^3$, and $V_B = 8.0 \text{ m}^3$. This way, you don't have to try to interpolate between the lines to read the value of P_B.

- Along an **isobar** (where **pressure** is constant), use the equation for the molar specific heat at constant **pressure**: $dQ = nc_P dT$.
- Since this is a **monatomic** ideal gas, $c_P = \frac{5}{2}R$. (See page 197.)
- You should get $Q_{BC} = \frac{5}{2}nR(T_C - T_B)$.
- You don't know temperature, so use the "trick" shown on page 204.
- You should get $nR(T_C - T_B) = P_C V_C - P_B V_B$ and $Q_{BC} = \frac{5}{2}(P_C V_C - P_B V_B)$.
- Recall that $V_B = 8.0 \text{ m}^3$, $P_B = 1 \text{ kPa}$, $V_C = 1.0 \text{ m}^3$, and $P_C = 1 \text{ kPa}$.
- The heat change is $Q_{BC} = -17{,}500 \text{ J} = -17.5 \text{ kJ}$. Since Q_{BC} is **negative**, the gas **releases** heat.
- Apply the **first law** of thermodynamics: $\Delta U_{BC} = Q_{BC} - W_{BC}$.
- The internal energy change is $\Delta U_{BC} = -10{,}500 \text{ J} = -10.5 \text{ kJ}$. Note that $-17{,}500 - (-7{,}000) = -17{,}500 + 7{,}000 = -10{,}500$.

(C) Path CA is an **isochor**: The **volume** is constant ($V_A = V_C$).

- Since the volume doesn't change, **no** work is done along path CA: $W_{CA} = 0$.
- Along an **isochor** (where **volume** is constant), use the equation for the molar specific heat at constant **volume**: $dQ = nc_V dT$.
- Since this is a **monatomic** ideal gas, $c_V = \frac{3}{2}R$. (See page 197.)
- You should get $Q_{CA} = \frac{3}{2}nR(T_A - T_C)$.
- You don't know temperature, so use the "trick" shown on page 204.
- You should get $nR(T_A - T_C) = P_A V_A - P_C V_C$ and $Q_{CA} = \frac{3}{2}(P_A V_A - P_C V_C)$.
- Read the initial and final values directly from the graph: $V_C = 1.0 \text{ m}^3$, $P_C = 1 \text{ kPa}$, $V_A = 1.0 \text{ m}^3$, and $P_A = 32 \text{ kPa}$.
- The heat change is $Q_{CA} = 46{,}500 \text{ J} = 46.5 \text{ kJ}$. Since Q_{CA} is **positive**, the gas **absorbs** heat.
- Apply the **first law** of thermodynamics: $\Delta U_{CA} = Q_{CA} - W_{CA}$.
- The internal energy change is $\Delta U_{CA} = 46{,}500 \text{ J} = 46.5 \text{ kJ}$.

(D) Use the equations $W_{net} = W_{AB} + W_{BC} + W_{CA}$, $Q_{net} = Q_{AB} + Q_{BC} + Q_{CA}$, and $\Delta U_{net} = \Delta U_{AB} + \Delta U_{BC} + \Delta U_{CA}$.

- The net work is $W_{net} = 29{,}000 \text{ J} = 29 \text{ kJ}$. The **net** work is **positive** because the path is **clockwise**. (Note: If you proceed to find the area using geometry, note that the top path is a **curve**, so it is **not** a triangle. **Don't** use the formula for a triangle.)
- The net heat change is $Q_{net} = 29{,}000 \text{ J} = 29 \text{ kJ}$.
- The net internal energy change is **zero**: $\Delta U_{net} = 0$.
- Note: $\Delta U_{net} = 0$ and $Q_{net} = W_{net}$ for **any complete** cycle (**closed** path).

80. Perform the integral for entropy change for this process: $\Delta S = \int_{Q=Q_0}^{Q} \frac{dQ}{T}$. (The word "slowly" suggests that the process is approximately quasistatic and reversible.)

- The temperature remains constant at $T = 0°\text{C}$ for this phase change. Since T is constant, you may pull it out of the integral.
- You should get $\Delta S = \frac{Q}{T}$.
- Use the equation for the heat associated with a phase transition: $Q = mL$. (See Chapter 15.) Melting is **endothermic**: The ice cube **absorbs** energy as it melts into ice water. Therefore, Q is **positive**.
- You should get $\Delta S = \frac{mL}{T}$.
- Convert the temperature to Kelvin: $T = 273$ K.
- Convert the mass to kilograms: $m = 0.910$ kg.
- The entropy of the system increases by $\Delta S = 1100$ J/K. If you're not using a calculator, it may help to note that $91 = 7 \times 13$ and $273 = 3 \times 7 \times 13$. Also note that $\frac{3.3}{3} = 1.1$.

---

81. Perform the integral for entropy change for this process: $\Delta S = \int_{Q=Q_0}^{Q} \frac{dQ}{T}$.

- Since T is **not** constant in this problem (unlike the previous problem), you may **not** pull it out of the integral. Instead, express dQ in terms of dT.
- Use the equation for specific heat capacity: $dQ = mc_w dT$.
- You should get $\Delta S = mc_w \int_{T=T_0}^{T} \frac{dT}{T}$.
- Recall from calculus that $\int \frac{dx}{x} = \ln(x)$. Also recall that $\ln(b) - \ln(a) = \ln\left(\frac{b}{a}\right)$. (For a quick review of **logarithms**, see Volume 2, Chapter 17.)
- You should get $\Delta S = mc_w \ln\left(\frac{T}{T_0}\right)$.
- Convert the temperatures to Kelvin: $T_0 = 360$ K and $T = 300$ K.
- Convert the mass to kilograms: $m = 0.500$ kg.
- The entropy of the system decreases by $\Delta S = -2100\ln(1.2)$ J/K. Note that $\frac{360}{300} = 1.2$. If you use a calculator, $\Delta S = -383$ J/K.

---

82. Study the example on page 207.

- Perform the integral for entropy change for this reversible process: $\Delta S = \int_{Q=Q_0}^{Q} \frac{dQ}{T}$.
- For an **isobaric** process, **pressure** remains constant. Express the heat in terms of the molar specific heat capacity at constant pressure: $dQ = nc_P dT$. You should get:

$$\Delta S = nc_P \int_{T=T_0}^{T} \frac{dT}{T}$$

- Recall from calculus that $\int \frac{dx}{x} = \ln(x)$. Also recall that $\ln(b) - \ln(a) = \ln\left(\frac{b}{a}\right)$. (For a quick review of **logarithms**, see Volume 2, Chapter 17.)
- You should get $\Delta S = nc_P \ln\left(\frac{T}{T_0}\right)$.
- You don't know temperature, so use the "trick" shown on page 204. It's a little different here: For an ideal gas, $PV = nRT$ such that $T = \frac{PV}{nR}$ and $T_0 = \frac{PV_0}{nR}$. (For this **isobar, pressure** is constant: $P_0 = P$.) Divide the two equations: $\frac{P}{nR}$ cancels out and you get $\frac{T}{T_0} = \frac{V}{V_0}$. (Alternatively, you could apply **Charles's law** - see Chapter 16 - to the **isobar**: $\frac{V_0}{T_0} = \frac{V}{T}$, which can be rewritten as $\frac{T}{T_0} = \frac{V}{V_0}$.)
- You should get $\Delta S = nc_P \ln\left(\frac{V}{V_0}\right)$.
- Since this is a **monatomic** ideal gas, $c_P = \frac{5}{2}R$. (See page 197.)
- Recall that the universal gas constant is $R = 8.314 \frac{\text{J}}{\text{mol}\cdot\text{K}} \approx \frac{25}{3} \frac{\text{J}}{\text{mol}\cdot\text{K}}$.
- The problem states that the volume of the gas doubles: $\frac{V}{V_0} = 2$.
- Note that $n = 6.0$ mol.
- The entropy of the system increases by $\Delta S \approx 125 \ln(2)$ J/K. If you use a calculator, this works out to $\Delta S = 87$ J/K.

Chapter 19: Heat Engines

83. Study the example on pages 225-227.

(A) Consult the notes on page 224.

- **Adiabat** ab: $Q_{ab} = 0$ along the adiabat. Apply the **first law** of thermodynamics: $\Delta U_{ab} = Q_{ab} - W_{ab} = -W_{ab}$. Use $U = \frac{3}{2}nRT$. You should get: $W_{ab} = \frac{3}{2}nR(T_a - T_b)$.
- **Isochor** bc: Volume is constant, so no work is done: $W_{bc} = 0$.
- **Adiabat** cd: $Q_{cd} = 0$ along the adiabat. Apply the **first law** of thermodynamics: $\Delta U_{cd} = Q_{cd} - W_{cd} = -W_{cd}$. Use $U = \frac{3}{2}nRT$. You should get: $W_{cd} = \frac{3}{2}nR(T_c - T_d)$.
- **Isochor** da: Volume is constant, so no work is done: $W_{da} = 0$.

(B) Consult the notes on page 224.

- **Adiabat** ab: $Q_{ab} = 0$ along the adiabat.
- **Isochor** bc: Volume is constant ($V_c = V_b$). Use the specific heat at constant **volume**. Note that $c_V = \frac{3}{2}R$ (see page 224). You should get $Q_{bc} = \frac{3}{2}nR(T_c - T_b)$.
- **Adiabat** cd: $Q_{cd} = 0$ along the adiabat.
- **Isochor** da: Volume is constant ($V_a = V_d$). Use the specific heat at constant **volume**. Note that $c_V = \frac{3}{2}R$ (see page 224). You should get $Q_{da} = \frac{3}{2}nR(T_a - T_d)$.

(C) Use the equation $W_{out} = W_{ab} + W_{bc} + W_{cd} + W_{da}$.

- The work output is $W_{out} = \frac{3}{2}nR(T_a - T_b + T_c - T_d)$.

(D) Part (L) shows that T_c is the **highest** temperature and T_a is the **lowest** temperature.

- The heat **input** is the **positive** heat change.
- The positive heat change is $Q_{in} = Q_{bc} = \frac{3}{2}nR(T_c - T_b)$ since $T_c > T_b$.

(E) The heat **output** (or exhaust) is the **negative** heat change.

- The negative heat change is $Q_{out} = Q_{da} = \frac{3}{2}nR(T_a - T_d)$ since $T_a < T_d$.

(F) Write the ideal gas law in ratio form for the two adiabats. Although these are adiabats, **don't** use the equations for adiabats yet – we will use them in part (G). Just use the ideal gas law for now.

- **Adiabat** ab: The ideal gas law is $\frac{P_a V_a}{T_a} = \frac{P_b V_b}{T_b}$.
- **Adiabat** cd: The ideal gas law is $\frac{P_c V_c}{T_c} = \frac{P_d V_d}{T_d}$.
- Solve for P_b and P_c in these two equations. Note that P_b is on the right, whereas P_c is on the left, in their respective equations. Tip: **Cross multiply** first.
- You should get $P_b = \frac{V_a}{V_b}\frac{T_b}{Ta} P_a$ and $P_c = \frac{V_d}{V_c}\frac{T_c}{T_d} P_d$.
- Note that $V_d = V_a$ and $V_c = V_b$. Therefore, $\frac{V_d}{V_c} = \frac{V_a}{V_b}$ and $P_c = \frac{V_a}{V_b}\frac{T_c}{T_d} P_d$.
- Substitute the compression ratio, $r = \frac{V_a}{V_b}$, into the equations for P_b and P_c.
- You should get $P_b = r\frac{T_b}{T_a} P_a$ and $P_c = r\frac{T_c}{T_d} P_d$.

(G) Now use the equations for the adiabats. See page 221.

- **Adiabat** ab: For an ideal gas along the adiabat, $P_a V_a^\gamma = P_b V_b^\gamma$.
- **Adiabat** cd: For an ideal gas along the adiabat, $P_c V_c^\gamma = P_d V_d^\gamma$.
- Solve for P_b and P_c in these two equations. Note that P_b is on the right, whereas P_c is on the left, in their respective equations.
- You should get $P_b = \frac{P_a V_a^\gamma}{V_b^\gamma} = P_a \left(\frac{V_a}{V_b}\right)^\gamma$ and $P_c = \frac{P_d V_d^\gamma}{V_c^\gamma} = P_d \left(\frac{V_d}{V_c}\right)^\gamma$.
- Note that $V_d = V_a$ and $V_c = V_b$. Therefore, $\frac{V_d}{V_c} = \frac{V_a}{V_b}$ and $P_c = P_d \left(\frac{V_a}{V_b}\right)^\gamma$.
- Substitute the compression ratio, $r = \frac{V_a}{V_b}$, into the equations for P_b and P_c. Also, note that $\gamma = \frac{5}{3}$ for a monatomic ideal gas (see page 197).
- You should get $P_b = r^{5/3}\, P_a$ and $P_c = r^{5/3}\, P_d$.

(H) Set the right-hand sides of $P_b = r\frac{T_b}{T_a} P_a$ and $P_b = r^{5/3}\, P_a$ equal to one another. Similarly, set the right-hand sides of $P_c = r\frac{T_c}{T_d} P_d$ and $P_c = r^{5/3}\, P_d$ equal to one another.

- You should get $r\frac{T_b}{T_a} P_a = r^{5/3}\, P_a$ and $r\frac{T_c}{T_d} P_d = r^{5/3}\, P_d$.

- P_a and P_d both cancel out. You should get $r\frac{T_b}{T_a} = r^{5/3}$ and $r\frac{T_c}{T_d} = r^{5/3}$.
- Multiply both sides of the first equation by T_a and the second equation by T_d.
- Divide both sides of both equations by r. Note that $\frac{r^{5/3}}{r} = r^{5/3-1} = r^{2/3}$ according to the rule $\frac{x^m}{x^n} = x^{m-n}$. Also note that $\frac{5}{3} - 1 = \frac{5}{3} - \frac{3}{3} = \frac{5-3}{3} = \frac{2}{3}$.
- You should get $T_b = r^{2/3}T_a$ and $T_c = r^{2/3}T_d$.

(I) Use the equation $e = 1 + \frac{Q_{out}}{Q_{in}}$. Recall that we found Q_{in} and Q_{out} in parts (D) and (E).

- The $\frac{3}{2}nR$'s cancel out. You should get: $e = 1 + \frac{T_a - T_d}{T_c - T_b}$. It is convenient to rewrite this as $e = 1 - \frac{T_d - T_a}{T_c - T_b}$, using $T_a - T_d = -(T_d - T_a)$.
- Use the equations from part (H), $T_b = r^{2/3}T_a$ and $T_c = r^{2/3}T_d$, to eliminate T_b and T_d from the efficiency equation. Note that $T_d = \frac{T_c}{r^{2/3}}$. You should get:

$$e = 1 - \frac{T_d - T_a}{T_c - T_b} = 1 - \frac{\frac{T_c}{r^{2/3}} - T_a}{T_c - r^{2/3}T_a}$$

- **Factor** $r^{2/3}$ out of the denominator. Note that $r^{2/3}\left(\frac{T_c}{r^{2/3}} - T_a\right) = T_c - r^{2/3}T_a$.
- You should get:

$$e = 1 - \frac{\frac{T_c}{r^{2/3}} - T_a}{r^{2/3}\left(\frac{T_c}{r^{2/3}} - T_a\right)}$$

- The $\left(\frac{T_c}{r^{2/3}} - T_a\right)$'s cancel out.
- The efficiency is $e = 1 - \left(\frac{1}{r}\right)^{2/3}$. It's the same as $e = 1 - \frac{1}{r^{2/3}}$.

(J) Use the equation $T_b = r^{2/3}T_a$ that we found in part (H).

- Solve for $r^{2/3}$ in this equation.
- You should get $r^{2/3} = \frac{T_b}{T_a}$, which equates to $\frac{1}{r^{2/3}} = \frac{T_a}{T_b}$. Plug this into $e = 1 - \frac{1}{r^{2/3}}$.
- The efficiency is $e = 1 - \frac{T_a}{T_b}$.

(K) Use the equation $T_c = r^{2/3}T_d$ that we found in part (H).

- Solve for $r^{2/3}$ in this equation.
- You should get $r^{2/3} = \frac{T_c}{T_d}$, which equates to $\frac{1}{r^{2/3}} = \frac{T_d}{T_c}$. Plug this into $e = 1 - \frac{1}{r^{2/3}}$.
- The efficiency is $e = 1 - \frac{T_d}{T_c}$.

(L) Study the *P*-*V* diagram. Clearly, $V_a > V_b$, which means that $\frac{V_a}{V_b} > 1$. Since $r = \frac{V_a}{V_b}$, it follows that $r > 1$. Study the temperature equations, knowing that $r > 1$.

- From $T_b = r^{2/3}T_a$, it follows that $T_b > T_a$ since $r^{2/3} > 1$. If you enter 1.01^(2/3) on your calculator, you will get approximately 1.0066556, which is greater than 1.

- From $T_c = r^{2/3}T_d$, it similarly follows that $T_c > T_d$.
- Along isochor bc, apply Gay-Lussac's law (Chapter 16), since **volume** is constant (or just cancel out volume in the ideal gas law: $\frac{P_b V_b}{T_b} = \frac{P_c V_c}{T_c}$) to get $\frac{P_b}{T_b} = \frac{P_c}{T_c}$. Solve for T_c. Tip: First **cross multiply**. You should get $T_c = \frac{P_c}{P_b} T_b$.
- Study the *P-V* diagram. Clearly, $P_c > P_b$, which means that $\frac{P_c}{P_b} > 1$.
- It follows that $T_c > T_b$.
- If you do the same thing along isochor da, you should get $\frac{P_d}{T_d} = \frac{P_a}{T_a}$ and $T_d = \frac{P_d}{P_a} T_a$.
- Since $P_d > P_a$ on the *P-V* diagram, $\frac{P_d}{P_a} > 1$ such that $T_d > T_a$.
- Put these inequalities together: $T_b > T_a$, $T_c > T_d$, $T_c > T_b$, and $T_d > T_a$.
- The highest temperature is $T_h = T_c$ and the lowest temperature is $T_\ell = T_a$.

(M) The efficiency of a Carnot heat engine operating between the same extreme temperatures would be $e_C = 1 - \frac{T_\ell}{T_h}$ (see page 220).

- Compare this with the result from part (J), which was $e = 1 - \frac{T_a}{T_b}$.
- Since $T_h = T_c$ and $T_\ell = T_a$, we can write the Carnot efficiency as $e_C = 1 - \frac{T_a}{T_c}$.
- Since $T_c > T_b$, it follows that $\frac{T_a}{T_c} < \frac{T_a}{T_b}$. A larger denominator makes a smaller fraction.
- Since $\frac{T_a}{T_c} < \frac{T_a}{T_b}$, we're subtracting a smaller number from one in $1 - \frac{T_a}{T_c}$ than we are in $1 - \frac{T_a}{T_b}$, which means that $1 - \frac{T_a}{T_c} > 1 - \frac{T_a}{T_b}$, showing that the Carnot heat engine is more efficient than an Otto heat engine operating between the same extreme temperatures.

---

84. Study the example on pages 225-227.

(A) Consult the notes on page 224.

- **Isochor** ab: Volume is constant, so no work is done: $W_{ab} = 0$.
- **Isobar** bc: Pressure is constant: $W_{bc} = \int_{V=V_b}^{V_c} P_b \, dV = P_b \int_{V=V_b}^{V_c} dV = P_b(V_c - V_b)$. Use the ideal gas law: $P_b V_b = nRT_b$ and $P_c V_c = nRT_c$. Since $P_c = P_b$, we can write $W_{bc} = P_b(V_c - V_b) = P_b V_c - P_b V_b = P_c V_c - P_b V_b = nR(T_c - T_b)$.
- **Isochor** cd: Volume is constant, so no work is done: $W_{cd} = 0$.
- **Isobar** da: Pressure is constant: $W_{da} = \int_{V=V_d}^{V_a} P_a \, dV = P_a \int_{V=V_d}^{V_a} dV = P_a(V_a - V_d)$. Use the ideal gas law: $P_a V_a = nRT_a$ and $P_d V_d = nRT_d$. Since $P_a = P_d$, we can write $W_{da} = P_a(V_a - V_d) = P_a V_a - P_a V_d = P_a V_a - P_d V_d = nR(T_a - T_d)$.

(B) Consult the notes on page 224.

- **Isochor** ab: Volume is constant ($V_b = V_a$). Use the specific heat at constant **volume**. Note that $c_V = \frac{3}{2}R$ (see page 224). You should get $Q_{ab} = \frac{3}{2}nR(T_b - T_a)$.

- **Isobar** bc: Pressure is constant ($P_c = P_b$). Use the specific heat at constant **pressure**. Note that $c_P = \frac{5}{2}R$ (see page 224). You should get $Q_{bc} = \frac{5}{2}nR(T_c - T_b)$.
- **Isochor** cd: Volume is constant ($V_d = V_c$). Use the specific heat at constant **volume**. Note that $c_V = \frac{3}{2}R$ (see page 224). You should get $Q_{cd} = \frac{3}{2}nR(T_d - T_c)$.
- **Isobar** da: Pressure is constant ($P_a = P_d$). Use the specific heat at constant **pressure**. Note that $c_P = \frac{5}{2}R$ (see page 224). You should get $Q_{da} = \frac{5}{2}nR(T_a - T_d)$.

(C) Use the equation $W_{out} = W_{ab} + W_{bc} + W_{cd} + W_{da}$.

- The work output is $W_{out} = nR(T_a - T_b + T_c - T_d)$.

(D) The heat **input** is the **positive** heat change.

- The positive heat changes are Q_{ab} and Q_{bc}, corresponding to increasing pressure and increasing volume, respectively. Note: In part (M), we rank the temperatures to make it more clear which heat changes are positive or negative.
- The heat input is $Q_{in} = Q_{ab} + Q_{bc}$.
- You should get $Q_{in} = \frac{3}{2}nR(T_b - T_a) + \frac{5}{2}nR(T_c - T_b)$.
- **Distribute** the 3 and 5 to write this as $Q_{in} = \frac{nR}{2}(3T_b - 3T_a) + \frac{nR}{2}(5T_c - 5T_b)$.
- **Factor** out the $\frac{nR}{2}$ to write this as $Q_{in} = \frac{nR}{2}(3T_b - 3T_a + 5T_c - 5T_b)$.
- **Combine like terms.** Note that $3T_b - 5T_b = -2T_b$.
- The heat input is $Q_{in} = \frac{nR}{2}(-3T_a - 2T_b + 5T_c)$.

(E) The heat **output** (or exhaust) is the **negative** heat change.

- The negative heat changes are Q_{cd} and Q_{da}.
- The heat output is $Q_{out} = Q_{cd} + Q_{da}$.
- You should get $Q_{out} = \frac{3}{2}nR(T_d - T_c) + \frac{5}{2}nR(T_a - T_d)$.
- **Distribute** the 3 and 5 to write this as $Q_{out} = \frac{nR}{2}(3T_d - 3T_c) + \frac{nR}{2}(5T_a - 5T_d)$.
- **Factor** out the $\frac{nR}{2}$ to write this as $Q_{out} = \frac{nR}{2}(3T_d - 3T_c + 5T_a - 5T_d)$.
- **Combine like terms.** Note that $3T_d - 5T_d = -2T_d$.
- The heat output is $Q_{out} = \frac{nR}{2}(5T_a - 3T_c - 2T_d)$.

(F) Write the ideal gas law for the two **isobars**. Apply Charles's law (Chapter 16), since **pressure** is constant (or cancel out pressure in the ideal gas law: $\frac{P_0V_0}{T_0} = \frac{PV}{T}$) to get $\frac{V_0}{T_0} = \frac{V}{T}$.

- **Isobar** bc: Charles's law is $\frac{V_b}{T_b} = \frac{V_c}{T_c}$.
- **Isobar** da: Charles's law is $\frac{V_d}{T_d} = \frac{V_a}{T_a}$.
- Solve for T_b and T_d in these two equations. Tip: **Cross multiply** first.
- You should get $T_b = \frac{V_b}{V_c}T_c$ and $T_d = \frac{V_d}{V_a}T_a$.
- Note that $V_b = V_a$ and $V_d = V_c$. Therefore, $\frac{V_b}{V_c} = \frac{V_a}{V_c}$ and $\frac{V_d}{V_a} = \frac{V_c}{V_a}$.

- Substitute the compression ratio, $r = \frac{V_c}{V_a}$, into the equations for T_b and T_d.
- You should get $T_b = \frac{T_c}{r}$ and $T_d = rT_a$. Note that $\frac{V_b}{V_c} = \frac{V_a}{V_c} = \frac{1}{r}$ and $\frac{V_d}{V_a} = \frac{V_c}{V_a} = r$.

(G) Use the equation $e = 1 + \frac{Q_{out}}{Q_{in}}$. Recall that we found Q_{in} and Q_{out} in parts (D) and (E).

- The $\frac{nR}{2}$'s cancel out. You should get: $e = 1 + \frac{5T_a - 3T_c - 2T_d}{-3T_a - 2T_b + 5T_c}$. It is convenient to rewrite this as $e = 1 - \frac{-5T_a + 3T_c + 2T_d}{-3T_a - 2T_b + 5T_c}$, using $5T_a - 3T_c - 2T_d = -(-5T_a + 3T_c + 2T_d)$. We will also reorder the terms: $e = 1 - \frac{3T_c - 5T_a + 2T_d}{5T_c - 3T_a - 2T_b}$.
- Use the equations from part (F), $T_b = \frac{T_c}{r}$ and $T_d = rT_a$, to eliminate T_b and T_d from the efficiency equation. You should get:

$$e = 1 - \frac{3T_c - 5T_a + 2rT_a}{5T_c - 3T_a - \frac{2T_c}{r}}$$

- **Combine like terms**. Note that $-5T_a + 2rT_a = (-5 + 2r)T_a = (2r - 5)T_a$ and that $5T_c - \frac{2T_c}{r} = \left(5 - \frac{2}{r}\right)T_c = \left(\frac{5r}{r} - \frac{2}{r}\right)T_c = \left(\frac{5r-2}{r}\right)T_c$.
- The efficiency is:

$$e = 1 - \frac{3T_c + (2r - 5)T_a}{\left(\frac{5r - 2}{r}\right)T_c - 3T_a}$$

(H) Divide the equation $T_b = \frac{T_c}{r}$ by $T_d = rT_a$.

- You should get $\frac{T_b}{T_d} = \frac{T_c}{r} \div rT_a$. To divide by rT_a, multiply by its **reciprocal**. Note that the reciprocal of rT_a is $\frac{1}{rT_a}$. You should get $\frac{T_b}{T_d} = \frac{T_c}{r} \times \frac{1}{rT_a}$. Note that $rr = r^2$.
- Rearrange this. Multiply both sides by r^2. You should get $\frac{T_c}{T_a} = r^2 \frac{T_b}{T_d}$.

(I) Write the ideal gas law for the two **isochors**. Apply Gay-Lussac's law (Chapter 16), since **volume** is constant (or cancel out volume in the ideal gas law: $\frac{P_0 V_0}{T_0} = \frac{PV}{T}$) to get $\frac{P_0}{T_0} = \frac{P}{T}$.

- **Isochor** ab: Gay-Lussac's law is $\frac{P_a}{T_a} = \frac{P_b}{T_b}$.
- **Isochor** cd: Gay-Lussac's law is $\frac{P_c}{T_c} = \frac{P_d}{T_d}$.
- Solve for T_b and T_d in these two equations. Tip: **Cross multiply** first.
- You should get $T_b = \frac{P_b}{P_a} T_a$ and $T_d = \frac{P_d}{P_c} T_c$.
- Note that $P_b = P_c$ and $P_a = P_d$. Therefore, $\frac{P_d}{P_c} = \frac{P_a}{P_b}$ and $T_d = \frac{P_a}{P_b} T_c$.
- Substitute the pressure ratio, $r_p = \frac{P_b}{P_a}$, into the equations for T_b and T_d.
- You should get $T_b = r_p T_a$ and $T_d = \frac{T_c}{r_p}$. Note that $\frac{P_d}{P_c} = \frac{P_a}{P_b} = \frac{1}{r_p}$.
- Note: $r_p = \frac{P_b}{P_a}$ and $r = \frac{V_c}{V_a}$ are two **different** ratios. Don't get these confused.

(J) Divide the equation $T_d = \frac{T_c}{r_p}$ by $T_b = r_p T_a$.

- You should get $\frac{T_d}{T_b} = \frac{T_c}{r_p} \div r_p T_a$. To divide by $r_p T_a$, multiply by its **reciprocal**. Note that the reciprocal of $r_p T_a$ is $\frac{1}{r_p T_a}$. You should get $\frac{T_d}{T_b} = \frac{T_c}{r_p} \times \frac{1}{r_p T_a}$. Note that $r_p r_p = r_p^2$.
- Rearrange this. Multiply both sides by r_p^2. You should get $\frac{T_c}{T_a} = r_p^2 \frac{T_d}{T_b}$.
- Note that these are r_p's, and **<u>not</u>** r's. Recall that $r_p = \frac{P_b}{P_a}$ and $r = \frac{V_c}{V_a}$.

(K) Since $\frac{T_c}{T_a} = r^2 \frac{T_b}{T_d}$ and $\frac{T_c}{T_a} = r_p^2 \frac{T_d}{T_b}$, we can set the right-hand sides equal to one another.

- You should get $r^2 \frac{T_b}{T_d} = r_p^2 \frac{T_d}{T_b}$. **Cross multiply** to get $r^2 T_b^2 = r_p^2 T_d^2$. Note, for example, that $T_b T_b = T_b^2$. Now squareroot both sides of the equation to get $rT_b = r_p T_d$. Divide both sides of the equation by T_d to get $\frac{T_b}{T_d} = \frac{r_p}{r}$.
- Plug $\frac{T_b}{T_d} = \frac{r_p}{r}$ into the equation $\frac{T_c}{T_a} = r^2 \frac{T_b}{T_d}$. Note that $r^2 \left(\frac{r_p}{r}\right) = rr_p$.
- You should get $\frac{T_c}{T_a} = rr_p$. Multiply both sides by T_a to get $T_c = rr_p T_a$.

(L) Plug $T_c = rr_p T_a$ into the equation for efficiency from part (G). You should get:

$$e = 1 - \frac{3rr_p T_a + (2r - 5)T_a}{\left(\frac{5r - 2}{r}\right) rr_p T_a - 3T_a}$$

- T_a cancels out. Note that $\left(\frac{5r-2}{r}\right) rr_p = (5r - 2)r_p = 5rr_p - 2r_p$. You should get:

$$e = 1 - \frac{3rr_p + 2r - 5}{5rr_p - 2r_p - 3}$$

(M) Study the *P-V* diagram. Clearly, $V_c > V_a$ and $P_b > P_a$, which means that $r = \frac{V_c}{V_a} > 1$ and $r_p = \frac{P_b}{P_a} > 1$. Study the temperature equations, knowing that $r_p > 1$ and $r > 1$.

- From $T_b = \frac{T_c}{r}$ (which is the same as $T_c = rT_b$), it follows that $T_c > T_b$.
- From $T_d = rT_a$, it follows that $T_d > T_a$.
- From $T_b = r_p T_a$, it follows that $T_b > T_a$.
- From $T_d = \frac{T_c}{r_p}$ (which is the same as $T_c = r_p T_d$), it follows that $T_c > T_d$.
- From $T_c = rr_p T_a$, it follows that $T_c > T_a$.
- Since $T_c > T_a$, $T_c > T_b$, and $T_c > T_d$, the highest temperature is $T_h = T_c$.
- Since $T_b > T_a$, $T_c > T_a$, and $T_d > T_a$, the lowest temperature is $T_\ell = T_a$.

(N) Plug $r = 2$ and $r_p = 4$ into the equation from part (L). You should get:

- $e = 1 - \frac{24+4-5}{40-8-3} = 1 - \frac{23}{29} = \frac{6}{29} = 0.21 = 21\%$ to two significant figures. (Multiply by 100% to express your answer as a percentage.) The efficiency of this heat engine is 21%.
- The efficiency of a Carnot heat engine operating between the same extreme temperatures would be $e_C = 1 - \frac{T_\ell}{T_h}$ (see page 220).

- Since $T_h = T_c$ and $T_\ell = T_a$, we can write the Carnot efficiency as $e_C = 1 - \frac{T_a}{T_c}$.
- Use the equation $T_c = rr_pT_a$ from part (K) to write this as $e_C = 1 - \frac{1}{rr_p}$.
- You should get: $e = 1 - \frac{1}{8} = \frac{7}{8} = 0.88 = 88\%$ to two significant figures. (Multiply by 100% to express your answer as a percentage.)
- The efficiency of the Carnot heat engine is 88% to two significant figures.
- Note that the Carnot heat engine is more efficient (as must be the case).

Chapter 20: Light Waves

85. Identify the known quantities in SI units.
    - The wavelength in air is $\lambda_a = 4.00 \times 10^{-7}$ m since $1 \text{ nm} = 1 \times 10^{-9}$ m.
    - The speed of light in air is about the same as in vacuum: $v_a = 3.00 \times 10^8$ m/s.

(A) Use the equation for wave speed. Solve for frequency: $f = \frac{v_a}{\lambda_a}$.

- The frequency is $f = 7.50 \times 10^{14}$ Hz. Note that $\frac{10^8}{10^{-7}} = 10^{8-(-7)} = 10^{8+7} = 10^{15}$ according to the rule $\frac{x^m}{x^n} = x^{m-n}$ and that $0.75 \times 10^{15} = 7.5 \times 10^{14}$.

(B) The speed of light in the medium is $v_m = 1.80 \times 10^8$ m/s.

- Use the equation for wave speed. Solve for wavelength: $\lambda_m = \frac{v_m}{f}$.
- The **frequency** is the **same** as for part (A).
- The wavelength is $\lambda_m = 2.40 \times 10^{-7}$ m, which is the same as $\lambda_m = 240$ nm.

---

86. Study the similar example on page 241.
    - The **power** (P) output of the star is **fixed**.
    - The intensity (I) depends on the distance (r) from the star.
    - Intensity is $I = \frac{P}{A}$, where $A = 4\pi r^2$ is the surface area of a sphere of radius r. Combine these equations to get $I = \frac{P}{4\pi r^2}$.
    - Write an equation for intensity for each planet's orbital radius.
    - You should have $I_1 = \frac{P}{4\pi r_1^2}$ and $I_2 = \frac{P}{4\pi r_2^2}$.
    - Solve for power in both equations. You should get $P = I_1 4\pi r_1^2 = I_2 4\pi r_2^2$.
    - Divide both sides of the equation by $4\pi r_2^2$. You should get $I_2 = \left(\frac{r_1}{r_2}\right)^2 I_1$.
    - According to the problem, $r_2 = 3r_1$, which means that $\frac{r_1}{r_2} = \frac{1}{3}$. This is equivalent to $\frac{r_2}{r_1} = 3$, but we need $\frac{r_1}{r_2}$, **not** $\frac{r_2}{r_1}$. Plug $\frac{r_1}{r_2} = \frac{1}{3}$ into the previous equation.
    - Note that $\frac{1}{3^2} = \frac{1}{9}$. You should get $I_2 = \frac{I_1}{9}$.
    - The intensity at the more distant planet's orbital radius is $I_2 = 200 \frac{\text{W}}{\text{m}^2}$.

Chapter 21: Reflection and Refraction

87. Study the examples on pages 247-249.

(A) Extend the incident ray (i) until it reaches the boundary between the water and glass.

- Since the boundary between the water and glass is horizontal, the normal (N) is a **vertical** line (since the normal must be **perpendicular** to the boundary).
- Draw the reflected ray coming back into the water, such that $\theta_r = \theta_i$.
- The refracted ray, going from water to glass, is **slowing down** (it's slower in glass).
- Bend the refracted ray **towards the normal** since it is slowing down.
- See the diagram below on the left.

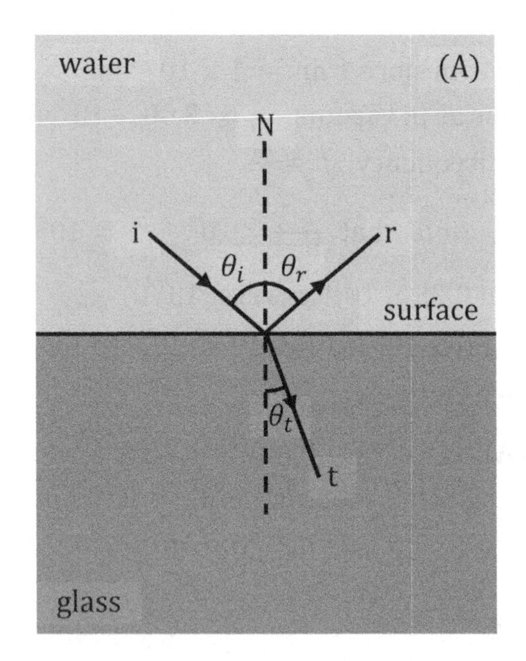

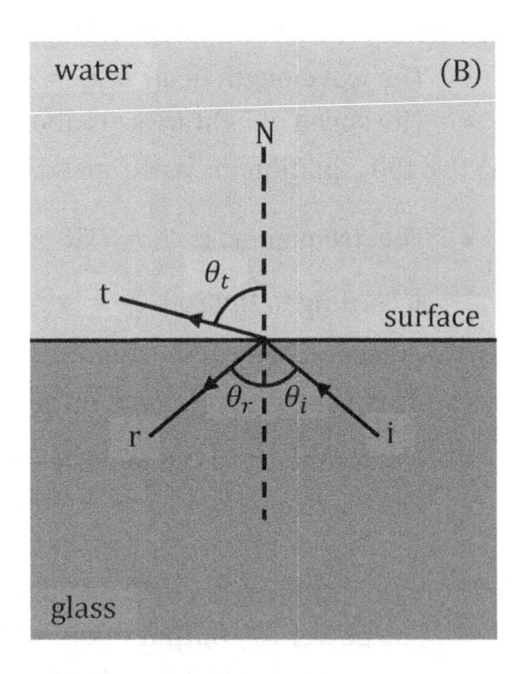

(B) Extend the incident ray (i) until it reaches the boundary between the glass and water.

- Since the boundary between the glass and water is horizontal, the normal (N) is a **vertical** line (since the normal must be **perpendicular** to the boundary).
- Draw the reflected ray coming back into the glass, such that $\theta_r = \theta_i$.
- The refracted ray, going from glass to water, is **speeding up** (it's faster in water).
- Bend the refracted ray **away from the normal** since it is speeding up.
- See the diagram above on the right.

(C) Extend the incident ray (i_1) until it reaches the boundary between the air and glass.

- Draw the normal **perpendicular** to the boundary (or surface or interface) between the air and glass. The normal is perpendicular to the side of the triangle.
- Draw the reflected ray coming back into the air, such that $\theta_{1r} = \theta_{1i}$.
- The refracted ray, going from air to glass, is **slowing down** (it's slower in glass).
- Bend the refracted ray **towards the normal** since it is slowing down.
- The refracted ray will reach a second boundary between glass and air.

- When the refracted ray reaches the right side of the triangle (it's a **prism**), a second reflection and refraction will occur.[1]
- The first refracted ray (t_1) serves as the second incident ray (i_2): $t_1 = i_2$. That is, the original ray of refraction inside the prism is incident upon a second surface. So the ray that is t_1 for the first surface is the same ray that is i_2 for the second surface.
- Draw a new normal **perpendicular** to the boundary (or surface or interface) between the glass and air. This normal is perpendicular to the right side of the triangle. You will have two normals: N_1 and N_2. One is normal to the left side of the triangle, while the other is normal to the right side of the triangle.
- Draw the second reflected ray coming back into the glass, such that $\theta_{2r} = \theta_{2i}$.
- The second refracted ray, going from glass to air, is **speeding up** (it's faster in air).
- Bend the second refracted ray **away from the normal** (N_2) since it is speeding up.
- **Note**: The first refracted ray (t_1) going from air to glass slows down and bends towards the first normal (N_1), whereas the second refracted ray (t_2) going from glass to air speeds up and bends away from the second normal (N_2).
- See the diagram below. **Note**: The first refracted ray (t_1) is not necessarily horizontal as shown below. It could be angled slightly upward (but it's definitely angled lower than the first incident ray), or it could be angled slightly downward (but it definitely does **<u>not</u>** angle below the first normal). Without knowing the exact index of refraction of the glass, we can't determine the exact angles.

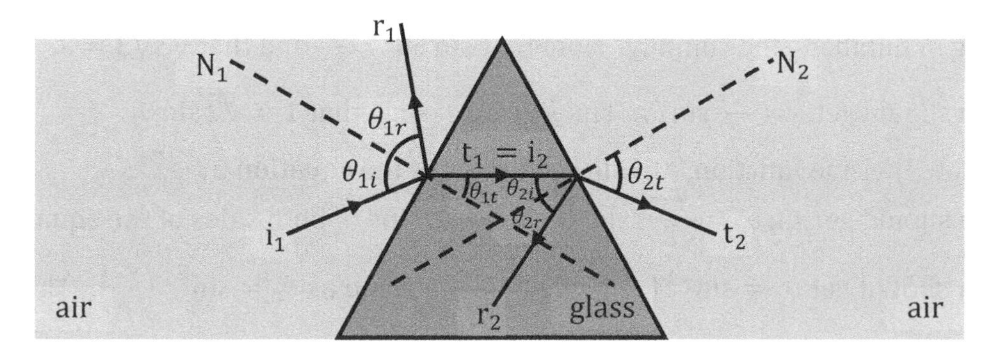

(D) Extend the incident ray (i) until it reaches the boundary between the two media.

- Since the boundary between the two media is a circle, the normal (N) is along a **radius** (since the normal must be **perpendicular** to the boundary, and since the radius is perpendicular to the circumference of the circle). Since the normal is along a radius, first draw the center of the circle (C), and then draw the normal from the center of the circle through the point where the incident ray (i) meets the circle.
- Draw the reflected ray coming back into the medium with $n_1 = 1.6$, such that $\theta_r = \theta_i$.

[1] The second reflected ray (the one formed from the first refracted ray) will refract and reflect yet again when it strikes the bottom surface. However, the question only asked us to draw two reflections and refractions.

- The refracted ray, going from $n_1 = 1.6$ to $n_2 = 1.3$, is **speeding up** (it's **faster** in the medium with the **smaller** index of refraction[2]).
- Bend the refracted ray **away from the normal** since it is speeding up.
- See the diagram below.

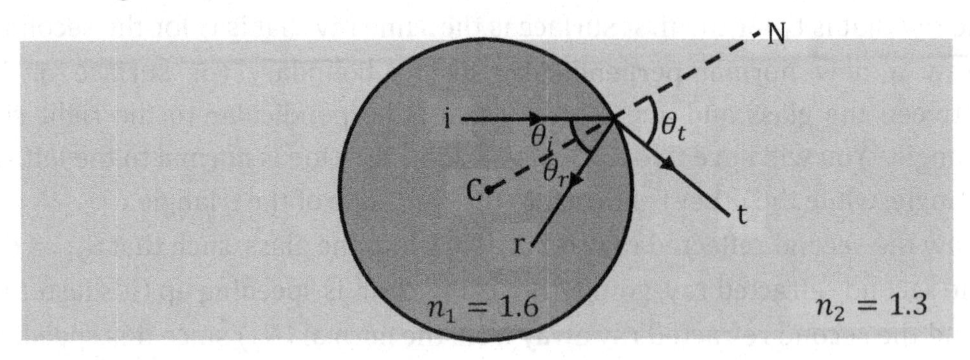

Chapter 22: Snell's Law

88. Identify the known quantities in appropriate units.
  - The incident angle is $\theta_i = 60°$.
  - The index of refraction of the banana juice (the incident medium) is $n_i = \sqrt{3}$.
  - The index of refraction of the glass (the refracting medium) is $n_t = \frac{3\sqrt{2}}{2}$.

(A) Apply Snell's law: $n_i \sin\theta_i = n_t \sin\theta_t$.

- Plug in numbers and simplify. Note that $\sin 60° = \frac{\sqrt{3}}{2}$ and that $\sqrt{3}\sqrt{3} = 3$.
- You should get $\frac{3}{2} = \frac{3\sqrt{2}}{2}\sin\theta_t$. The $\frac{3}{2}$'s cancel such that $1 = \sqrt{2}\sin\theta_t$.
- Isolate the sine function: Divide both sides of the equation by $\sqrt{2}$.
- You should get $\sin\theta_t = \frac{1}{\sqrt{2}}$. Take the inverse sine of both sides of the equation.
- You should get $\theta_t = \sin^{-1}\left(\frac{1}{\sqrt{2}}\right)$, which is the same as $\theta_t = \sin^{-1}\left(\frac{\sqrt{2}}{2}\right)$. These are the same because $\frac{1}{\sqrt{2}} = \frac{1}{\sqrt{2}}\frac{\sqrt{2}}{\sqrt{2}} = \frac{\sqrt{2}}{2}$ if you **rationalize** the denominator.
- The angle of refraction is $\theta_t = 45°$ because $\sin(45°) = \frac{\sqrt{2}}{2}$. See page 252.

(B) Use the equation $v_i = \frac{c}{n_i}$. The speed of light in **vacuum** is $c = 3.00 \times 10^8$ m/s.

- Note that $\frac{3}{\sqrt{3}} = \sqrt{3}$ because $\sqrt{3}\sqrt{3} = 3$.
- The speed of light in the banana juice is $v_i = \sqrt{3} \times 10^8$ m/s. If you use a calculator, this works out to $v_i = 1.7 \times 10^8$ m/s to two significant figures.

[2] That's because $n = \frac{c}{v}$. A **smaller** speed of light in the medium (v) makes n **larger**.

(C) Use the equation $v_t = \frac{c}{n_t}$. The speed of light in **vacuum** is $c = 3.00 \times 10^8$ m/s.

- Note that $\frac{3}{\frac{3\sqrt{2}}{2}} = 3 \times \frac{2}{3\sqrt{2}}$ (to divide by a fraction, multiply by its **reciprocal**). Also note that $\frac{2}{\sqrt{2}} = \sqrt{2}$ because $\sqrt{2}\sqrt{2} = 2$.
- The speed of light in the glass is $v_t = \sqrt{2} \times 10^8$ m/s. If you use a calculator, this works out to $v_t = 1.4 \times 10^8$ m/s to two significant figures.

---

89. Identify the known quantities in appropriate units.

- The incident angle is $\theta_i = 60°$.
- The speed of light in the first (incident) medium is $v_i = \sqrt{6} \times 10^8$ m/s.
- The speed of light in the second (refracting) medium is $v_t = \sqrt{2} \times 10^8$ m/s.

(A) Use the equation $n_i = \frac{c}{v_i}$. The speed of light in **vacuum** is $c = 3.00 \times 10^8$ m/s.

- Note that $\frac{3}{\sqrt{6}} = \frac{3}{\sqrt{6}}\frac{\sqrt{6}}{\sqrt{6}} = \frac{3\sqrt{6}}{6} = \frac{\sqrt{6}}{2}$ if you **rationalize** the denominator since $\sqrt{6}\sqrt{6} = 6$.
- The index of refraction of the first medium is $n_i = \frac{\sqrt{6}}{2}$. If you use a calculator, this works out to $n_i = 1.2$ to two significant figures.

(B) Use the equation $n_t = \frac{c}{v_t}$. The speed of light in **vacuum** is $c = 3.00 \times 10^8$ m/s.

- Note that $\frac{3}{\sqrt{2}} = \frac{3}{\sqrt{2}}\frac{\sqrt{2}}{\sqrt{2}} = \frac{3\sqrt{2}}{2}$ if you **rationalize** the denominator since $\sqrt{2}\sqrt{2} = 2$.
- The index of refraction of the second medium is $n_t = \frac{3\sqrt{2}}{2}$. If you use a calculator, this works out to $n_t = 2.1$ to two significant figures.

(C) Apply Snell's law: $n_i \sin\theta_i = n_t \sin\theta_t$. Use your answers to parts (A) and (B).

- Plug in numbers and simplify. Note that $\sin 60° = \frac{\sqrt{3}}{2}$.
- You should get $\frac{\sqrt{18}}{4} = \frac{3\sqrt{2}}{2}\sin\theta_t$. Note that $\frac{\sqrt{6}}{2}\frac{\sqrt{3}}{2} = \frac{\sqrt{18}}{4}$.
- Note that $\sqrt{18} = \sqrt{(2)(9)} = \sqrt{2}\sqrt{9} = 3\sqrt{2}$, such that $\frac{\sqrt{18}}{4} = \frac{3\sqrt{2}}{4}$ and $\frac{3\sqrt{2}}{4} = \frac{3\sqrt{2}}{2}\sin\theta_t$.
- Divide both sides of the equation by $3\sqrt{2}$. You should get $\frac{1}{4} = \frac{1}{2}\sin\theta_t$.
- Isolate the sine function: Multiply both sides of the equation by 2.
- You should get $\sin\theta_t = \frac{1}{2}$. Take the inverse sine of both sides of the equation.
- You should get $\theta_t = \sin^{-1}\left(\frac{1}{2}\right)$.
- The angle of refraction is $\theta_t = 30°$ because $\sin(30°) = \frac{1}{2}$. See page 252.

90. First draw a ray diagram following the technique from Chapter 21.
- The boundary is vertical, so the normal is **horizontal** (**perpendicular** to the boundary).
- The refracted ray is **slowing down** because $n_2 > n_1$.
- Bend the refracted ray **towards the normal** since it is slowing down.
- The refracted ray reaches a second boundary at the third medium.
- The first refracted ray (t_1) serves as the second incident ray (i_2): $t_1 = i_2$.
- Draw a new normal **perpendicular** to the boundary.
- The second refracted ray is **slowing down** because $n_3 > n_2$.
- Bend the second refracted ray **towards the normal** since it is slowing down.

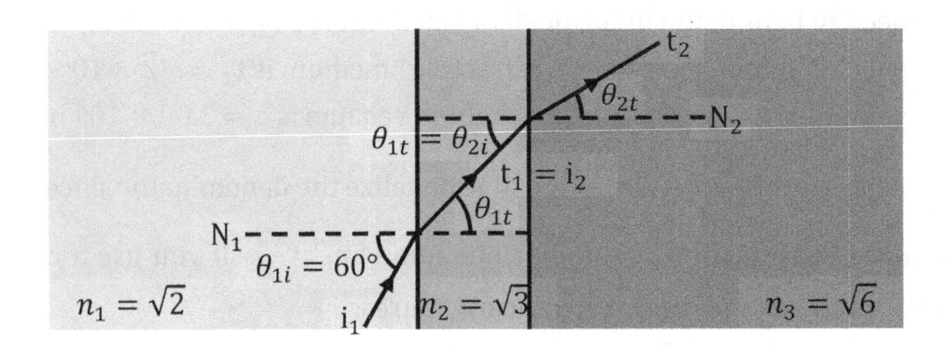

(A) Identify the known quantities in appropriate units.
- The first incident angle is $\theta_{1i} = 60°$.
- The indices of refraction are $n_{1i} = \sqrt{2}$, $n_{1t} = n_{2i} = \sqrt{3}$, and $n_{2t} = \sqrt{6}$. The reason that $n_{1t} = n_{2i}$ is that the first refracted ray is also incident upon the third medium.
- Apply Snell's law: $n_{1i} \sin\theta_{1i} = n_{1t} \sin\theta_{1t}$.
- Plug in numbers and simplify. Note that $\sin 60° = \frac{\sqrt{3}}{2}$ and that $\sqrt{2}\sqrt{3} = \sqrt{6}$.
- You should get $\frac{\sqrt{6}}{2} = \sqrt{3}\sin\theta_{1t}$. Isolate the sine function: Divide both sides of the equation by $\sqrt{2}$. Note that $\frac{\sqrt{6}}{\sqrt{3}} = \sqrt{\frac{6}{3}} = \sqrt{2}$. You should get $\sin\theta_{1t} = \frac{\sqrt{2}}{2}$.
- Take the inverse sine of both sides of the equation. You should get $\theta_{1t} = \sin^{-1}\left(\frac{1}{\sqrt{2}}\right)$.
- The **first** angle of refraction is $\theta_{1t} = 45°$ because $\sin(45°) = \frac{\sqrt{2}}{2}$. See page 252.
- Now apply Snell's law to the **second** refraction: $n_{2i} \sin\theta_{2i} = n_{2t} \sin\theta_{2t}$.
- Since the first refracted ray serves as the second incident ray, $n_{1t} = n_{2i} = \sqrt{3}$ and $\theta_{1t} = \theta_{2i} = 45°$. (Here, we are applying the rule from geometry that **alternate interior angles** are equal: In the diagram above, θ_{1t} and θ_{2i} are alternate interior angles because N_1 and N_2 are parallel – they are both horizontal. See page 270.)
- Recall that $n_{2t} = \sqrt{6}$. Plug in numbers and simplify.
- You should get $\frac{\sqrt{6}}{2} = \sqrt{6}\sin\theta_{2t}$. Note that $\sin 45° = \frac{\sqrt{2}}{2}$ and that $\sqrt{3}\sqrt{2} = \sqrt{6}$.

- Isolate the sine function: Divide both sides of the equation by $\sqrt{6}$.
- You should get $\sin\theta_{2t} = \frac{1}{2}$. Take the inverse sine of both sides of the equation.
- You should get $\theta_{2t} = \sin^{-1}\left(\frac{1}{2}\right)$.
- The **second** angle of refraction is $\theta_{2t} = 30°$ because $\sin(30°) = \frac{1}{2}$. See page 252.

(B) If you remove the second medium, there is only one refraction and the diagram looks like the picture below.

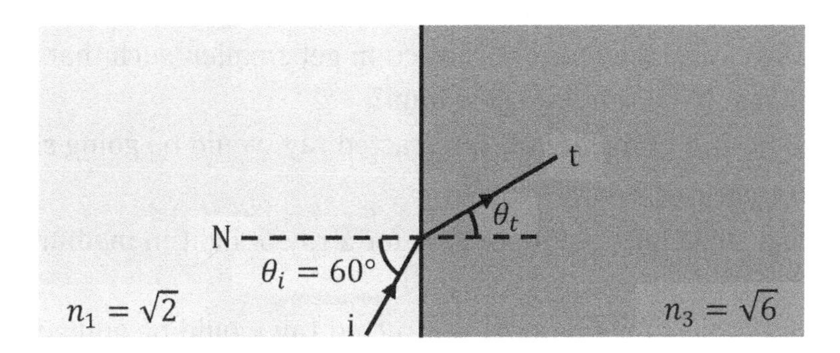

- Apply Snell's law: $n_i \sin\theta_i = n_t \sin\theta_t$. Note that $\theta_i = 60°$, $n_i = \sqrt{2}$, and $n_t = \sqrt{6}$.
- Plug in numbers and simplify. Note that $\sin 60° = \frac{\sqrt{3}}{2}$.
- You should get $\frac{\sqrt{6}}{2} = \sqrt{6}\sin\theta_t$. Note that $\sqrt{2}\sqrt{3} = \sqrt{6}$.
- Isolate the sine function: Divide both sides of the equation by $\sqrt{6}$.
- You should get $\sin\theta_t = \frac{1}{2}$. Take the inverse sine of both sides of the equation.
- You should get $\theta_t = \sin^{-1}\left(\frac{1}{2}\right)$.
- The angle of refraction is $\theta_t = 30°$ because $\sin(30°) = \frac{1}{2}$. See page 252.

(C) In parts (A) and (B) both, we found that the final angle of refraction was $\theta_{2t} = 30°$.

- Evidently, the second medium has **<u>no</u>** impact on the final angle of refraction.
- Let's see why that is. For the first refraction, $n_{1i}\sin\theta_{1i} = n_{1t}\sin\theta_{1t}$. For the second refraction, $n_{2i}\sin\theta_{2i} = n_{2t}\sin\theta_{2t}$.
- The ray in the second medium serves both as the first refracted ray and the second incident ray: $n_{2i} = n_{1t}$ and $\theta_{2i} = \theta_{1t}$. Plug these into the second equation to get $n_{1t}\sin\theta_{1t} = n_{2t}\sin\theta_{2t}$.
- The two equations are now $n_{1i}\sin\theta_{1i} = n_{1t}\sin\theta_{1t}$ and $n_{1t}\sin\theta_{1t} = n_{2t}\sin\theta_{2t}$.
- Since $n_{1t}\sin\theta_{1t}$ equals $n_{1i}\sin\theta_{1i}$ and since $n_{1t}\sin\theta_{1t}$ also equals $n_{2t}\sin\theta_{2t}$, it follows that $n_{1i}\sin\theta_{1i} = n_{2t}\sin\theta_{2t}$. (It's basically, if A = B and B = C, then A = C.)
- The equation $n_{1i}\sin\theta_{1i} = n_{2t}\sin\theta_{2t}$ is exactly what we did in part (B), if you look at the numbers. Thus, we see that the term $n_{1t}\sin\theta_{1t}$ is effectively the middleman in a way that has **<u>no</u>** effect on θ_{2t}.

Chapter 23: Total Internal Reflection

91. Total internal reflection is only possible if the refracted ray would travel **faster** than the incident ray.

- Does light travel **faster** in a medium with a higher or lower index or refraction?
- According to the equation $n = \frac{c}{v}$, a faster speed (meaning a larger v) of light in the medium results in a **smaller** index of refraction. (Light travels fastest through a vacuum, which has the minimum possible index of refraction of exactly 1.)
- In which case would the index of refraction get smaller, such that light would travel faster in the would-be refracting medium?

(A) Going from medium A to medium B, a refracted ray would be going **slower** because the index of refraction **increases** from $\sqrt{3}$ to $\sqrt{6}$.

- Total internal reflection is **not** possible for a ray of light in medium A incident upon medium B.

(B) Going from medium B to medium A, a refracted ray would be going **faster** because the index of refraction **decreases** from $\sqrt{6}$ to $\sqrt{3}$.

- Total internal reflection **is** possible for a ray of light in medium B incident upon medium A.
- Apply Snell's law: $n_i \sin\theta_i = n_t \sin\theta_t$.
- Set $\theta_i = \theta_c$ and $\theta_t = 90°$. Note that $n_i = \sqrt{6}$ and $n_t = \sqrt{3}$.
- Plug in numbers and simplify. Note that $\sin 90° = 1$.
- You should get $\sqrt{6}\sin\theta_c = \sqrt{3}$.
- Isolate the sine function: Divide both sides of the equation by $\sqrt{6}$.
- You should get $\sin\theta_c = \frac{\sqrt{2}}{2}$. Note that $\frac{\sqrt{3}}{\sqrt{6}} = \sqrt{\frac{3}{6}} = \frac{1}{\sqrt{2}} = \frac{1}{\sqrt{2}}\frac{\sqrt{2}}{\sqrt{2}} = \frac{\sqrt{2}}{2}$ since $\sqrt{2}\sqrt{2} = 2$.
- Take the inverse sine of both sides of the equation. You should get $\theta_c = \sin^{-1}\left(\frac{\sqrt{2}}{2}\right)$. Note that this is the same as $\theta_c = \sin^{-1}\left(\frac{1}{\sqrt{2}}\right)$ because $\frac{1}{\sqrt{2}} = \frac{\sqrt{2}}{2}$.
- The critical angle for total internal reflection is $\theta_c = 45°$ because $\sin(45°) = \frac{\sqrt{2}}{2}$. See page 260.

Chapter 24: Dispersion and Scattering

92. From the diagram, $\theta_{1i} = 60°$, $n_1 = 2$, $n_2 = \sqrt{6}$, and $n_3 = \sqrt{3}$.

(A) Apply Snell's law to the first refraction: $n_1 \sin\theta_{1i} = n_2 \sin\theta_{1t}$. Note that $\sin 60° = \frac{\sqrt{3}}{2}$.

- You should get $\sqrt{3} = \sqrt{6}\sin\theta_{1t}$. Isolate the sine function.
- Divide both sides of the equation by $\sqrt{6}$. You should get $\sin\theta_{1t} = \frac{\sqrt{2}}{2}$. Note that

$\frac{\sqrt{3}}{\sqrt{6}} = \sqrt{\frac{3}{6}} = \sqrt{\frac{1}{2}} = \frac{1}{\sqrt{2}} = \frac{1}{\sqrt{2}}\frac{\sqrt{2}}{\sqrt{2}} = \frac{\sqrt{2}}{2}$ because $\sqrt{2}\sqrt{2} = 2$.

- Take the inverse sine of both sides of the equation.
- You should get $\theta_{1t} = \sin^{-1}\left(\frac{\sqrt{2}}{2}\right)$.
- The **first** angle of refraction is $\theta_{1t} = 45°$ because $\sin 45° = \frac{\sqrt{2}}{2}$.

(B) Note: Although the first refracted ray (t_1) serves as the second incident ray (i_2), the angles θ_{1t} and θ_{2i} are **not** equal (because they are relative to two different normals).

- Use the triangle shown in the problem. The three angles add up to 180°.
- You should get $A + \alpha + \beta = 180°$. Plug in the prism angle ($A = 75°$).
- You should get $\alpha + \beta = 105°$.
- Study the diagram given in the problem: α and θ_{1t} are **complementary**.
- This means that $\alpha + \theta_{1t} = 90°$. Plug in $\theta_{1t} = 45°$, which we found previously.
- You should get $\alpha = 45°$. Plug this into $\alpha + \beta = 105°$.
- You should get $\beta = 60°$. Note that β and θ_{2i} are **complementary**.
- This means that $\beta + \theta_{2i} = 90°$. Plug in $\beta = 60°$.
- You should get $\theta_{2i} = 30°$.
- Apply Snell's law to the second refraction: $n_2 \sin\theta_{2i} = n_3 \sin\theta_{2t}$. Use $\theta_{2i} = 30°$, $n_2 = \sqrt{6}$, and $n_3 = \sqrt{3}$. Note that $\sin 30° = \frac{1}{2}$.
- You should get $\frac{\sqrt{6}}{2} = \sqrt{3}\sin\theta_{2t}$. Isolate the sine function: Divide both sides by $\sqrt{3}$.
- You should get $\sin\theta_{2t} = \frac{\sqrt{2}}{2}$. Note that $\frac{\sqrt{6}}{\sqrt{3}} = \sqrt{\frac{6}{3}} = \sqrt{2}$.
- Take the inverse sine of both sides of the equation.
- You should get $\theta_{2t} = \sin^{-1}\left(\frac{\sqrt{2}}{2}\right)$.
- The **second** angle of refraction is $\theta_{2t} = 45°$ because $\sin 45° = \frac{\sqrt{2}}{2}$.

(C) Recall that $\theta_{2i} = 30°$ and $\theta_{2t} = 45°$.

- If there were no refraction at all, θ_{2i} and θ_{2t} would be equal.
- The difference between these two angles shows how much the second refracted ray is bent compared to the second incident ray (which is the same as the first refracted ray).
- The angle between the two rays is $\theta_{2t} - \theta_{2i} = 15°$.

Chapter 25: Thin Lenses

93. Identify the given information. **Note:** It's okay to leave the distances in the units given.
    - The object distance is $p = 200$ cm. The first object distance is always positive.
    - The focal length is $f = 40$ cm. Focal length is **positive** for a **convex** lens.
    - The object height is $h_o = 8.0$ mm. The first object height is always positive. **Note:** It's okay for h_o to be in millimeters (it just needs to have the same units as h_i).

(A) Use the equation $\frac{1}{p} + \frac{1}{q} = \frac{1}{f}$. Plug in numbers.

- You should get $\frac{1}{200} + \frac{1}{q} = \frac{1}{40}$. First isolate $\frac{1}{q}$. Subtract $\frac{1}{200}$ from both sides.
- You should get $\frac{1}{q} = \frac{1}{40} - \frac{1}{200}$. Find a **common denominator**.
- You should get $\frac{1}{q} = \frac{4}{200}$. Take the reciprocal of both sides and simplify.
- The image forms at $q = 50$ cm.

(B) Combine the two magnification equations together: $\frac{h_i}{h_o} = -\frac{q}{p}$.

- Multiply both sides of the equation by h_o. You should get $h_i = -\frac{q}{p} h_o$.
- Be careful with the signs.
- The image height is $h_i = -2.0$ mm. Note that the centimeters (cm) **cancel** out.

(C) Use the formula $M = \frac{h_i}{h_o}$ or $M = -\frac{q}{p}$.

- The magnification is $M = -\frac{1}{4} \times$, which is the same as $M = -0.25 \times$.

(D) Is the image upright or inverted? Review the sign conventions on page 277.

- Examine the **sign** of the **magnification.**
- Since M is **negative,** the image is **inverted.**

(E) Is the image real or virtual? Review the sign conventions on page 277.

- Examine the **sign** of the **image distance.**
- Since q is **positive,** the image is **real.**

---

94. Identify the given information. **Note:** It's okay to leave the distances in the units given.
    - The object distance is $p = 192$ cm. The first object distance is always positive.
    - The focal length is $f = -64$ cm. Focal length is **negative** for a **concave** lens.
    - The object height is $h_o = 24$ mm. The first object height is always positive. **Note:** It's okay for h_o to be in millimeters (it just needs to have the same units as h_i).

(A) Use the equation $\frac{1}{p} + \frac{1}{q} = \frac{1}{f}$. Plug in numbers.

- You should get $\frac{1}{192} + \frac{1}{q} = \frac{1}{-64}$. First isolate $\frac{1}{q}$. Subtract $\frac{1}{192}$ from both sides.
- You should get $\frac{1}{q} = -\frac{1}{64} - \frac{1}{192}$. Find a **common denominator.**
- You should get $\frac{1}{q} = -\frac{4}{192}$. Take the reciprocal of both sides and simplify.

- The image forms at $q = -48$ cm.

(B) Combine the two magnification equations together: $\frac{h_i}{h_o} = -\frac{q}{p}$.

- Multiply both sides of the equation by h_o. You should get $h_i = -\frac{q}{p} h_o$.
- Be careful with the minus signs. (Two minus signs make a plus sign.)
- The image height is $h_i = 6.0$ mm. Note that the centimeters (cm) **cancel** out.

(C) Use the formula $M = \frac{h_i}{h_o}$ or $M = -\frac{q}{p}$.

- The magnification is $M = \frac{1}{4} \times$, which is the same as $M = 0.25 \times$.

(D) Is the image upright or inverted? Review the sign conventions on page 277.

- Examine the **sign** of the **magnification**.
- Since M is **positive**, the image is **upright**.

(E) Is the image real or virtual? Review the sign conventions on page 277.

- Examine the **sign** of the **image distance**.
- Since q is **negative**, the image is **virtual**.

---

95. Identify the given information. **Note**: It's okay to leave the distances in the units given.

- The object distance is $p = 30$ cm. The first object distance is always positive.
- The focal length is $f = 90$ cm. Focal length is **positive** for a **convex** lens.
- The object height is $h_o = 16$ mm. The first object height is always positive. **Note**: It's okay for h_o to be in millimeters (it just needs to have the same units as h_i).

(A) Use the equation $\frac{1}{p} + \frac{1}{q} = \frac{1}{f}$. Plug in numbers.

- You should get $\frac{1}{30} + \frac{1}{q} = \frac{1}{90}$. First isolate $\frac{1}{q}$. Subtract $\frac{1}{30}$ from both sides.
- You should get $\frac{1}{q} = \frac{1}{90} - \frac{1}{30}$. Find a **common denominator**.
- You should get $\frac{1}{q} = \frac{-2}{90}$. Take the reciprocal of both sides and simplify.
- The image forms at $q = -45$ cm.

(B) Combine the two magnification equations together: $\frac{h_i}{h_o} = -\frac{q}{p}$.

- Multiply both sides of the equation by h_o. You should get $h_i = -\frac{q}{p} h_o$.
- Be careful with the minus signs. (Two minus signs make a plus sign.)
- The image height is $h_i = 24$ mm. Note that the centimeters (cm) **cancel** out.

(C) Use the formula $M = \frac{h_i}{h_o}$ or $M = -\frac{q}{p}$.

- The magnification is $M = \frac{3}{2} \times$, which is the same as $M = 1.5 \times$.

(D) Is the image upright or inverted? Review the sign conventions on page 277.

- Examine the **sign** of the **magnification**.
- Since M is **positive**, the image is **upright**.

(E) Is the image real or virtual? Review the sign conventions on page 277.

- Examine the **sign** of the **image distance**.
- Since q is **negative**, the image is **virtual**.

---

96. Identify the given information. **Note**: It's okay to leave the distances in the units given.

- The first object distance is $p_1 = 48$ cm. The first object distance is always positive.
- The focal lengths are $f_1 = -24$ cm and $f_2 = 32$ cm. Focal length is **negative** for the **concave** lens and **positive** for the **convex** lens.
- The distance between the centers of the lenses is $L = 80$ cm.
- The first object height is $h_{1o} = 36$ mm. The first object height is always positive. **Note**: It's okay for h_{1o} to be in millimeters (as long as all of the h's are consistent).

(A) Use the equation $\frac{1}{p_1} + \frac{1}{q_1} = \frac{1}{f_1}$. Plug in numbers.

- You should get $\frac{1}{48} + \frac{1}{q_1} = \frac{1}{-24}$. First isolate $\frac{1}{q_1}$. Subtract $\frac{1}{48}$ from both sides.
- You should get $\frac{1}{q_1} = -\frac{1}{24} - \frac{1}{48}$. Find a **common denominator**.
- You should get $\frac{1}{q_1} = -\frac{3}{48}$. Take the reciprocal of both sides and simplify.
- The first image forms at $q_1 = -16$ cm. You're <u>**not**</u> finished yet.
- Treat the image of the first lens as the object for the second lens.
- The image of the first lens forms 16 cm from the first lens and on the **same** side (since q_1 is **negative**) of the lens as the original object. Since the lenses are 80 cm apart, the object distance for the second lens is 96 cm. (Draw a picture if it helps.)
- Alternatively, use equation $p_2 = L - q_1$. Note that $80 - (-16) = 80 + 16$.
- The object distance for the second lens is $p_2 = 96$ cm.
- Now use equation $\frac{1}{p_2} + \frac{1}{q_2} = \frac{1}{f_2}$. Plug in numbers.
- You should get $\frac{1}{96} + \frac{1}{q_2} = \frac{1}{32}$. First isolate $\frac{1}{q_2}$. Subtract $\frac{1}{96}$ from both sides.
- You should get $\frac{1}{q_2} = \frac{1}{32} - \frac{1}{96}$. Find a **common denominator**.
- You should get $\frac{1}{q_2} = \frac{2}{96}$. Take the reciprocal of both sides and simplify.
- The **final** image forms at $q_2 = 48$ cm.

(B) First use the equations $M_1 = -\frac{q_1}{p_1}$ and $M_2 = -\frac{q_2}{p_2}$.

- You should get $M_1 = \frac{1}{3}\times$ and $M_2 = -\frac{1}{2}\times$.
- Now use the formula $M = M_1 M_2$.
- The **overall** magnification is $M = -\frac{1}{6}\times$. If you use a calculator, $M = 0.17\times$.

(C) Use the equation $M = \frac{h_{2i}}{h_{1o}}$. Solve for h_{2i}. Multiply both sides of the equation by h_{1o}.

- You should get $h_{2i} = M h_{1o}$. Note that $h_{1o} = 36$ mm.
- The **final** image height is $h_{2i} = -6.0$ mm. Note that the centimeters (cm) **cancel** out.

(D) Is the final image upright or inverted? Review the sign conventions on page 277.

- Examine the **sign** of the overall **magnification.**
- Since M is **negative**, the final image is **inverted.**

(E) Is the final image real or virtual? Review the sign conventions on page 277.

- Examine the **sign** of the second **image distance.**
- Since q_2 is **positive**, the final image is **real.**

---

97. Study the two similar examples on page 284.

(A) It doesn't actually matter whether the lens is convex or concave. You **don't** need to use an equation with focal length to solve the problem.

- Apply the two magnification equations.
- Set the two magnification equations equal to one another: $\frac{h_i}{h_o} = -\frac{q}{p}$.
- Study the sign conventions on page 277.
- p and h_o are both positive.
- Therefore, according to $\frac{h_i}{h_o} = -\frac{q}{p}$, the image height ($h_i$) and image distance ($q$) must have **opposite signs.**
- According to page 277, an image that is **real** would have a **positive** image distance (q), and an image that is **upright** would also have a **positive** image height (h_i).
- However, since we already reasoned from $\frac{h_i}{h_o} = -\frac{q}{p}$ that h_i and q must have **opposite signs**, it **isn't** possible for the image of a single lens to be both real and upright (because h_i and q would both be positive).

(B) This is just like part (A).

- According to page 277, an image that is **virtual** would have a **negative** image distance (q), and an image that is **inverted** would also have a **negative** image height (h_i).
- We already showed in part (A) that h_i and q must have **opposite signs.** Therefore, it **isn't** possible for the image of a single lens to be both virtual and inverted (because h_i and q would both be negative).

---

98. Follow the directions on page 279. Study the two examples on pages 287-290.

- Begin by drawing a **convex** lens and the **optic axis** (see page 276).
- Draw and label **two** foci (F) the same distance from each side of the lens.
- Draw and label an object (O) **twice** as far as the focus from the lens.
- Draw the first ray from the top of the object **parallel** to the optic axis until it reaches the lens. Since the lens is **convex**, the refracted ray will converge to the **far focus.**
- Draw the second ray straight through the center of the lens (virtually undeflected).
- Draw the third ray from the top of the object through the **near focus** (since that focus hasn't yet been used) until it reaches the lens. Since the lens is convex, the refracted ray will then travel **parallel** to the optic axis.

- Draw and label the image where the three **output** rays intersect. (The output rays – the rays leaving the lens – intersect at the image. Be careful not to look at the input rays – the rays entering the lens – which intersect at the object.)
- You can see the completed diagram below.
- **How to check your answer with math**: In this problem, the object distance is twice the focal length ($p = 2f$). Therefore, $\frac{1}{p} + \frac{1}{q} = \frac{1}{f}$ becomes $\frac{1}{2f} + \frac{1}{q} = \frac{1}{f}$. Subtract $\frac{1}{2f}$ from both sides to get $\frac{1}{q} = \frac{1}{f} - \frac{1}{2f}$. Make a common denominator: $\frac{1}{q} = \frac{2}{2f} - \frac{1}{2f}$. You should get $\frac{1}{q} = \frac{1}{2f}$. Invert both sides to get $q = 2f$. Since $f > 0$ for a convex lens, $q > 0$.
- **How to use this to check your answer**: (1) We found that $q > 0$. For a lens, $q > 0$ if the image is on the **opposite** side of the lens compared to the object. (2) Since $q = 2f$ and $p = 2f$, the image should be about as far from the lens as the object is. (3) Since $M = -\frac{q}{p} = -\frac{2f}{2f} = -1\times$, the image should be about as tall as the object, and since $M < 0$, the image should be **inverted**. You can verify these details in the diagram below.

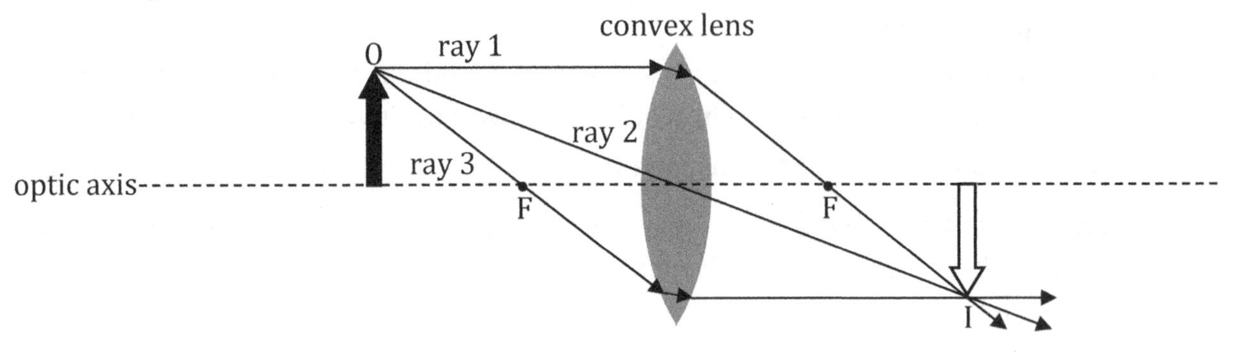

99. Follow the directions on page 279. Study the two examples on pages 287-290.
    - Begin by drawing a **convex** lens and the **optic axis** (see page 276).
    - Draw and label <u>two</u> foci (F) the same distance from each side of the lens.
    - Draw and label an object (O) <u>half</u> as far as the focus from the lens.
    - Draw the first ray from the top of the object **parallel** to the optic axis until it reaches the lens. Since the lens is **convex**, the refracted ray will converge to the **<u>far</u> focus**.
    - Draw the second ray straight through the center of the lens (virtually undeflected).
    - Draw the third ray from the **<u>near</u> focus** (since that focus hasn't yet been used) through the top of the object until it reaches the lens. Since the lens is convex, the refracted ray will then travel **parallel** to the optic axis.
    - Draw and label the image where the three **output** rays (the ones leaving the lens) intersect. Since the output rays don't intersect to the right of the lens, you must **extrapolate backwards to the left of the lens.**
    - You can see the completed diagram on the following page.

- **How to check your answer with math**: In this problem, the object distance is one-half the focal length ($p = \frac{f}{2}$). Therefore, $\frac{1}{p} + \frac{1}{q} = \frac{1}{f}$ becomes $\frac{2}{f} + \frac{1}{q} = \frac{1}{f}$. Subtract $\frac{2}{f}$ from both sides to get $\frac{1}{q} = \frac{1}{f} - \frac{2}{f}$. You should get $\frac{1}{q} = -\frac{1}{f}$. Invert both sides to get $q = -f$. Since $f > 0$ for a convex lens, $q < 0$.
- **How to use this to check your answer**: (1) We found that $q < 0$. For a lens, $q < 0$ if the image is on the **same** side of the lens as the object. (2) Since $q = -f$, the image should form very close to the left focus. (3) Since $M = -\frac{q}{p} = -\frac{(-f)}{f/2} = 2\times$, the image should be about twice tall as the object, and since $M > 0$, the image should be **upright**. You can verify these details in the diagram below.

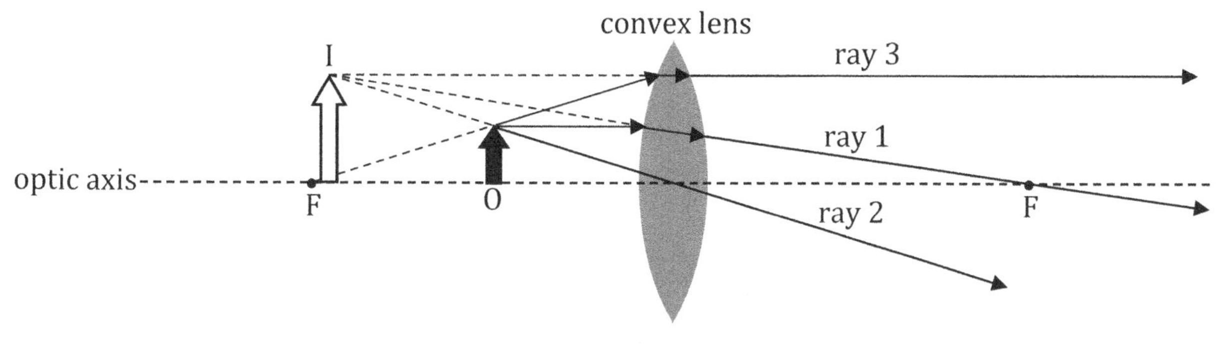

100. Follow the directions on page 279. Study the two examples on pages 287-290.
- Begin by drawing a **concave** lens and the **optic axis** (see page 276).
- Draw and label <u>two</u> foci (F) the same distance from each side of the lens.
- Draw and label an object (O) <u>half</u> as far as the focus from the lens.
- Draw the first ray from the top of the object **parallel** to the optic axis until it reaches the lens. Since the lens is **concave**, the refracted ray will diverge from the **<u>near</u> focus**. Use a dashed line to show that this ray diverges from the near focus.
- Draw the second ray straight through the center of the lens (virtually undeflected).
- Draw the third ray from the top of the object towards the **<u>far</u> focus** (since that focus hasn't yet been used) until it reaches the lens. Since the lens is concave, the refracted ray will then travel **parallel** to the optic axis before it actually reaches the far focus.
- Draw and label the image where the three **output** rays (the ones leaving the lens) intersect. Since the output rays don't intersect to the right of the lens, you must **extrapolate backwards to the left of the lens**. Add a dashed line to extrapolate ray 3 to the left in order to help you locate the image.
- You can see the completed diagram on the following page.

- **How to check your answer with math**: In this problem, the object distance is one-half the focal length ($p = -\frac{f}{2}$). (There is a minus sign here because $f < 0$ for a concave lens, while p must be positive.) Therefore, $\frac{1}{p} + \frac{1}{q} = \frac{1}{f}$ becomes $-\frac{2}{f} + \frac{1}{q} = \frac{1}{f}$. Add $\frac{2}{f}$ to both sides to get $\frac{1}{q} = \frac{1}{f} + \frac{2}{f}$. You should get $\frac{1}{q} = \frac{3}{f}$. Invert both sides to get $q = \frac{f}{3}$. Since $f < 0$ for a concave lens, $q < 0$.
- **How to use this to check your answer**: (1) We found that $q < 0$. For a lens, $q < 0$ if the image is on the **same** side of the lens as the object. (2) Since $q = \frac{f}{3}$, the image should be about one-third of the distance from the lens to the focus. Also, since $p = -\frac{f}{2}$, and since $\frac{1}{3}$ is smaller than $\frac{1}{2}$, the image should be a little closer to the lens than the object is. (3) Since $M = -\frac{q}{p} = -\frac{(f/3)}{(-f/2)} = \frac{2}{3}\times$, the image should be about two-thirds the size of the object, and since $M > 0$, the image should be **upright**. You can verify these details in the diagram below.

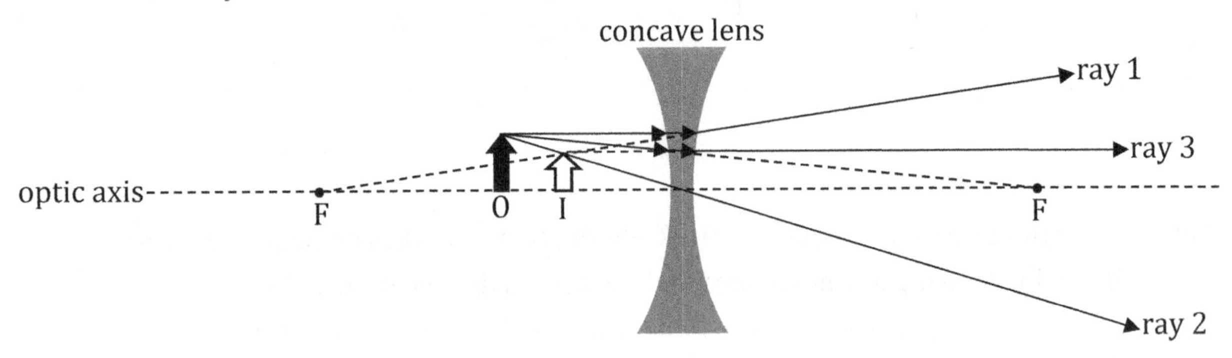

Chapter 26: Spherical Mirrors

101. Identify the given information. **Note**: It's okay to leave the distances in the units given.
- The object distance is $p = 21$ cm. The first object distance is always positive.
- The radius of curvature is $R = 28$ cm. It is **positive** for a **concave** mirror.
- The object height is $h_o = 5.0$ mm. The first object height is always positive. **Note**: It's okay for h_o to be in millimeters (it just needs to have the same units as h_i).

(A) First find the focal length. Use the equation $R = 2f$. Solve for f.
- The focal length is $f = 14$ cm. Focal length is **positive** for a **concave** mirror.
- Use the equation $\frac{1}{p} + \frac{1}{q} = \frac{1}{f}$. Plug in numbers.
- You should get $\frac{1}{21} + \frac{1}{q} = \frac{1}{14}$. First isolate $\frac{1}{q}$. Subtract $\frac{1}{21}$ from both sides.
- You should get $\frac{1}{q} = \frac{1}{14} - \frac{1}{21}$. Find a **common denominator**.
- You should get $\frac{1}{q} = \frac{1}{42}$. Take the reciprocal of both sides.

- The image forms at $q = 42$ cm.

(B) Combine the two magnification equations together: $\frac{h_i}{h_o} = -\frac{q}{p}$.

- Multiply both sides of the equation by h_o. You should get $h_i = -\frac{q}{p} h_o$.
- Be careful with the signs.
- The image height is $h_i = -10$ mm. Note that the centimeters (cm) **cancel** out.

(C) Use the formula $M = \frac{h_i}{h_o}$ or $M = -\frac{q}{p}$.

- The magnification is $M = -2.0 \times$.

(D) Is the image upright or inverted? Review the sign conventions on page 301.

- Examine the **sign** of the **magnification**.
- Since M is **negative**, the image is **inverted**.

(E) Is the image real or virtual? Review the sign conventions on page 301.

- Examine the **sign** of the **image distance**.
- Since q is **positive**, the image is **real**.

---

102. Identify the given information. **Note:** It's okay to leave the distances in the units given.

- The object distance is $p = 80$ cm. The first object distance is always positive.
- The radius of curvature is $R = -40$ cm. It is **negative** for a **convex** mirror.
- The object height is $h_o = 15$ mm. The first object height is always positive. **Note:** It's okay for h_o to be in millimeters (it just needs to have the same units as h_i).

(A) First find the focal length. Use the equation $R = 2f$. Solve for f.

- The focal length is $f = -20$ cm. Focal length is **negative** for a **convex** mirror.
- Use the equation $\frac{1}{p} + \frac{1}{q} = \frac{1}{f}$. Plug in numbers.
- You should get $\frac{1}{80} + \frac{1}{q} = \frac{1}{-20}$. First isolate $\frac{1}{q}$. Subtract $\frac{1}{80}$ from both sides.
- You should get $\frac{1}{q} = -\frac{1}{20} - \frac{1}{80}$. Find a **common denominator**.
- You should get $\frac{1}{q} = -\frac{5}{80}$. Take the reciprocal of both sides and simplify.
- The image forms at $q = -16$ cm.

(B) Combine the two magnification equations together: $\frac{h_i}{h_o} = -\frac{q}{p}$.

- Multiply both sides of the equation by h_o. You should get $h_i = -\frac{q}{p} h_o$.
- Be careful with the minus signs. (Two minus signs make a plus sign.)
- The image height is $h_i = 3.0$ mm. Note that the centimeters (cm) **cancel** out.

(C) Use the formula $M = \frac{h_i}{h_o}$ or $M = -\frac{q}{p}$.

- The magnification is $M = \frac{1}{5} \times$, which is the same as $M = 0.20 \times$.

(D) Is the image upright or inverted? Review the sign conventions on page 301.

- Examine the **sign** of the **magnification**.

- Since M is **positive**, the image is **upright**.

(E) Is the image real or virtual? Review the sign conventions on page 301.

- Examine the **sign** of the **image distance**.
- Since q is **negative**, the image is **virtual**.

---

103. Identify the given information. **Note**: It's okay to leave the distances in the units given.

- The object distance is $p = 27$ cm. The first object distance is always positive.
- The radius of curvature is $R = 108$ cm. It is **positive** for a **concave** mirror.
- The object height is $h_o = 7.0$ mm. The first object height is always positive. **Note**: It's okay for h_o to be in millimeters (it just needs to have the same units as h_i).

(A) First find the focal length. Use the equation $R = 2f$. Solve for f.

- The focal length is $f = 54$ cm. Focal length is **positive** for a **concave** mirror.
- Use the equation $\frac{1}{p} + \frac{1}{q} = \frac{1}{f}$. Plug in numbers.
- You should get $\frac{1}{27} + \frac{1}{q} = \frac{1}{54}$. First isolate $\frac{1}{q}$. Subtract $\frac{1}{27}$ from both sides.
- You should get $\frac{1}{q} = \frac{1}{54} - \frac{1}{27}$. Find a **common denominator**.
- You should get $\frac{1}{q} = \frac{-1}{54}$. Take the reciprocal of both sides.
- The image forms at $q = -54$ cm.

(B) Combine the two magnification equations together: $\frac{h_i}{h_o} = -\frac{q}{p}$.

- Multiply both sides of the equation by h_o. You should get $h_i = -\frac{q}{p} h_o$.
- Be careful with the minus signs. (Two minus signs make a plus sign.)
- The image height is $h_i = 14$ mm. Note that the centimeters (cm) **cancel** out.

(C) Use the formula $M = \frac{h_i}{h_o}$ or $M = -\frac{q}{p}$.

- The magnification is $M = 2.0 \times$.

(D) Is the image upright or inverted? Review the sign conventions on page 301.

- Examine the **sign** of the **magnification**.
- Since M is **positive**, the image is **upright**.

(E) Is the image real or virtual? Review the sign conventions on page 301.

- Examine the **sign** of the **image distance**.
- Since q is **negative**, the image is **virtual**.

104. Follow the directions on page 302. Study the two examples on pages 307-310.

- Begin by drawing a **concave** mirror and the **optic axis** (see page 300).
- Draw and label a focus (F) in front of the concave mirror, then add a center of curvature (C) that is <u>twice</u> (since $R = 2f$) as far as the focus from the mirror, and draw and label an object (O) that is <u>three times</u> as far as the focus from the mirror.
- Draw the first ray from the top of the object **parallel** to the optic axis until it reaches the mirror. The reflected ray will pass through the **focus**.
- Draw the second ray through the center of curvature. Since this ray is perpendicular to the mirror (because this ray travels along a radius, and a radius is perpendicular to the surface of a sphere), it reflects back on itself.
- Draw the third ray from the top of the object through the **focus** until it reaches the mirror. The reflected ray will then travel **parallel** to the optic axis.
- Draw and label the image where the three **output** rays intersect. (The output rays – the reflected rays leaving the mirror – intersect at the image. Be careful not to look at the input rays – the rays incident upon the mirror – which intersect at the object.)
- You can see the completed diagram below.
- **How to check your answer with math**: In this problem, the object distance is three times the focal length ($p = 3f$). Therefore, $\frac{1}{p} + \frac{1}{q} = \frac{1}{f}$ becomes $\frac{1}{3f} + \frac{1}{q} = \frac{1}{f}$. Subtract $\frac{1}{3f}$ from both sides to get $\frac{1}{q} = \frac{1}{f} - \frac{1}{3f}$. Make a common denominator: $\frac{1}{q} = \frac{3}{3f} - \frac{1}{3f}$. You should get $\frac{1}{q} = \frac{2}{3f}$. Invert both sides to get $q = \frac{3f}{2}$. Since $f > 0$ for a concave mirror, $q > 0$.
- **How to use this to check your answer**: (1) We found that $q > 0$. For a mirror, $q > 0$ if the image is on the **same** side of the mirror as the object. (2) Since $q = \frac{3f}{2}$ and $R = 2f$, the image should be about halfway between the focus and the center of curvature. (3) Since $M = -\frac{q}{p} = -\frac{3f/2}{3f} = -\frac{1}{2}\times$, the image should be about half as tall as the object, and since $M < 0$, the image should be **inverted**. You can verify these details in the diagram below.

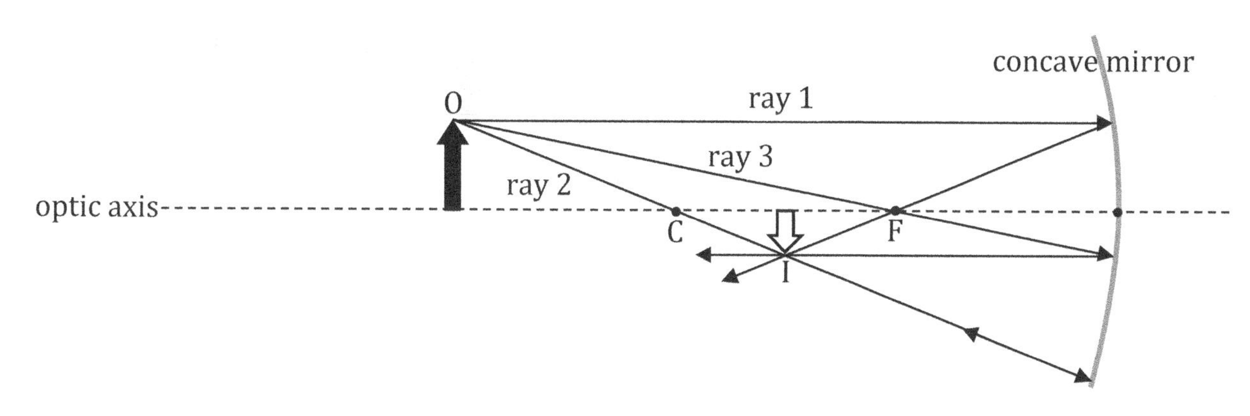

105. Follow the directions on page 302. Study the two examples on pages 307-310.

- Begin by drawing a **concave** mirror and the **optic axis** (see page 300).
- Draw and label a focus (F) in front of the concave mirror, then add a center of curvature (C) that is **twice** (since $R = 2f$) as far as the focus from the mirror, and draw and label an object (O) that is **halfway** between the focus and the mirror.
- Draw the first ray from the top of the object **parallel** to the optic axis until it reaches the mirror. The reflected ray will pass through the **focus.**
- Draw the second ray from the top of the object directly away from point C. Since this ray is perpendicular to the mirror (because this ray travels along a radius, and a radius is perpendicular to the surface of a sphere), it reflects back on itself.
- Draw the third ray from the top of the object away from the **focus** until it reaches the mirror. The reflected ray will then travel **parallel** to the optic axis.
- Draw and label the image where the three **output** rays intersect (the ones reflected by the mirror). Since the output rays don't intersect to the left of the mirror, you must **extrapolate backwards to the right of the mirror.**
- You can see the completed diagram below.
- **How to check your answer with math**: In this problem, the object distance is one-half the focal length ($p = \frac{f}{2}$). Therefore, $\frac{1}{p} + \frac{1}{q} = \frac{1}{f}$ becomes $\frac{2}{f} + \frac{1}{q} = \frac{1}{f}$. Subtract $\frac{2}{f}$ from both sides to get $\frac{1}{q} = \frac{1}{f} - \frac{2}{f}$. You should get $\frac{1}{q} = -\frac{1}{f}$. Invert both sides to get $q = -f$. Since $f > 0$ for a concave mirror, $q < 0$.
- **How to use this to check your answer**: (1) We found that $q < 0$. For a mirror, $q < 0$ if the image is on the **opposite** side of the mirror compared to the object. (2) Since $q = -f$, the image should form behind the mirror about as far from the mirror as the focus is in front of the mirror. (3) Since $M = -\frac{q}{p} = -\frac{(-f)}{f/2} = 2\,\times$, the image should be about twice tall as the object, and since $M > 0$, the image should be **upright**. You can verify these details in the diagram below.

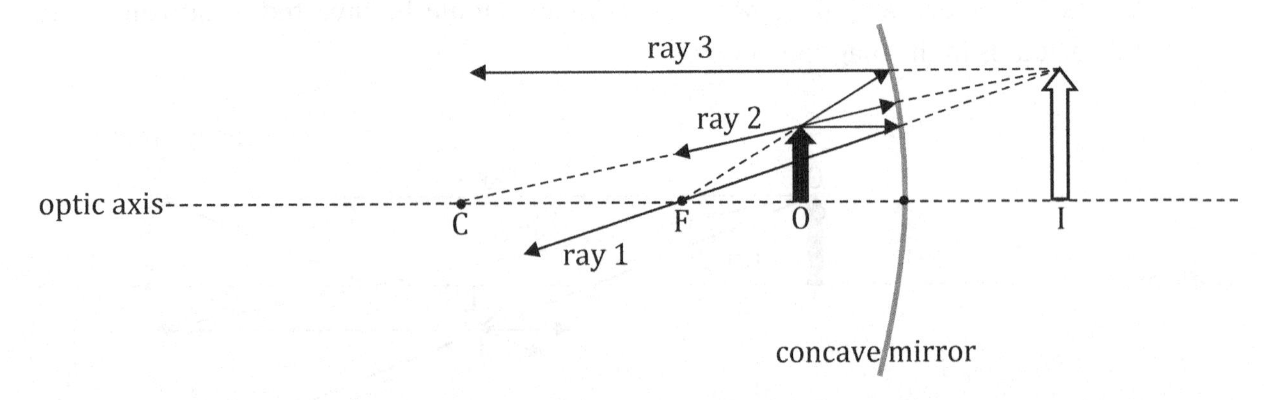

106. Follow the directions on page 302. Study the two examples on pages 307-310.

- Begin by drawing a **convex** mirror and the **optic axis** (see page 300).
- Draw and label a focus (F) **behind** the convex mirror, then add a center of curvature (C) that is **twice** (since $R = 2f$) as far as the focus from the mirror, and draw and label an object (O) that is just as far **in front** of the mirror as the focus is behind the mirror.
- Draw the first ray from the top of the object **parallel** to the optic axis until it reaches the mirror. The reflected ray will diverge from the **focus**. Use a dashed line to show that this ray diverges from the focus.
- Draw the second ray towards the center of curvature. Since this ray is perpendicular to the mirror (because this ray travels along a radius, and a radius is perpendicular to the surface of a sphere), it reflects back on itself.
- Draw the third ray from the top of the object towards the **focus** until it reaches the mirror. The reflected ray will then travel **parallel** to the optic axis.
- Draw and label the image where the three **output** rays intersect (the ones reflected by the mirror). Since the output rays don't intersect to the left of the mirror, you must **extrapolate backwards to the right of the mirror**.
- You can see the completed diagram below.
- **How to check your answer with math**: In this problem, the object distance is the same as the focal length ($p = -f$), apart from a minus sign (since $f < 0$ for a convex mirror, while p must be positive). Therefore, $\frac{1}{p} + \frac{1}{q} = \frac{1}{f}$ becomes $-\frac{1}{f} + \frac{1}{q} = \frac{1}{f}$. Add $\frac{1}{f}$ to both sides to get $\frac{1}{q} = \frac{1}{f} + \frac{1}{f}$. You should get $\frac{1}{q} = \frac{2}{f}$. Invert both sides to get $q = \frac{f}{2}$. Since $f < 0$ for a convex mirror, $q < 0$.
- **How to use this to check your answer**: (1) We found that $q < 0$. For a mirror, $q < 0$ if the image is on the **opposite** side of the mirror compared to the object. (2) Since $q = \frac{f}{2}$, the image should be about one-half of the distance from the mirror to the focus. (3) Since $M = -\frac{q}{p} = -\frac{(f/2)}{(-f)} = \frac{1}{2}\times$, the image should be about one-half the size of the object, and since $M > 0$, the image should be **upright**. You can verify these details in the diagram below.

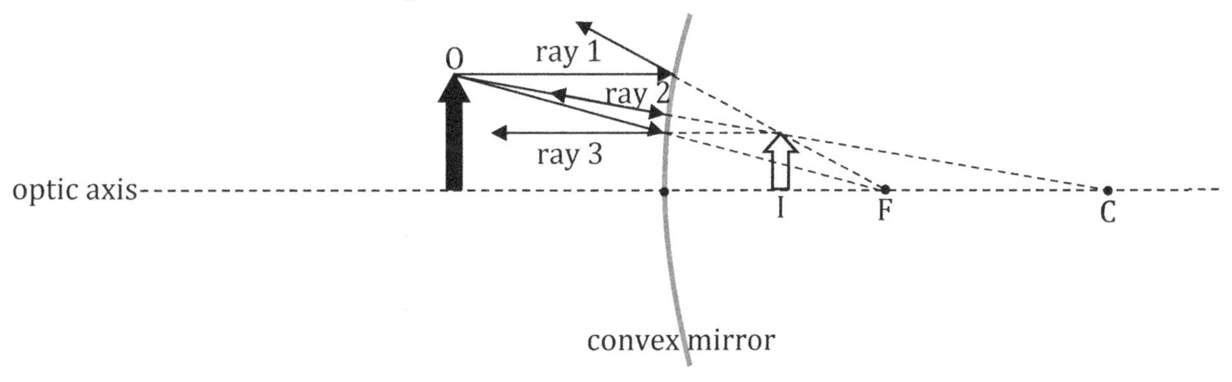

Chapter 27: Single-slit Diffraction

107. Identify the given information in consistent units.

- The slit width is $a = 8.0 \times 10^{-6}$ m (since 1 mm = 10^{-3} m and $0.0080 = 8.0 \times 10^{-3}$).
- The distance from the slit to the screen is $L = 1.50$ m (since 1 cm = 10^{-2} m).
- The question specifies the 10th dark fringe from the optic axis: $m = 10$.
- The distance from the 10th dark fringe to the optic axis is $x_{10} = \frac{\sqrt{3}}{2}$ m.
- The problem is asking for the wavelength (λ).
- First apply the equation $\tan\theta_m = \frac{x_m}{L}$. You should get $\theta_{10} = \tan^{-1}\left(\frac{x_{10}}{L}\right)$.
- The angle is $\theta_{10} = 30°$. Note that $\frac{\sqrt{3}}{2} \div 1.5 = \frac{\sqrt{3}}{2} \div \frac{3}{2} = \frac{\sqrt{3}}{2} \times \frac{2}{3} = \frac{\sqrt{3}}{3}$ and $\tan 30° = \frac{\sqrt{3}}{3}$.
- Now apply the equation $a \sin\theta_m = m\lambda$. Solve for λ. You should get $\lambda = \frac{a \sin\theta_{10}}{10}$.
- The wavelength is $\lambda = 400$ nm, which is the same as $\lambda = 4.0 \times 10^{-7}$ m.

---

108. Identify the given information in consistent units.

- The slit width is $a = 2.6 \times 10^{-5}$ m (since 1 mm = 10^{-3} m and $0.026 = 2.6 \times 10^{-2}$).
- The distance from the slit to the screen is $L = 0.80$ m (since 1 cm = 10^{-2} m).
- The question specifies the 3rd dark fringe from the optic axis: $m = 3$.
- The distance from the 3rd dark fringe to the optic axis is $x_3 = 0.060$ m.
- The problem is asking for the wavelength (λ).
- Use the equations $\tan\theta_m = \frac{x_m}{L}$ and $a \sin\theta_m = m\lambda$.
- If you're not using a calculator, note that $\tan\theta_m \approx \sin\theta_m$ since θ_m is small.
- This means that $\frac{x_m}{L} \approx \frac{m\lambda}{a}$ (since $\tan\theta_m = \frac{x_m}{L}$ and $\sin\theta_m = \frac{m\lambda}{a}$).
- Solve for λ. First **cross multiply**. You should get $\lambda \approx \frac{a x_m}{mL}$.
- The wavelength is $\lambda \approx 650$ nm, which is the same as $\lambda \approx 6.5 \times 10^{-7}$ m.
- If you use a calculator, you may follow the method from Problem 107. Using a calculator, you should get $\theta_m = 4.289°$ and $\lambda = 648$ nm $= 6.5 \times 10^{-7}$ m.

---

109. Let the equations $a \sin\theta_m = m\lambda$ and $\tan\theta_m = \frac{x_m}{L}$ serve as your guide.

(A) Solve for θ_m in terms of a.

- You should get $\theta_m = \sin^{-1}\left(\frac{m\lambda}{a}\right)$.
- A narrower slit has a smaller value of a, which means a **larger** value of $\frac{m\lambda}{a}$ (since a is in the denominator). This results in a larger value of θ_m and a larger value of x_m (since $\tan\theta_m = \frac{x_m}{L}$).
- The distance to each dark spot from the optic axis will be **longer** (since θ_m and x_m will be bigger). The bright bands (fringes) will also become wider.

(B) Solve for θ_m in terms of λ (since the value of the **wavelength**, λ, depends on the **color**).

- You should get $\theta_m = \sin^{-1}\left(\frac{m\lambda}{a}\right)$.
- Green light has a **shorter** wavelength (λ) than red light (see Chapter 20).[1] A shorter wavelength color has a smaller value of λ. This results in a smaller value of θ_m and a smaller value of x_m (since $\tan\theta_m = \frac{x_m}{L}$).
- The distance to each dark spot from the optic axis will be **shorter** (since θ_m and x_m will be smaller). The bright bands (fringes) will also become narrower.

(C) Solve for θ_m in each equation L.

- You should get $\theta_m = \sin^{-1}\left(\frac{m\lambda}{a}\right)$ and $\theta_m = \tan^{-1}\left(\frac{x_m}{L}\right)$.
- From the equation $\theta_m = \sin^{-1}\left(\frac{m\lambda}{a}\right)$, we see that θ_m **doesn't change** because the wavelength (λ) and slit width (a) aren't changed in this problem.
- From the equation $\theta_m = \tan^{-1}\left(\frac{x_m}{L}\right)$, we see that the ratio $\frac{x_m}{L}$ must be constant in order for θ_m to remain constant. Moving the screen farther from the slit results in a larger value of L, which means a **larger** value of x_m (so that $\frac{x_m}{L}$ remains constant).
- The distance to each dark spot from the optic axis will be **longer** (x_m will be bigger). The bright bands (fringes) will also become wider. However, the angles to the dark fringes will be **unchanged** (since θ_m is constant).

Chapter 28: Diffraction Grating

110. Convert the grating constant from lines/cm to lines/m.

- Multiply by 100 to convert from lines/cm to lines/m. (If we were converting from cm to m, we would divide by 100, but we're converting from $\frac{1}{\text{cm}}$ to $\frac{1}{\text{m}}$, so we multiply by 100.)
- The grating **constant** is $\frac{1}{d} = 500{,}000 \frac{\text{lines}}{\text{m}}$. Note that $5000 \frac{\text{lines}}{\text{cm}} = 5000 \frac{\text{lines}}{\text{cm}} \times \frac{100 \text{ cm}}{1 \text{ m}}$.
- Find the **reciprocal** of the grating constant to determine the grating spacing.
- The grating **spacing** is $d = \frac{1}{500{,}000}$ m $= 0.0000020$ m $= 2.0 \times 10^{-6}$ m.
- The question specifies the 2nd bright fringe from the optic axis: $m = 2$.
- The corresponding angle is $\theta_2 = 30°$.
- Use the equation $d\sin\theta_m = m\lambda$. Solve for λ. You should get $\lambda = \frac{d\sin\theta_2}{2}$.
- The wavelength is $\lambda = 500$ nm, which is the same as $\lambda = 5.0 \times 10^{-7}$ m.

111. Study the example on page 325.

- Multiply by 100 to convert from lines/cm to lines/m. (If we were converting from cm to m, we would divide by 100, but we're converting from $\frac{1}{\text{cm}}$ to $\frac{1}{\text{m}}$, so we multiply by 100.)

[1] In ROY G. BIV, red (R) has longer wavelength while violet (V) has shorter wavelength. Green (G) thus has shorter wavelength than red (R).

- The grating **constant** is $\frac{1}{d} = 800{,}000 \frac{\text{lines}}{\text{m}}$. Note that $8000 \frac{\text{lines}}{\text{cm}} = 8000 \frac{\text{lines}}{\text{cm}} \times \frac{100 \text{ cm}}{1 \text{ m}}$.
- Find the **reciprocal** of the grating constant to determine the grating spacing.
- The grating **spacing** is $d = \frac{1}{800{,}000}$ m $= 0.00000125$ m $= 1.25 \times 10^{-6}$ m.
- The wavelength is $\lambda = 4.50 \times 10^{-7}$ m (since 1 nm $= 10^{-9}$ m and $450 = 4.50 \times 10^2$).
- To determine the number of observable fringes, set $\theta_{max} = 90°$.
- Use the equation $d \sin\theta_m = m\lambda$ with $\theta_{max} = 90°$. Solve for m_{max}.
- You should get $m_{max} = \frac{d}{\lambda}$. Recall from trig that $\sin 90° = 1$.
- You should get $m_{max} = \frac{25}{9}$, which is between 2 and 3. Since m_{max} must be an **integer**, $m_{max} = 2$ (always round **down** to find m_{max} - since anything higher than the value of m_{max} is unobservable, rounding up would be incorrect).
- The total number of observable fringes equals $2m_{max} + 1$.
- There are 5 observable fringes.

Chapter 29: Double-slit Interference

112. Study the example on page 331.

(A) Read page 328 and study the diagrams.

- There are $N - 2$ weak bright fringes between each pair of main bright fringes, where N is the number of slits.
- In the given diagram, there are 2 weak bright fringes between each pair of main bright fringes.
- Set $N - 2$ equal to 2 and solve for N.
- There are $N = 4$ slits.

(B) Identify the relevant given information.

- The wavelength is $\lambda = 6.00 \times 10^{-7}$ m (since 1 nm $= 10^{-9}$ m and $600 = 6.00 \times 10^2$).
- The distance from the slit to the screen is $L = 0.80$ m (since 1 cm $= 10^{-2}$ m).
- The given diagram shows that $x_2 = 4.0$ mm $= 0.0040$ m for the 2nd (main) bright fringe. The corresponding value of m is $m = 2$.
- For the 2nd **bright** fringe, use the formulas $d \sin\theta_m = m\lambda$ and $\tan\theta_m = \frac{x_m}{L}$.
- If you're not using a calculator, it is convenient to approximate $\sin\theta_m \approx \tan\theta_m$ (since the angles are small in this problem). Solve for $\sin\theta_m$ to get $\sin\theta_m = \frac{m\lambda}{d}$. When you make the small angle approximation, you should get $\frac{m\lambda}{d} \approx \frac{x_m}{L}$. Solve for d. First **cross multiply**. You should get $d \approx \frac{2\lambda L}{x_2}$.
- The slit spacing is $d \approx 0.00024$ m $= 0.24$ mm. If you use a calculator, you should get $\theta_2 = \tan^{-1}\left(\frac{x_2}{L}\right) = 0.2865°$ and $d = 0.00024$ m $= 0.24$ mm.

(C) Identify the relevant given information.

- The wavelength is $\lambda = 6.00 \times 10^{-7}$ m (since 1 nm $= 10^{-9}$ m and $600 = 6.00 \times 10^2$).
- The distance from the slit to the screen is $L = 0.80$ m (since 1 cm $= 10^{-2}$ m).
- The given diagram shows that $x_1 = 6.0$ mm $= 0.0060$ m for the 1st **missing order** (the first gap where it seems like there "should" be a main bright fringe where there isn't one). The corresponding value of n is $n = 1$. **Note**: We're using the index n for missing orders and the index m for bright or dark fringes. The value $x_1 = 6.0$ mm corresponds to the **first** missing order ($n = 1$).
- For the 1st **missing order**, use the formulas $a \sin\theta_n = n\lambda$ and $\tan\theta_n = \frac{x_n}{L}$.
- If you're not using a calculator, it is convenient to approximate $\sin\theta_n \approx \tan\theta_n$ (since the angles are small in this problem). Solve for $\sin\theta_n$ to get $\sin\theta_n = \frac{n\lambda}{a}$. When you make the small angle approximation, you should get $\frac{n\lambda}{a} \approx \frac{x_n}{L}$. Solve for a. First **cross multiply**. You should get $a \approx \frac{\lambda L}{x_1}$ (using $n = 1$).
- The slit width is $a \approx 0.000080$ m $= 0.080$ mm. If you use a calculator, you should get[2] $\theta_1 = \tan^{-1}\left(\frac{x_1}{L}\right) = 0.4297°$ and $a = 0.000080$ m $= 0.080$ mm.

Chapter 30: Polarization

113. Since the reflected ray is **completely polarized**, you may apply Snell's law for **Brewster's angle**: $n_i \tan\theta_B = n_t$.

- You should get $\sqrt{2}\tan\theta_B = \sqrt{6}$.
- Isolate the tangent function. Divide both sides of the equation by $\sqrt{2}$.
- You should get $\tan\theta_B = \sqrt{3}$. Note that $\frac{\sqrt{6}}{\sqrt{2}} = \sqrt{\frac{6}{2}} = \sqrt{3}$.
- Take the inverse tangent of both sides. You should get $\theta_B = \tan^{-1}(\sqrt{3})$.
- **Brewster's angle** is $\theta_B = 60°$ because $\tan(60°) = \sqrt{3}$.
- The **incident** angle equals Brewster's angle $\theta_i = \theta_B = 60°$.
- The **reflected** angle equals the incident angle: $\theta_r = \theta_i = 60°$ (the law of **reflection**).
- For Brewster's angle, the **refracted** angle makes a 90° angle with the reflected angle.

[2] Note that there are two different kinds of angles. In part (B), we worked with θ_2 of the interference pattern (related to the slit spacing, d, and the second bright fringe), whereas in part (C) we are working with θ_1 of the diffraction pattern (related to the slit width, a, and the first missing order). The reason that θ_1 is greater than θ_2 is that these are two different types of angles (one has subscript $m = 2$, the other has subscript $n = 1$). In fact, if you study the diagram, for this problem you can see that there should be a "third" main bright fringe where there is instead the first missing order, such that θ_3 (with $m = 3$) of the interference pattern is **equal** to θ_1 (with $n = 1$) of the diffraction pattern. (If you want to make it easier to tell the difference between interference and diffraction angles, you could write the equation for the missing orders as $a \sin\varphi_m = m\lambda$, using φ instead of θ. Then, for this problem, $\theta_3 = \varphi_1$. However, almost every textbook uses the same symbol, θ_m, in both formulas.)

See the diagram on page 335. Note that $\theta_r + 90° + \theta_t = 180°$.

- Solve for the refracted angle. You should get $\theta_t = 90° - \theta_r$. Plug in $\theta_r = 60°$.
- The **refracted** angle is $\theta_t = 30°$. Note that $\theta_t = 30°$ is the angle of refraction, whereas $\theta_r = 60°$ is the angle of reflection.

---

114. Since the light is initially **unpolarized**, the first polarizer simply reduces the intensity by **half**. After passing through the first polarizer, the light becomes polarized. Apply Malus's law for the second and third polarizers. Study the second example on page 338.

(A) The initial intensity is $I_0 = 960 \frac{\text{W}}{\text{m}^2}$.

- Apply the equation $I_1 = \frac{I_0}{2}$.
- The light has an intensity of $I_1 = 480 \frac{\text{W}}{\text{m}^2}$ after passing through the first polarizer.
- Apply Malus's law for the second polarizer: $I_2 = I_1 \cos^2 \theta_{12}$, where θ_{12} is the angle between the first two polarizers. Note that $\theta_{12} = 75° - 45° = 30°$.
- The light has an intensity of $I_2 = 360 \frac{\text{W}}{\text{m}^2}$ after passing through the second polarizer.
- Note that cosine is **squared**. Recall from trig that $\cos 30° = \frac{\sqrt{3}}{2}$. Note that $\left(\frac{\sqrt{3}}{2}\right)^2 = \frac{3}{4}$.
- Apply Malus's law for the third polarizer: $I_3 = I_2 \cos^2 \theta_{23}$, where θ_{23} is the angle between the last two polarizers. Note that $\theta_{23} = 135° - 75° = 60°$.
- The light has an intensity of $I_3 = 90 \frac{\text{W}}{\text{m}^2}$ after passing through the third polarizer. Note that cosine is **squared**. Recall from trig that $\cos 60° = \frac{1}{2}$.

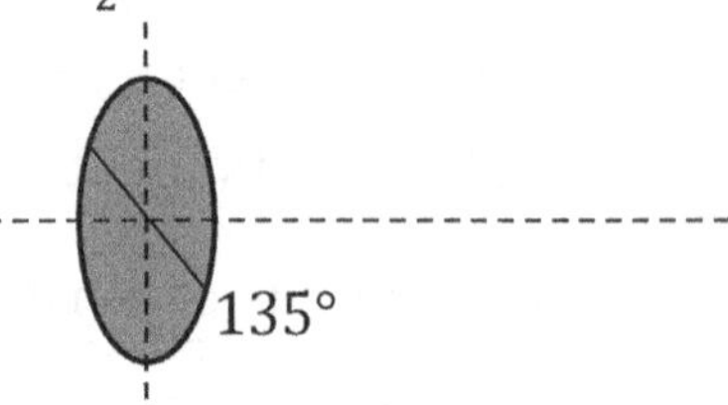

(B) Start over. The initial intensity is $I_0 = 960 \frac{\text{W}}{\text{m}^2}$. The new setup is sketched above.

- Apply the equation $I_1 = \frac{I_0}{2}$.
- The light has an intensity of $I_1 = 480 \frac{\text{W}}{\text{m}^2}$ after passing through the first polarizer.
- Apply Malus's law for the last polarizer: $I_3 = I_1 \cos^2 \theta_{13}$, where θ_{13} is the angle between the first and last polarizers. Note that $\theta_{13} = 135° - 45° = 90°$.
- The light has an intensity of $I_3 = 0$ after passing through the last polarizer. Recall from trig that $\cos 90° = 0$.
- No light makes it through the last polarizer when the middle polarizer is removed because the remaining polarizers are **perpendicular** to one another and $\cos 90° = 0$.

WAS THIS BOOK HELPFUL?

A great deal of effort and thought was put into this book, such as:

- Breaking down the solutions to help make physics easier to understand.
- Careful selection of examples and problems for their instructional value.
- Multiple stages of proofreading, editing, and formatting.
- Two physics instructors worked out the solution to every problem to help check all of the final answers.
- Dozens of actual physics students provided valuable feedback.

If you appreciate the effort that went into making this book possible, there is a simple way that you could show it:

<u>Please take a moment to post an honest review.</u>

For example, you can review this book at Amazon.com or BN.com (for Barnes & Noble).

Even a short review can be helpful and will be much appreciated. If you're not sure what to write, following are a few ideas, though it's best to describe what's important to you.

- Were you able to understand the explanations?
- Did you appreciate the list of symbols and units?
- Was it easy to find the equations you needed?
- How much did you learn from reading through the examples?
- Did the hints and intermediate answers section help you solve the problems?
- Would you recommend this book to others? If so, why?

<u>Are you an international student?</u>

If so, please leave a review at Amazon.co.uk (United Kingdom), Amazon.ca (Canada), Amazon.in (India), Amazon.com.au (Australia), or the Amazon website for your country.

The physics curriculum in the United States is somewhat different from the physics curriculum in other countries. International students who are considering this book may like to know how well this book may fit their needs.

THE SOLUTIONS MANUAL

The solution to every problem in this workbook can be found in the following book:

100 Instructive Calculus-based Physics Examples

Fully Solved Problems with Explanations

Volume 3: Waves, Fluids, Sound, Heat, and Light

Chris McMullen, Ph.D.

ISBN: 978-1-941691-21-2

If you would prefer to see every problem worked out completely, along with explanations, you can find such solutions in the book shown below. (The workbook you are currently reading has hints, intermediate answers, and explanations. The book described above contains full step-by-step solutions.)

ABOUT THE AUTHOR

Chris McMullen is a physics instructor at Northwestern State University of Louisiana and also an author of academic books. Whether in the classroom or as a writer, Dr. McMullen loves sharing knowledge and the art of motivating and engaging students.

He earned his Ph.D. in phenomenological high-energy physics (particle physics) from Oklahoma State University in 2002. Originally from California, Dr. McMullen earned his Master's degree from California State University, Northridge, where his thesis was in the field of electron spin resonance.

As a physics teacher, Dr. McMullen observed that many students lack fluency in fundamental math skills. In an effort to help students of all ages and levels master basic math skills, he published a series of math workbooks on arithmetic, fractions, algebra, and trigonometry called the Improve Your Math Fluency Series. Dr. McMullen has also published a variety of science books, including introductions to basic astronomy and chemistry concepts in addition to physics textbooks.

Dr. McMullen is very passionate about teaching. Many students and observers have been impressed with the transformation that occurs when he walks into the classroom, and the interactive engaged discussions that he leads during class time. Dr. McMullen is well-known for drawing monkeys and using them in his physics examples and problems, applying his creativity to inspire students. A stressed-out student is likely to be told to throw some bananas at monkeys, smile, and think happy physics thoughts.

Author, Chris McMullen, Ph.D.

PHYSICS

The learning continues at Dr. McMullen's physics blog:

www.monkeyphysicsblog.wordpress.com

More physics books written by Chris McMullen, Ph.D.:

- An Introduction to Basic Astronomy Concepts (with Space Photos)
- The Observational Astronomy Skywatcher Notebook
- An Advanced Introduction to Calculus-based Physics
- Essential Calculus-based Physics Study Guide Workbook
- Essential Trig-based Physics Study Guide Workbook
- 100 Instructive Calculus-based Physics Examples
- 100 Instructive Trig-based Physics Examples
- Creative Physics Problems
- A Guide to Thermal Physics
- A Research Oriented Laboratory Manual for First-year Physics

SCIENCE

Dr. McMullen has published a variety of **science** books, including:

- Basic astronomy concepts
- Basic chemistry concepts
- Balancing chemical reactions
- Creative physics problems
- Calculus-based physics textbook
- Calculus-based physics workbooks
- Trig-based physics workbooks

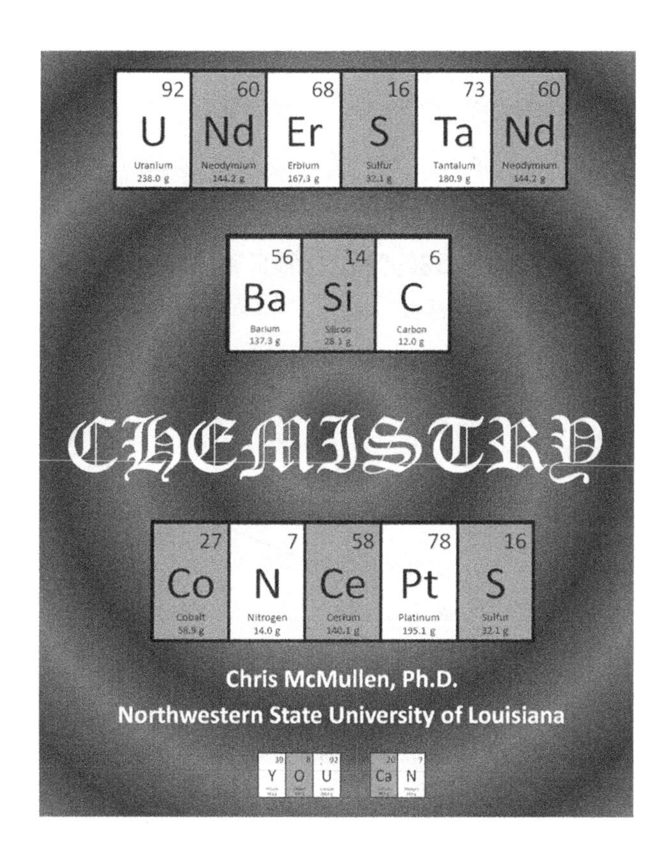

MATH

This series of math workbooks is geared toward practicing essential math skills:

- Algebra and trigonometry
- Fractions, decimals, and percents
- Long division
- Multiplication and division
- Addition and subtraction

www.improveyourmathfluency.com

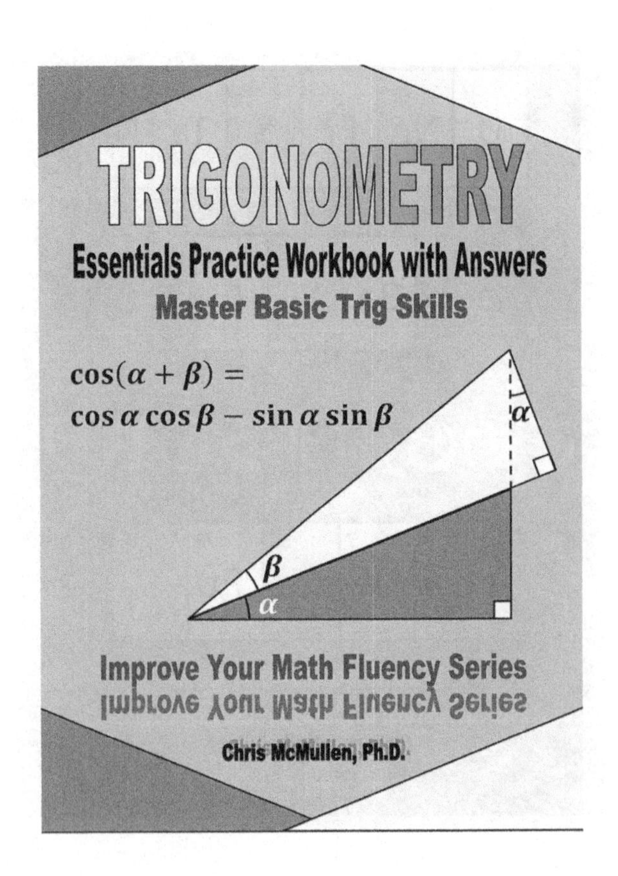

PUZZLES

The author of this book, Chris McMullen, enjoys solving puzzles. His favorite puzzle is Kakuro (kind of like a cross between crossword puzzles and Sudoku). He once taught a three-week summer course on puzzles. If you enjoy mathematical pattern puzzles, you might appreciate:

300+ Mathematical Pattern Puzzles

Number Pattern Recognition & Reasoning

- pattern recognition
- visual discrimination
- analytical skills
- logic and reasoning
- analogies
- mathematics

VErBAl ReAcTiONS

Chris McMullen has coauthored several word scramble books. This includes a cool idea called **VErBAl ReAcTiONS**. A VErBAl ReAcTiON expresses word scrambles so that they look like chemical reactions. Here is an example:

$$2\,\mathrm{C} + \mathrm{U} + 2\,\mathrm{S} + \mathrm{Es} \rightarrow \mathrm{S\ U\ C\ C\ Es\ S}$$

The left side of the reaction indicates that the answer has 2 C's, 1 U, 2 S's, and 1 Es. Rearrange CCUSSEs to form SUCCEsS.

Each answer to a **VErBAl ReAcTiON** is not merely a word, it's a chemical word. A chemical word is made up not of letters, but of elements of the periodic table. In this case, SUCCEsS is made up of sulfur (S), uranium (U), carbon (C), and Einsteinium (Es).

Another example of a chemical word is GeNiUS. It's made up of germanium (Ge), nickel (Ni), uranium (U), and sulfur (S).

If you enjoy anagrams and like science or math, these puzzles are tailor-made for you.

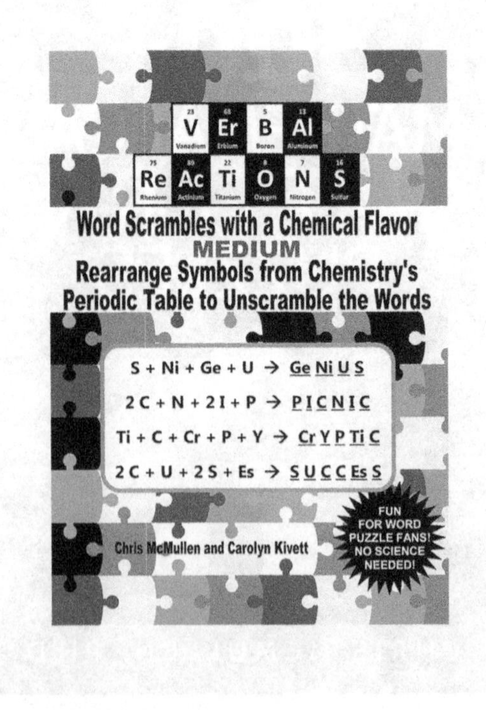

BALANCING CHEMICAL REACTIONS

$$2\ C_2H_6 + 7\ O_2 \rightarrow 4\ CO_2 + 6\ H_2O$$

Balancing chemical reactions isn't just chemistry practice.

These are also **fun puzzles** for math and science lovers.

Balancing Chemical Equations Worksheets

Over 200 Reactions to Balance

Chemistry Essentials Practice Workbook with Answers

Chris McMullen, Ph.D.

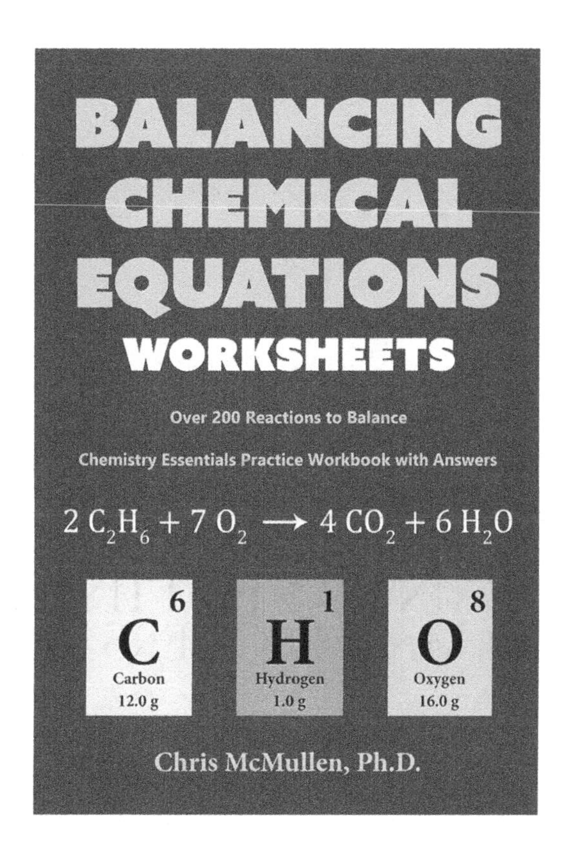

CURSIVE HANDWRITING

for... MATH LOVERS

Would you like to learn how to write in cursive?

Do you enjoy math?

This cool writing workbook lets you practice writing math terms with cursive handwriting. Unfortunately, you can't find many writing books oriented around math.

Cursive Handwriting for Math Lovers

by Julie Harper and Chris McMullen, Ph.D.

Made in the USA
Monee, IL
31 March 2024

56130981R00247